General and Inorganic Chemistry Made Easy

VCH

General and Inorganic Chemistry Made Easy

J. Nentwig
M. Kreuder
K. Morgenstern

VCH

Joachim Nentwig
Ahornweg 11
D-3043 Schneverdingen
Germany

Manfred Kreuder
Karl Morgenstern
Bayer AG, Werk Uerdingen
D-4150 Krefeld-Uerdingen

Library of Congress Cataloging-in-Publication Data

Nentwig, J. (Joachim)
General and inorganic chemistry made easy / by Joachim Nentwig, Manfred Kreuder, Karl Morgenstern
p. cm. -- (Chemistry made easy)
Includes index
ISBN 1-56081-502-7. -- ISBN 3-527-28348-X (Weinheim). -- ISBN 1-56081-549-3 (set)
1. Chemistry. 2. Chemistry, Inrganic. I. Kreuder, Manfred. II. Morgenstern, Karl. III. Title. IV. Series.
QD33.N445 1991 91-42954
5470--dc20 CIP

Printed in the United States of America

ISBN: 1-56081-502-7 VCH Publishers, Inc.
ISBN: 3-527-28348-X VCH Verlagsgesellschaft

Printing history:

10 9 8 7 6 5 4 3 2 1

VCH Publishers, Inc.
220 East 23rd Street
Suite 909
New York, N.Y. 10010

VCH Verlagsgesellschaft mbH
P.O. Box 10 11 61
D-6940 Weinheim
Federal Republic of Germany

VCH Publishers (UK) Ltd.
8 Wellington Court
Cambridge CB1 1HW
United Kingdom

PREFACE

In order to make the technique of programmed learning and our subject matter optimally compatible, we used a mixture of "linear programming" (after Skinner) and "branched programming" (after Crowder). Mixed programs of this type have been proven particularly effective. The term *hybrid program* was proposed for them at DIDACTA, an international conference on education and educational methods in Switzerland.

For us, programming the material meant the creation of an eticing stimulus and an educational aid for the reader. A programmed text requires the reader to complete one step before the next can be made, leading to an intensive absorption of the material. The reader is drawn into a question-and-answer game where the solution to every question must be found, thereby enabling the user to take control of the learning performance. The places the reader in a classroom-like situation, with the added advantage of solely determining the rate of progress. A well-programmed text is, to a large extent, capable of replacing a tutor. Therefore, this text is of particular value wherever the opportunity to attend classes or lectures does not exist (or was missed), and in cases where additional self initiative is desired.

We ascribe the huge successes of the original German editions of *General and Inorganic Chemistry Made Easy* and *Organic Chemistry Made Easy* to the fact that all the programs were thoroughly tested in classrooms and in face-to-face tests before they were published. Thus, they are not merely "desk programs," but are instead instructional and learning aids that have proven their value in practice.

This edition was revised, updated, and rearranged. Naturally, we would be grateful for any critical comments or suggestions.

January 1992

J.N.
M.K.
K.M.

Contents

How to use this book

For those interested in acquiring a thorough grounding in the fundamentals of chemistry, this book provides an effective and simple means. The method of teaching utilized here is based on the proven approach of

programmed instruction.

The material we cover is presented in the form of 29 separate Programs. Although at the outset relatively light demands are made on the student, the situation changes as the student begins to advance through the book. Because our method is rather a specialized one, we outline below details of the techniques employed.

The material in each Program has been divided up into numerous small units or *frames*. At the end of every frame you, the student, are asked to answer one or more questions. The intention is to check whether you have really understood the content of the particular frame. Such questions may involve the setting up of a formula or an equation, the calculation of some quantity, the insertion of missing words into sentences and so on. After answering, you are advised to proceed to another frame where the correct answer is to be found or where any mistakes you may have made are analyzed. Only at this stage should you consider moving on to learn something new. By following this procedure you will be conducted step by step through all of the Programs.

To illustrate our approach more fully, we now present two very simple sample frames. These demonstrate the two basic ways in which questions will be posed. We mention in passing that you will continually have need of a pen (or pencil) and paper as you work your way through each Program.

---

1 Chemistry may be studied by an effective technique known as programmed instruction. The material covered in this volume is presented in the form of 25 separate Programs.

Now write out the completed version of the following sentence: We shall be studying chemistry here by means of .... instruction.

Continue your reading at 11 .

You would then be required to write down: "We shall be studying chemistry here by means of *programmed* instruction."

Please do *not* write your answer within the actual frame itself. That would make it impossible for you to revise the material at some later stage! Be sure to write all of your answers and notes on separate sheets of paper and not directly into the book.

After writing down your answer, turn to the next frame indicated; in the above case you would turn to frame number [11]. There you will be able to check your answer. (Do please make sure that in turning up the relevant frame you do not jump over into some other Program!)

The route to be followed through any given Program is indicated not by the page number but rather by the frame number located at the top left hand corner of each frame. Thus, you would check your answer here against the correct solution given in frame [11]. Filled-in words are always *italicized* in our solutions.

---

[11] We shall be studying chemistry here by means of *programmed* instruction.

It has been demonstrated that programmed instruction is a more effective technique than learning from a conventional textbook.

A Program can facilitate learning by covering the material in small stages, by enabling the student to check each answer instantly and by permitting the student to work at his own individual pace throughout.

What is the most effective way of learning?

- By making use of a conventional textbook → [3]
- By making use of a Program → [18]
- There is no difference between the two ways → [34]
- I cannot answer this question → [54]

The student should now decide in favor of one of the four possible options and then turn to the frame indicated for the option chosen.

If the student had chosen frames [3] or [34], he would discover that his answer was *not correct.* No one should feel too bad about that: additional help is always provided elsewhere as indicated. The correct answer is, of course, given in frame [18].

Whenever you feel rather uncertain about how to answer a particular question and would appreciate having some further explanation on it, you are advised to choose a frame such as 54 . It is always wiser to follow this route than simply to guess at an answer.

During the coming weeks you will begin to realize that learning by means of programmed instruction is not only much easier and more enjoyable than conventional textbook learning, but also that you can progress through the material more rapidly. However, you should not look upon programmed instruction as a miracle method to attain your goal without any real work on your part being necessary! You will certainly *not* be able to learn chemistry "in your sleep" by this means. It is only by conscientiously working your way through each and every Program that success can be achieved.

Below we give some tips which should help to speed you on your way:

1. When commencing work on a Program you should be in a relaxed state and have sufficient time to complete the Program. It is best to work through a Program in *one* sitting if possible and then to re-work it again several times over the next few days. Of course, if you do have to interrupt working your way through a Program, simply jot down the number of the last frame you studied.

2. The individual frames are intended to be *worked* through and not merely *read* through. Do not leave a frame until you are quite certain you have grasped its content. A Program is not designed in such a way that you can work backwards through it again! Check all your answers very carefully against those given in the next frame and correct any errors you may have made. Do not get upset if you make mistakes the first time you work through a Program. However, it would be a great pity if you did not learn from the mistakes you make!

3. Absolute *honesty* is called for at those points where you are asked to select one out of several possible answers to a question. We recognize, of course, that it would be only too easy to cheat here. Our advice, however, is always to make a serious attempt to arrive at the correct answer. Cheating will certainly *not bring* the kind of *success* you could otherwise expect. From time to time you will have the opportunity of choosing as your answer some such phrase as "I do not understand the question" or, simply, "I do not know". Never hesitate to select this answer if you genuinely feel uncertain. You will then be guided to a frame where additional help is available.

4. The continual checking of your progress and the choices you have to make in answering questions serve to introduce a certain excitement into the learning process. The learning process is thereby transformed into something resembling the playing of a game. This effect is an intended one and is highly desirable, even though there is a potential danger that students may thereby be encouraged to work through Programs too fast and perhaps even start guessing at answers. Be on your guard against this possible disadvantage! You should always endeavor to work slowly, to remain calm and to maximize your concentration.

Success in learning requires *active participation* on the part of the student, *immediate checking* of all answers and progression through the material at an *individually paced speed.* Programmed instruction satisfies all of these requirements.

Program 1

Introduction

1 Chemistry is one of several natural sciences. Other sciences which fall within this general category include botany, zoology, mineralogy and physics. As the fields of chemistry and physics are rather closely inter-related, they have a number of features in common. However, we wish to make clear from the outset that a fundamental distinction exists between chemical and physical processes. Let's try to explain this distinction by considering first the behavior of water.

Water, as we all know, exists not only as liquid water but also as ice and steam. It can thus occur as a solid, a liquid or a gas.

This could be expressed rather more scientifically by saying that water occurs in three different *states*, namely the solid, the liquid or the gaseous state.

Now try to answer the following questions:

What is water called in the solid state?
What is water called in the gaseous state?

Using a separate sheet of paper write your answer down as a complete sentence, and then check it with that given in [8] .

2 You have quite correctly appreciated that the second experiment involves a physical process (the freezing of water to form ice), for only a change of state occurs here.

However, the first experiment (the evaporation of water) does *not* involve a chemical process but rather a physical one. Steam is merely another *form* of water, for, upon cooling down, steam is converted back into liquid *water*.

Continue reading now at [7] .

3 You are neither concentrating nor working carefully enough. Your answer reveals that you have not fully absorbed the material covered this far. You are therefore advised to start again at [1] . Do not allow yourself to be distracted, and do not proceed with undue haste, but rather work your way steadily through the program.

4 You have correctly appreciated that the substance glass is not changed when two wine glasses are tapped together to produce a ringing sound. However, the second experiment also involves a physical process, for when the glass is smashed by a powerful blow the glass itself remains unchanged. Even glass fragments or finely ground glass powder are after all still glass. There is *no* actual change of substance here and so the process involved (Experiment 2) *cannot* be a chemical one.

Continue your reading at 13 .

5 You have quite rightly appreciated that the light gray powder is *not* a chemical compound.

Grinding, however, is a *physical* process, as in both the crushing and the mixing operations the substances themselves *remain unchanged* and only their state changes.

It's a pity that you are not yet able to decide with certainty whether a given process is a chemical or a physical one. It would be advisable to repeat this program as soon as possible!

For the present, however, go on to 29 .

6 Sulfur dioxide is a chemical compound which is formed from the combination of *sulfur* and *oxygen.*

The *chemical compound* sulfur dioxide is produced by a *chemical process.* A conflagration results when sulfur and oxygen combine together. Conflagrations or the evolution of heat frequently accompany the formation of chemical compounds.

This brings us now to an important new concept: the *chemical compound.*

If two substances combine together in a chemical process, there results a ....

Jot down the missing words and then go on to 35 .

7 Your answer is quite correct.

The freezing and evaporation of water are indeed *physical* processes as only *changes of state* occur.

On the other hand, *chemical* processes are those in which the substances involved .....

Write out the full definition and then check your answer with that in 17 .

8 Water in the solid state is *ice* whereas water in the gaseous state is *steam.*

However, it is not only water which occurs in these three *states* (solid, liquid and gaseous).

Iron may be melted down, i.e. converted from the solid to the liquid state.

Gasoline becomes vaporized in the carburettor of an automobile. This represents conversion from the .... state to the ...... state.

Jot down the missing words and check your answer with that in 16 .

9 You have obviously not read the description of the experiment very carefully. Otherwise, you would have known that the mixture was not heated to glowing in the flame but that the glowing started quite suddenly by itself after the initial warming of the mixture. In fact, the glowing became more intense after the flame had been removed. Thus, the great amount of heat generated here was produced within the mixture itself.

The resulting gray mass is clearly one single substance. For this reason it is not possible to recognize any of the original iron or sulfur under a microscope. Moreover, one is not able to remove any free iron from the mass with a magnet. Any attempt to extract sulfur from the gray mass by means of carbon disulfide would also be doomed to failure.

Repeat now from 38 .

10

1. The melting of sugar is a physical process. The change of state is from solid to liquid.

2. Strong heating causes the sugar to burn and form charcoal. This involves a change (transformation) of substance and therefore represents a chemical process.

3. The rusting of iron produces a substance which has quite different properties from those of the original iron. A change (transformation) of substance and therefore a chemical process is involved.

4. a) The melting of wax involves only a change of state and is thus a physical process.

b) The burning of wax involves a change (transformation) of substance and is thus a chemical process.

Consider examples of other changes and processes encountered in everyday life and decide whether they are chemical or physical in nature.

Continue your reading at 19 .

---

11 It's just a well that you didn't guess here but that you honestly admitted that certain points remain unclear to you.

The dissolving of sugar in water is a physical process.

The sugar undergoes no change (or transformation) of substance in this process, which implies that no new substance having completely different properties is formed. By means of another simple physical process (evaporation of the water) one can readily recover the original sugar.

Compare these observations with the definition we gave for the formation of a chemical compound, and then go back to 30 .

---

12 If two or more substances combine together to form a *new substance having completely different properties*, the new substance is said to be a *chemical compound*. (The original substances, incidentally, may themselves be chemical compounds in that they can be formed from yet other substances.)

To clarify matters let's suppose that we perform the following experiment:

4 g (grams) of sulfur and 7 g of iron filings are ground up together in a mortar, i.e. crushed and mixed. A light gray powder results from the admixture of the yellow sulfur and the gray iron filings. If one now examines this mixture through a microscope one can recognize *unchanged* particles of sulfur and iron adjacent to one another.

Which of the following statements is correct?

- Grinding is a physical process;
  the light gray powder is a chemical compound → 43

- Grinding is a physical process;
  the light gray powder is *not* a chemical compound → 29

- Grinding is a chemical process;
  the light gray powder is a chemical compound → [36]

- Grinding is a chemical process;
  the light gray powder is *not* a chemical compound → [5]

---

[13] Yes, two physical processes are involved. When the glasses are tapped together to produce the ringing sound, the substance is not changed in any way. Even if we break the glass, the substance will not be changed, for glass fragments or finely ground glass powder still remain glass. There is clearly no actual change (transformation) of substance here.

Let's consider yet a third example:

Sulfur is a yellow substance which can easily be reduced to powder owing to its brittle nature.

Experiment 1. Some sulfur is carefully heated in a porcelain spoon. After a short while the sulfur melts.

Experiment 2. The same sulfur is heated more strongly. The molten sulfur begins to burn with a blue flame. Pungent gases are produced as a result of the combustion process. After some time has elapsed all the sulfur will have disappeared. Upon cooling the gases produced during the combustion, no sulfur is deposited.

What types of process are involved in these two experiments?

- Two physical processes → [39]
- Two chemical processes → [21]
- The first experiment involves a physical process whereas the second involves a chemical one → [27]
- The first experiment involves a chemical process whereas the second involves a physical one → [3]

---

[14] Correct! The evaporation or freezing of water changes *only* its state.

Proceed now to [15] .

---

15 Physical processes are those in which the substance remains unchanged and only a change in its state and associated properties occurs.

Processes in which the substances themselves are changed (transformed) are chemical processes (coal → ash and smoke).

Definition: Chemical processes are those in which the substances involved are *changed* (transformed) *into different substances*.

Be sure to take good note of this definition too!

All *combustions* are examples of chemical processes. Think, for instance, of the burning not only of coal, but also of coal gas, wax candles, cigarettes, matches, etc.

Suppose now the following two experiments are performed:

1. Some water is heated until it boils. At 100 °C the boiling actually starts and steam is given off. After a while all the water has evaporated without a trace.

2. Some water is cooled down. At 0 °C it freezes and turns into ice.

What types of process are involved in these two experiments?

- Two chemical processes → 25
- Two physical processes → 7
- The first experiment involves a chemical process whereas the second involves a physical one → 2
- The first experiment involves a physical process whereas the second involves a chemical one → 31

16 Gasoline becomes vaporized in the carburettor of a car. This represents conversion from the *liquid* to the *gaseous* state.

In such a change of state the *substance* itself *remains unchanged*, as it is *merely* converted from one state to another. A change of state can readily be *reversed*.

Example: If some water in a beaker is boiled, steam will be given off (conversion from liquid to gas). After a short while, drops of water are formed on the cold lid of the beaker as the steam cools down (conversion from gas to liquid).

Using this example as your starting point, describe the sequence of changes of state which occur when water is converted from liquid → solid → liquid. Check your answer with that in 24 .

---

17 Chemical processes are those in which the substances involved are *changed* or (more accurately) *transformed into different substances.*

If you managed to complete this definition correctly you are coming along very nicely! If, on the other hand, you had difficulties or made mistakes, please now make sure that you commit this important concept to memory.

Write down the definition of a physical change and check it with that given in 26 .

---

18 Your answer is wrong! The dissolving of sugar in water is a *physical* process as the water contains the sugar in dissolved form. There is *no* actual *change (transformation) of substance* for either the sugar or the water. Even when in solution the sugar may be recognized from its taste.

The important concept of the "chemical compound" is obviously not yet fully clear to you. You should therefore repeat from 35 .

---

19 Do you still remember the chemical process involving the combustion of sulfur? You might like to try out the experiment yourself. If so, first buy a stick of sulfur from a drugstore for a few cents. This can be lit with a match, though care must be taken to carry out the experiment on a non-combustible base, e.g. on a porcelain plate. The sulfur burns away with a blue flame and a pungent gas is given off.

In the process the sulfur combines with the oxygen in the air to form a new substance, sulfur dioxide, which makes its appearance as a pungent gas. Sulfur and oxygen in the sulfur dioxide are now no longer free but are present in combined form. In fact, we could say that a chemical compound has been formed.

Jot down the missing words in the following sentence:

Sulfur dioxide is a chemical compound which is formed by the combination of .... and ....

Check your answer with that in 6 .

---

20 It's just as well that you didn't guess here but that you are instead seeking further clarification.

In the description of the experiment two things must surely have struck you, namely:

1. The glowing of the mass became more intense although the flame had actually been removed.
   This fact clearly points to a chemical process. Chemical processes often involve the liberation of heat or even a conflagration.

2. The gray mass which was formed had quite different properties from those of the original mixture.

Go back now to 38 and make a careful comparison of the properties of the gray mass with those of the original mixture.

---

21 You have correctly appreciated that the *combustion* of sulfur is a *chemical* process.

However, the melting of sulfur involves only a *change of state* (from solid to liquid) and not a transformation of substance. The melting of sulfur is thus a *physical* process.

Continue your reading at 37 .

---

22 *Sulfur* + *Oxygen* → Sulfur dioxide

Sulfur dioxide is a chemical compound which is formed by the combination of sulfur and oxygen.

What is meant by a chemical compound?

If two or more substances combine together to form a ...., the new substance is said to be ....

Write out the definition in full and check with that in 12 .

---

23 You doubtless came up with the right answer here: The coal itself does *not* remain unchanged but is in fact completely transformed. Accordingly, this change *cannot* possibly be physical in nature. How, then, did the definition of a physical change read?

Physical processes are ....
Complete this sentence and check your answer with that in 15 .

---

24 When water is cooled below 0 °C ice is formed (conversion from liquid to solid). If this ice is warmed up water is again formed (conversion from solid to liquid).

Changes of state are *physical processes* and have nothing to do with chemistry.

Other properties which may undergo change during a *physical process* include the heat content (hot/cold), the magnetism (magnetic/non-magnetic), the particle size (large/medium/small) and the state of solution (dissolved/undissolved).

Note that in all of the changes mentioned here the *substance* itself *remains unchanged*. This implies that if one were to reproduce the original conditions (of temperature, etc.) the substance would again assume its original state. This is actually a very important criterion for determining which processes are physical.

Now complete the sentence: Physical processes are those in which the substance remains unchanged and only its ...... changes.

Proceed to 32 .

---

25 Your answer is not correct.

Chemical processes are those in which the substances involved are changed (transformed) into *different* substances.

Consider for a moment the following question: Is water changed (transformed) into a *different* substance when it is either evaporated or frozen?

- Yes → 33
- No → 14

---

26 Physical processes are those in which the substance remains unchanged and only a change of its state and associated properties occurs.

Let's take a look at a second example.

Experiment 1: Two wine glasses are gently tapped together to produce a ringing sound.

Experiment 2: The two glasses are tapped together so violently that the glass is shattered. The broken glass fragments are ground down in a mortar until a powder is obtained.

What types of process are involved in these two experiments?

- Two physical processes → 13
- The first involves a physical process whereas the second involves a chemical one → 4
- Two chemical processes → 41
- The first involves a chemical process whereas the second involves a physical one → 34

---

27 The melting of sulfur is a physical process whereas the combustion of sulfur is a chemical process.

Your answer reveals that you have fully understood the material covered so far. For this reason you may now jump over to 19 .

Alternatively, if you still have nagging doubts about certain points, you could test yourself at 37 .

---

28 Both questions should have been answered with a "no", for neither the tapping nor the grinding of the glass can change it into a *different* substance.

Go back now to 26 and answer the question there again.

---

29 Quite correct! The grinding up of iron filings and sulfur powder is a physical process. The substances remain unchanged as only their state of subdivision changes.

The mixture of iron and sulfur particles, i.e. the light gray powder, is not a chemical compound. *No new substance* having different properties is formed. We may convince ourselves of this by separating the mixture again.

The separation can easily be achieved by two quite different approaches:

1. As iron is attracted by a magnet whereas sulfur is not, the iron could be removed by a magnet. The sulfur would then be left behind.

2. As sulfur is soluble in a certain solvent (carbon disulfide) whereas iron is not, the sulfur could be dissolved out of the mixture by carbon disulfide. After filtering such a solution the iron would be left behind on the filter. The sulfur could then be recovered by evaporating the solution which had passed through.

Caution!! Carbon disulfide is highly flammable and very poisonous!

Both of these separation methods are .... (physical/chemical?) processes. → 38

---

30 Your answer is correct. The gray product is a new chemical compound. The strong glowing, which becomes even more intense after removal of the flame, indicates that a chemical process is taking place and that a new compound is being formed. The resulting compound of sulfur and iron is known as *iron sulfide*. It has quite different properties from those of a mixture of sulfur and iron.

In contrast to a mixture, a chemical compound *cannot* be split up into its original components by a purely *physical* process.

In a further experiment suppose some sugar is dissolved in water. The water takes on a sweet taste, though otherwise it is not immediately obvious that the water contains any dissolved sugar. If we were to evaporate the water, a layer of sugar would be left behind.

Is a new chemical compound formed when sugar is dissolved in water?

- Yes → 18
- No → 42
- I don't know → 11

---

31 You have correctly appreciated that the first experiment (evaporation of water) involves a physical process, for evaporation entails merely a change of state.

The second experiment, however, involves *not* a chemical but rather a physical process, as the freezing of water to form ice also entails only a change of state! There is *no* actual change (transformation) of substance occurring here. Moreover, provided that sufficient heat is applied to the ice, it will melt and form *water* again.

Continue your reading at 7 .

---

32 Physical processes are those in which the substance remains unchanged and only its *state* changes.

Now let's reformulate this important notion in order to arrive at a *definition*. (Definitions are given to to make it easier for you to give specific answers to questions on important concepts.)

Definition: *Physical* processes are those in which the *substance* remains unchanged and only a *change of its state and associated properties occurs.*

Please make sure that you memorize this definition!

In addition to providing useful heat, the burning of coal produces ash and smoke. What do you suppose is actually happening here? Does the coal remain unchanged and is only a change of state involved?

The answer is given in [23], but you should consider this question very carefully before checking.

---

33 No! When water evaporates it is *not* changed (transformed) into a *different* substance but *only* its *state* is changed. The steam produced may be converted back into liquid water again merely by cooling it down.

Similarly, when water freezes and forms ice it is *not* changed (transformed) into a different substance but *only* its *state* is changed. Provided that sufficient heat is applied, the ice will melt and turn into water again.

Repeat from [16].

---

34 You have quite correctly appreciated that when glass is smashed by a powerful blow, the substance glass itself is not changed (transformed) in any way.

However, the first experiment also involves a physical process, for when wine glasses are tapped together to produce a ringing sound, the substance glass is *not* changed at all. Because there is *no* change (transformation) of substance, the process *cannot* be a chemical one!

Continue your reading at [13].

---

35 *Chemical compound*

For this most important concept we shall now formulate a general definition which should be easy to remember.

Definition: If two or more substances combine together to form a new substance having different properties, the new substance is said to be a *chemical compound.*

For example, rust is a chemical compound which is formed by the combination of iron and oxygen. It has quite different properties from those of the starting material iron.

Iron + Oxygen → Rust

How is the compound sulfur dioxide produced?

.... + ... → Sulfur dioxide

Complete the line above and check your answer at 22.

---

36 No! Grinding a substance, i.e. crushing and mixing it, produces only a change in its state (from large lumps to fine powder).

As long as sulfur and iron particles remain in the mixture in *unchanged* form, one cannot speak of a change (transformation) of substance.

Return to 12 and re-think your answer!

---

37 For each of the following experiments decide whether a chemical or a physical process is involved:

1. Some sugar is carefully heated up in a beaker. A clear, viscous, molten mass of liquid is formed.

2. The molten sugar is heated more strongly. It becomes dark, vapors are given off, and finally all that remains is a black cinder.

3. Some iron is allowed to rust away.

4. A wax candle is lit and in the burning process one can observe
   (a) melting of the wax by the heat of the flame;
   (b) burning of the wax at the wick of the candle.

Jot down the type of process involved in each of the experiments (1, 2, 3, 4a and 4b) and check your answers at 10.

---

38 Both methods of separation are *physical* processes.

(If you are still not too clear on this, read the supplementary material presented in 40 . An experiment is outlined there that you might like to try out yourself.)

Let's now suppose that a mixture of 4 g of sulfur and 7 g of iron filings is carefully heated in a test tube. Only a small flame is used and the heat is applied to one *specific* spot in the mixture. Before the whole of the mixture in the tube has been heated up, a glow begins to develop at the spot which is being heated. Even if the flame is now *entirely* removed from this spot, the glowing continues and in fact becomes ever more intense until it envelops all of the mixture. Upon cooling, the test tube is found to contain a uniform gray mass. Under a microscope one is no longer able to detect any of the original sulfur or iron particles. Moreover, it is not possible to extract from the gray mass either iron (with a magnet) or sulfur (with carbon disulfide).

What does the gray mass represent?

- A mixture of sulfur and iron which has been heated to glowing in a flame → 9
- A new chemical compound → 30
- I don't know → 20

---

39 You have quite correctly appreciated that the *melting* of sulfur is a *physical* process, as only a change of state occurs (from solid to liquid).

During the combustion of the sulfur, however, a pungent gas is produced. This gas is *not* vaporized sulfur, since the sulfur has been changed (transformed) into a new substance as a result of the combustion. (No sulfur would be deposited upon cooling down the combustion gases.) Accordingly, the *combustion* of sulfur is a *chemical* process.

Continue your reading at 37 .

---

40 Although both the first method of separation (removal of iron with a magnet) and the second method of separation (dissolving in carbon disulfide and then filtering) involve physical processes, it is perhaps rather more self-evident for the first method.

There is clearly no change (transformation) of substance when iron is extracted by means of a magnet and the sulfur is similarly unchanged after its separation.

Upon treating the mixture with carbon disulfide, the sulfur dissolves completely, just as salt does in water. There is, however, *no* change (transformation) of substance taking place. The proof of this may be seen in the fact that after the carbon disulfide has been evaporated off, the sulfur is left behind in an unchanged state.

If you find this procedure difficult to visualize, try carrying out the following experiment yourself:

Experiment	*Analogy*
1. Mix together some sand and salt.	Sulfur and iron are ground up together.
2. Add some water and stir (the salt dissolves but not the sand).	The sulfur is dissolved in carbon disulfide whereas the iron will not dissolve.
3. Now filter this using a coffee filter. The sand remains on the filter.	Upon filtering the iron remains behind on the filter.
4. Evaporate the filtered solution in a flat dish. The salt is recovered as the residue – taste it if you wish!	Carbon disulfide is then evaporated off from the filtered solution. The sulfur is recovered as a residue.

Continue your reading at 38 .

---

41 No! There are *not* two chemical processes involved here.

Consider for a moment the following questions:

1. Is glass changed (transformed) into a *different* substance when it is tapped and a ringing sound results?.......(Yes/No).

2. Is glass changed (transformed) into a *different* substance when it is ground up?.......(Yes/No).

Jot down your answers and then check them at 28 .

42 Your answer is quite correct.

The dissolving of sugar in water involves a physical process as no new chemical compound is formed.

The separation of the two substances, sugar and water, from one another is straightforward:
We either evaporate or boil off the water and the sugar is left behind. The steam given off in the process can be cooled down in a suitable apparatus and thereby changed back into water (condensed).

There are certain cases, however, where the distinction between a physical and a chemical process, or between a chemical compound and a mixture, is not quite so clearcut as in the examples we have cited here. Sometimes a great deal of apparatus is necessary and very sophisticated procedures have to be adopted to make the distinction. In all cases though, it is very important that each experiment to be carried out is properly prepared, performed with the utmost conscienciousness, and that all events occurring during the course of an experiment are carefully and accurately observed. Even the most insignificant observations can turn out to be of the greatest importance in chemical experiments!

To sum up then:

You have learnt and hopefully retained (!) three new definitions. By applying these definitions you should be able to distinguish between simple *physical* and *chemical processes* and decide whether or not a *chemical compound* is formed.

Mixtures and solutions may be separated by means of *physical processes* whereas *chemical compounds cannot* be separated in this way.

End of Program 1.

43 You have quite correctly understood that grinding is a physical process.

However, the light gray powder, consisting of a mixture of finely ground iron and sulfur particles, is *not* a chemical compound.

The reasons for this are as follows:

1. Chemical compounds are formed only in chemical processes and never in physical processes.

2. Although the iron and the sulfur are in close contact, they are not combined together. One is still able to identify yellow sulfur and gray iron particles (with a microscope if necessary).

Go on now to [29] .

Program 2

The Air

1 We all know that many substances burn, e.g. gasoline, coal, wood and paper. Phosphorus too is flammable.

Let's imagine now that the following two experiments are performed:

Experiment 1. A piece of phosphorus is put into a small dish floating on a water surface. The dish is covered by a gas jar as indicated in Figure 1. Quite spontaneously the phosphorus soon begins to smoulder and then burn. At the same time the water level in the gas jar is observed to rise. When the conflagration ceases – which occurs before all of the phosphorus has been burnt up – the water stops rising. (The conflagration will cease only if sufficient phosphorus has been used.) In addition to the unused phosphorus a white substance is now present in the dish. Approximately *one fifth* (1/5) of the original quantity of air in the gas jar is *missing.*

(Please do not perform this experiment yourself. Experiments with phosphorus are very dangerous!)

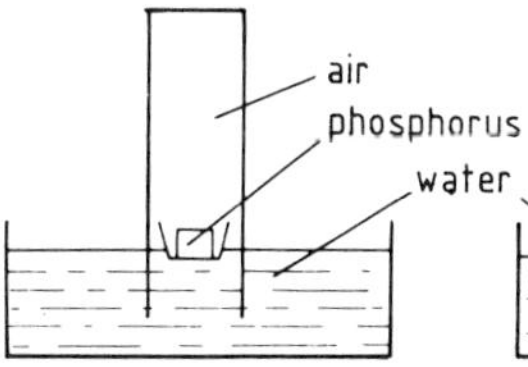

Fig. 1.

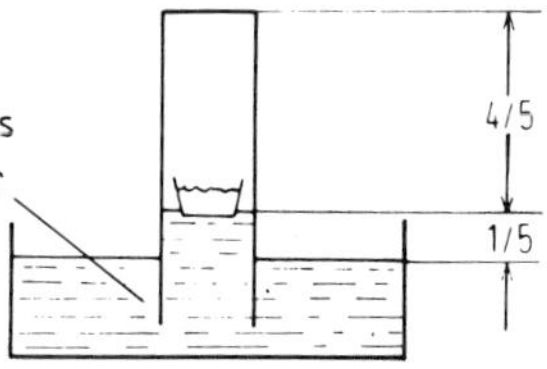

Fig. 2.

Experiment 2. Small pieces of copper wire are placed inside a glass tube and heated in a flame until they glow red hot. One liter of air (= 1000 cm^3) is blown in at one end of the tube. The gas which passes out at the other end of the tube is collected and it is found that it represents only about 800 cm^3. In other words, around 200 cm^3 is now missing.

How much of the air blown in is missing?

- About 1/3 → 8
- About 1/5 → 16
- I don't know → 24

2 Your answer is wrong. Just think again about what happens in the combustion of phosphorus: phosphorus combines with oxygen which it takes from the air. Something is thus added to the phosphorus, namely that amount of oxygen which is used up during the combustion process.

You should read 10 again.

3 Your answer is wrong. Air is not a uniform substance but rather a mixture of two main gases. *One* of these is involved in combustion and is thereby *used up* in the process. The *other* is *not* involved in combustion and so is *not used up*.

Read through 1 again.

4 Your answer is wrong. The oxidation does involve a *chemical reaction*, though the nitrogen does *not* participate. In fact, every sort of combustion or oxidation will cease in the presence of nitrogen. For instance, a burning candle would be extinguished by nitrogen.

Read 19 once again.

5 Your answer is quite correct.

The color of litmus changes from *blue* to *red* when the solution is acidified. This may easily be remembered with the aid of the mnemonic:

Acid makes Red

as both words end with the letter d.

By way of revision we return now to the subject of air and pose the question:

What is air?

- A mixture of two main gases → 13
- A uniform gas → 30
- A chemical compound of two gases → 21

6 Your answer is wrong. Please note carefully:

In any combustion, e.g. that of phosphorus, the air is reduced in volume by about 1/5. The remaining 4/5 will not support combustion. In fact, all combustion processes cease in this 4/5 of the air.

Jot down now the missing words in the sentence below:

Air consists of about 1/5 of .... and about 4/5 of ....

Continue your reading at 15 .

7 You have correctly appreciated that in the combustion only a *part* of the air is used up, though the actual amount you have given is wrong.

Go back and read 1 again.

8 Your answer is wrong. 1/3 of 1000 cm^3 is 333 cm^3.

In actual fact only about 200 cm^3 are missing! Recalculate your answer and then return to 1 .

9 Nitrogen *evaporates first* from liquid air and then the oxygen.

Let's study the liquefaction and evaporation of air in somewhat greater detail. When air is cooled down from +20 °C through 0 °C, −10 °C, −50 °C, −100 °C, −150 °C and so on, it becomes liquid at around −200 °C. If the light blue colored liquid air is now slowly warmed up from −200 °C to about −195 °C, it will begin to boil. The nitrogen (b.p. −196 °C) then gradually evaporates off. Thereafter we may suppose that the temperature is increased from −195 °C through −194 °C, −193 °C, etc. to reach −183 °C. In the process the color of the liquid air changes from light to deep blue, as the proportion of the colorless nitrogen is reduced and that of the deep blue colored oxygen is correspondingly increased. At −183 °C only liquid oxygen still remains, though this too will eventually completely evaporate at this temperature.

Write down now:

1. The color of liquid air.
2. The color and boiling point of liquid nitrogen.
3. The color and boiling point of liquid oxygen. → 22

10 Your answer is quite right.

A *chemical reaction* in which a substance *combines* with oxygen is called an oxidation.

If phosphorus is burnt it combines with the oxygen in the air. In this oxidation a chemical compound is formed consisting of phosphorus and oxygen, i.e. a phosphorus oxide. It is a solid white substance.

Is the phosphorus oxide which forms lighter or heavier than the original phosphorus?

- It is lighter → 34
- It is heavier → 26
- It is equally heavy → 2
- I don't know → 18

11 Your answer is wrong. Just think again for a moment about the difference between a chemical and a physical process. In a chemical process there is a transformation of substance whereas in a physical process there is only a change of state, e.g. a transition from liquid to solid, in the materials present.

The liquefaction of air involves only a change of state and this is therefore a .... process.

Jot down the missing word and then go to 31 .

12 Only *oxygen* can support combustion whereas *nitrogen* cannot.

In combustion processes oxides are formed.

Example:

Phosphorus + Oxygen produce a ......

Complete the sentence and check your answer at 29 .

13 Your answer is quite right.

Air is a *mixture* consisting of two main gases.
What are these two gases called?

Air is a mixture consisting mainly of .... and ....

Jot down the two names and then check them at 25.

---

14 The blue color of litmus dye turns to *red* in the presence of acids.

Now write out answers to the following questions:

1. To what temperature must air be cooled to liquefy it?

2. What is the color of liquid air?

3. Which component will evaporate first when liquid air is warmed up?

4. Which two gases are produced industrially in great quantities from liquid air? → 32

---

15 Your answer is quite correct.

Air is a mixture consisting of about 20% (1/5) oxygen and about 80% (4/5) nitrogen. Air is a colorless and odorless gas. One can also liquefy air.

In order to visualize the liquefaction of air more easily, consider the behavior of steam. When cooled down steam is deposited in the form of water, e.g. on the inner side of the lid on a beaker.

Air too becomes liquid when it is cooled down, though the air must be very intensely cooled – to about −200 °C (read as minus 200 degrees Celsius).

Freshly condensed liquid air is a bluish liquid mixture containing approximately .... times as much nitrogen as oxygen.

Jot down the missing word and then go to 23.

---

16 Your answer is quite right.

Approximately 1/5 of the air blown in is missing; that is to say, just as much as was missing in the first experiment!

If a burning candle is put into the air left over at the end of this experiment (about 1/5 of the original amount), it will be immediately extinguished.

We may therefore draw the following conclusions:

Air contains two quite different kinds of gas. One of these gases is involved in combustion and is used up in the combustion process. This gas makes up about 1/5 of the total in air. The other gas (forming about 4/5 of the original air) is not involved in combustion processes. On the contrary, it will actually extinguish any flame present.

What proportion of air gets used up in a combustion process?

- The air is completely used up. → 3
- About 1/5 of the air is used up. → 27
- About 4/5 of the air is used up. → 7

---

17 You have unfortunately mixed up the colors.

When *acids* are added, litmus changes its color from .... to ....

Jot down the missing words and then continue reading at 5 .

---

18 Think again about what happens when phosphorus burns: phosphorus combines with oxygen which it takes from the air. Something is thus added to the phosphorus, namely that quantity of oxygen removed from the air during the combustion process.

Read 10 again.

---

19 Compare the following facts:

Air consists of around 1/5 (or 20%) *oxygen* and around 4/5 (or 80%) *nitrogen.*

Oxygen supports combustion whereas nitrogen will extinguish any flame. It is thus not the air as such which supports combustion, but rather the oxygen contained in it. In any combustion process the oxygen will combine with the combustible material (e.g. with coal or phosphorus) and so will be removed from the air. Such a process is a type of *chemical reaction* known as an *oxidation.*

What then is an oxidation?

- A chemical reaction (process) in which a substance combines with oxygen → 10
- A chemical reaction (process) in which a substance combines with nitrogen → 4
- Every chemical reaction is an oxidation → 28

---

20 Liquid air has a temperature of around −200 °C.

When liquid air is warmed up it starts to evaporate. In the process the liquid nitrogen evaporates both sooner and faster than the liquid oxygen. The liquid air thus becomes increasingly impoverished in nitrogen, until finally only liquid oxygen remains. The remaining oxygen will then in turn also evaporate.

Write down the sequence in which the components of liquid air evaporate when it is warmed up.

→ 9

---

21 Your answer reveals that you have not fully understood the material covered this far.

Air is a mixture of two main gases, and is *not* a chemical compound. Although in a mixture the substances are in close contact, they are not changed in any way. Think, for instance, of the mixture of sulfur and iron, discussed earlier on.

Not only solid substances can be mixed, but also liquids or gases. Air is a .... of different gases.

Jot down the missing word and then continue reading at 13 .

---

22 Now check your answers:

1. Light blue
2. Colorless, −196 °C
3. Deep blue, −183 °C

As nitrogen evaporates from liquid air at a lower temperature than oxygen, the two substances may be collected separately. In fact, each can be obtained in a fairly pure state. (Exactly how pure will depend, of course, on the quality of the technical equipment.

In industry *large amounts* of oxygen and nitrogen are produced in this way from air. These gases are then filled under high pressure into steel bottles. Small amounts of gas may be drawn off from such a bottle as required. (Note that oxygen and nitrogen are *not* prepared from liquid air in the laboratory. The necessary apparatus would be too complicated and unwieldy.)

To summarize briefly, we now know that:

Air is a colorless, odorless gas; it is a mixture consisting of:

about 1/5 (20%) oxygen and

about 4/5 (80%) nitrogen.

Only .... can support combustion whereas .... cannot.

Write down the completed sentence and check your version with that given in [12] .

---

[23] Yes, *liquid* air contains approximately *four* times more nitrogen than oxygen, just as gaseous air does.

Is liquefaction of air a chemical or a physical process?

- A chemical process → [11]
- A physical process → [31]

---

[24] About 200 cm^3 are missing. What fraction of 1,000 cm^3 does this amount represent?

100 cm^3 represent one tenth (1/10) of 1,000 cm^3
250 cm^3 represent one quarter (1/4) of 1,000 cm^3
500 cm^3 represent one half (1/2) of 1,000 cm^3
200 cm^3 represent one .... (1/.....) of 1,000 cm^3

Supply the missing fraction and check your answer at [16] .

---

[25] Air is a mixture consisting mainly of *oxygen* and *nitrogen*.

Have you remembered the proportions of these two gases in air?

- About 2/5 oxygen and about 3/5 nitrogen → 6
- About 4/5 oxygen and about 1/5 nitrogen → 33
- About 1/5 oxygen and about 4/5 nitrogen → 15

---

26 Your answer is quite correct.

The product, i.e. the resulting substance, of the oxidation (phosphorus oxide) will be *heavier* than the original phosphorus because the latter has taken up oxygen.

If we now try to dissolve the combustion product of phosphorus, i.e. the phosphorus oxide, in water, we find that it dissolves readily. Moreover, water having phosphorus oxide dissolved in it is acidic, because an acid is formed in the process.

To test whether water having phosphorus oxide dissolved in it really is acidic, one should avoid using the tongue as this can be dangerous. One may use instead certain dyes which change their color in the presence of an acid. One example of such a dye is *litmus*; this has the property of changing from *blue* to *red* when acids are added to it.

If the combustion product of phosphorus, i.e. phosphorus oxide, is dissolved in water containing litmus, the color of this dye will change. What color change will occur?

- A change from blue to red → 5
- A change from red to blue → 17

---

27 Your answer is quite correct.

In the combustion process about 1/5 of the air is used up.

The gas used up in the combustion we call *oxygen*.

The gas not involved in any way in the combustion process we call *nitrogen*.

Air is a mixture of two gases and consists of about 1/5 (or 20%) .... and about 4/5 (or 80%) ....

Jot down the missing words. → 19

---

28 No, an oxidation is only a *certain type* of chemical reaction, namely a chemical reaction in which a substance combines with ....

Jot down the missing word. → 10

29 Phosphorus + Oxygen produce a *Phosphorus Oxide.*

If this phosphorus oxide is dissolved in water, an *acidic* solution is obtained. Blue litmus dye added to such a solution would change its color to ....

Jot down the color. → 14

30 Your answer reveals that you have not fully understood the material covered this far.

Read 1 again and pay particular attention to the experiments described there. It should then be clear to you that air is *not* a uniform gas.

31 Your answer is quite right.

The liquefaction of air involves only a change of state; it is thus a *physical* process.

If liquid air is poured on to a table, it evaporates in just the same way as a drop of water does from a hotplate. The table is some 200 °C warmer than the liquid air, as the latter has a temperature of about −.......... °C.

Jot down the temperature of liquid air, and then go on to 20 .

32 The correct answers are:

1. *−200 °C*
2. *Light blue*
3. *Nitrogen*
4. *Nitrogen* and *oxygen*

Did you still make some mistakes when repeating this exercise? If so, you should in your own interest now work through the Program again.

End of Program 2.

33 Your answer is wrong. Do not forget that in the combustion process about 1/5 of the air gets used up. The remaining 4/5 *cannot* support combustion; in fact it will extinguish any combustion.

You have it seems mixed up the proportions of the two main constituents of air. Accordingly, you should note down:

The composition of the air: about 1/5 of ....

about 4/5 of ....

Continue your reading at 15 .

34 Your answer is wrong. Consider again what happens when phosphorus burns: phosphorus combines with oxygen which it takes from the air. Something is thus added to the phosphorus, namely that quantity of oxygen removed from the air during the combustion.

Read through 10 again.

Program 3

Oxygen

1 We have already learnt quite a lot about oxygen, as it was frequently mentioned in the last Program during our discussion on air. Let's briefly recapitulate here what was said:

Air is a mixture of two main gases and consists of about 20% oxygen and about 80% nitrogen.

Oxygen will support combustion. Oxygen can be liquefied: liquid oxygen boils at –183 °C and has a blue color.

Oxygen is produced industrially by warming up liquid air, whereupon first the .... boils off and then the .... follows. These two substances are collected *one after the other*; in this way the two main constituents of air may be separated.

Jot down the color of liquid oxygen, and the sequence in which the two main components of liquid air boil off when the latter is warmed up. → 8

---

2 No, water is a *compound!*

Repeat now from 28 .

---

3 Your answer is quite right.

Electrolysis is a *chemical* process in which compounds are decomposed by means of direct electric current.

Water can thus be shown to be a *compound* of oxygen and hydrogen.

On the contrary, air is a .... of oxygen and nitrogen.

Jot down the missing word. → 12

---

4 Your answer is wrong. The mixture will not get any lighter. Nothing gets lost; rather the two substances mixed together, sulfur and iron, form the compound iron sulfide (without anything having been added or removed).

Read through 37 again.

3

5 Your answer is wrong. Oxygen, hydrogen and iron are elements. Air, however, is a *mixture* consisting mainly of the elements oxygen and nitrogen.

Go back to 34 and answer the question there again.

---

6 Oxygen is a *gas* which may be liquefied to form a *blue* liquid.

Oxygen can be produced as follows:

1. By heating .... (in the presence of manganese dioxide as a catalyst).
2. By separating off the .... from liquid air.
3. By electrolyzing .... in Hofmann's apparatus for the decomposition of water.

Write out the three procedures referred to here. → 25

---

7 Your answer is wrong. A combustion is not a method for preparing oxygen. On the contrary, oxygen is *used up* in such a process.

Read through 46 again.

---

8 Liquid oxygen is *deep blue*. First the *nitrogen* (b.p. −196 °C) boils off, and then the *oxygen* (b.p. −183 °C).

If you got anything wrong here, you should very thorougly work through Program 2 again. It is recommended that you become really familiar with this material before attempting to learn something new.

A second way of producing oxygen consists of decomposing water by means of an electric current. For this purpose one uses *Hofmann's apparatus for the decomposition of water* as shown in the Figure below.

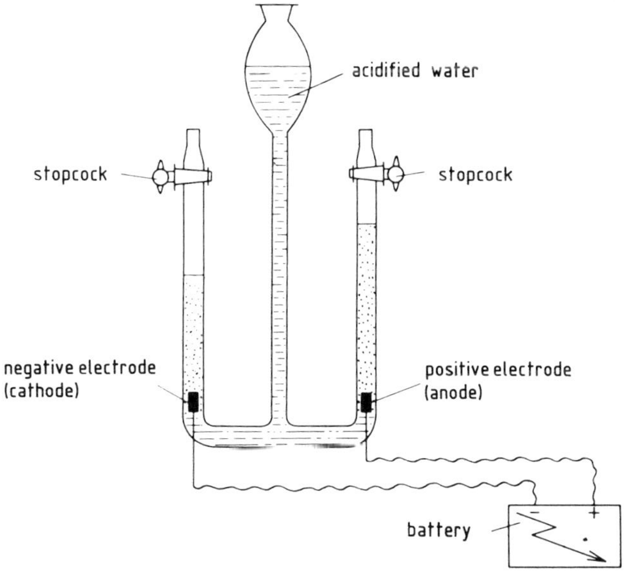

Make a sketch of this apparatus. Note that it consists of three glass tubes joined together at the bottom. The two side arms are closed off at the top with a .... and inside each of these at the bottom is mounted an .... Both of the electrodes are connected to the ....

Jot down the missing words, and then check them at 16 .

---

9 We have already mentioned that the production of oxygen by means of liquefying air and then separating off the nitrogen is unsuitable for the laboratory. This is because of the large and unwieldy apparatus required, which is suited only for industrial production.

Thus, your answer was wrong. → 30

---

10 Yes, a *compound* is decomposed by means of *chemical* processes.

How may water be decomposed into oxygen and hydrogen?

- By means of a physical process → 2
- By means of a chemical process → 18

---

11 *Physical* processes are those in which the substance remains unchanged and only a change of state and associated properties occurs.

In electrolysis, however, water is decomposed by means of direct current into two substances (oxygen and hydrogen) having quite different properties from those of water. The electrolysis thus brings about not a change of state, but rather a *transformation of substance.*

Processes in which the substances involved are transformed into different substances are *chemical* processes.

Electrolysis is therefore a .... process.

Jot down the missing word. → 3

12 *mixture*

Air is a *mixture* whereas water is a ....

Jot down the missing word. → 20

13 Your answer is quite right.

The mass will not change as no substance is added during the reaction. Nothing is taken away either.

The same principle can be observed in all chemical reactions. It is known as the *Law of Conservation of Mass* and it states that:

In any chemical process mass (i.e. substance) can be neither created nor destroyed.

Let's now briefly summarize what we have learnt so far.

Oxygen is a colorless, odorless and tasteless .... It supports combustion, but will not burn itself. One may detect oxygen by making use of the fact that a glowing taper thrust into it will flare up quite vigorously. Oxygen may be liquefied to form a .... colored liquid. The boiling point of liquid oxygen is −183 °C.

Jot down both of the missing words. → 6

14 Your answer is wrong. Pure oxygen does *not* act as a catalyst in any combustion process, as the oxygen becomes chemically bound to another substance and so gets used up in the process.

Combustion processes proceed somewhat slower in air because in air the oxygen is present only in dilute form, i.e. about 20% is oxygen. → 33

---

15 No, the row you have chosen contains only *compounds*. You should be looking here for *elements*.

Go back to 34 and consider again whether each of the substances listed there is an element or a compound. Remember that elements cannot be further decomposed by chemical processes whereas compounds can (because the latter are composed of elements).

---

16 *stopcock, electrode, battery*

In the apparatus for decomposing water, the electrodes used are in the form of small metal plates.

These are connected to the terminals of a battery (or to some other source of direct current), so that one plate becomes a *positive electrode* and the other plate a *negative electrode*. The positive electrode is referred to as the *anode* and the negative electrode as the *cathode*.

If direct current is passed through the acidified water in the apparatus, e.g. by connecting it up to a battery, a vigorous formation of gas is observed at each of the electrodes. The small gas bubbles formed rise up and collect beneath the stopcocks.

The gas which collects under one of the stopcocks (the one on the *right* in Fig. 3) is found to be pure oxygen whereas that beneath the other stopcock (on the *left* in our figure) turns out to be hydrogen (which will be discussed later).

Using these statements in conjunction with your sketch of the apparatus, complete the following sentences:

At the positive electrode (or ....) the gas formed is ....

At the negative electrode (or ....) the gas formed is .... → 24

---

17 Your answer is partially correct.

The electrolysis of water is also suitable for the preparation of oxygen in the *laboratory*. However, in industry the electrolysis of water is hardly ever used for the production of oxygen owing to the high costs of the electric current required. → 22

3

18 Your answer is quite correct.

Water is a compound. It can therefore be decomposed only by means of a *chemical* process, e.g. by electrolysis.

In the electrolysis of water in .... apparatus for the ...., oxygen is formed at the positive electrode (anode) and hydrogen at the negative electrode (cathode).

Jot down the missing words and check them at 26 .

---

19 Yes, mixtures may be separated by means of *physical* processes.
How may a compound be decomposed?

- By means of physical processes → 27
- By means of chemical processes → 10

---

20 *compound*

Air is a mixture of .... and ....

Water is a compound of .... and ....

Jot down the missing words. → 28

---

21 Your answer is partially correct.

The heating of potassium chlorate is also a suitable method for the preparation of oxygen in the *laboratory*. → 22

---

22 Your answer is quite correct.

The electrolysis of water (method 2) and the heating of potassium chlorate (method 3) produce small amounts of oxygen both rapidly and cheaply and are therefore suitable for the laboratory preparation of the gas.

These two methods, however, are not suitable for the industrial production of oxygen as they would be uneconomical.

The method of preparing oxygen from potassium chlorate has two notable disadvantages, namely that the oxygen comes off very unevenly and that the chlorate has to be heated very strongly. The reaction proceeds much faster if *manganese dioxide* is added to the potassium chlorate. Moreover, it proceeds more smoothly and the oxygen comes off sooner and at a lower temperature.

The manganese dioxide itself is not used up in this reaction and is present in the same state at the end of the reaction as at the beginning.

Substances which behave in this way and have the effect of speeding up a chemical process are called *catalysts*. The actual process involved is described as *catalysis*.

Now consider the question:

Is oxygen acting as a catalyst when a glowing taper thrust into it burns much faster than it would in air?

- Yes → [14]
- No → [33]

---

[23] Your answer is wrong. Phosphorus, hydrogen and sulfur are elements. But iron sulfide is a *compound!*

Phosphorus, hydrogen and sulfur cannot be further decomposed chemically because they are *elements*. Iron sulfide, however, because it is a *compound*, can be decomposed by means of .... processes. → [10]

---

[24] At the positive electrode (or *anode*) the gas formed is *oxygen*.
At the negative electrode (or *cathode*) the gas formed is *hydrogen*.

This type of decomposition of a substance by means of direct electric current is called *electrolysis*.

Is electrolysis a chemical process?

- Yes → [3]
- No, it is a physical process → [32]
- I don't know → [11]

---

[25] 1. By heating *postassium chlorate* (in the presence of manganese dioxide as a catalyst).

2. *By separating off the nitrogen* from liquid air.

3. By *electrolyzing water* in Hofmann's apparatus for the decomposition of water.

Method 2 is well suited for ....
Methods 1 and 3 are well suited for ....

Note down whether the methods referred to are well suited for the laboratory of the industrial production of oxygen. → 61

---

26 In *Hofmann's apparatus* for the *decomposition of water,* **oxygen** is formed at the positive electrode (anode) and **hydrogen** is formed at the negative electrode (cathode).

Oxygen and hydrogen *cannot* be further decomposed by means of any chemical processes. Such substances are called *chemical elements.*

Iron, sulfur and phosphorus also *cannot* be decomposed by means of chemical processes.

Iron, sulfur and phosphorus are therefore ....

Jot down the missing words and then go to 34 .

---

27 No, think for instance of the *compound* of iron and sulfur, iron sulfide! Neither iron nor sulfur can be separated from this compound by means of physical processes (e.g. by use of a magnet or by dissolving in carbon disulfide).

Compounds **cannot** be separated by means of physical processes.

Compounds can be decomposed only by means of .... processes.

Jot down the missing word. → 10

---

28 Air is a mixture of *oxygen* and *nitrogen.*
Water is a compound of *oxygen* and *hydrogen.*

The difference between a *mixture* and a *compound* is very important!

How can a mixture be separated?

- By means of physical processes → 19
- By means of chemical processes → 36
- I don't know → 44

---

29 Your answer is wrong. The *more* oxygen which is available to the flame in any combustion process, the *hotter* the flame will be.
Read through [62] again.

30 The following are methods for the production of oxygen:

1. Liquefaction of air; the nitrogen is separated off to leave behind *oxygen.*

2. Electrolysis of water in Hofmann's apparatus for the decomposition of water. At the positive electrode *oxygen* is formed and at the negative electrode hydrogen.

3. Heating of potassium chlorate; *oxygen* is evolved.

Which of these methods is suitable for use in the laboratory?

- Method 1 → [9]
- Method 2 → [21]
- Method 3 → [17]
- Methods 2 and 3 → [22]
- I don't know → [40]

31 Your answer is wrong. There are various types of oxide. We have already come across phosphorus *oxide* and iron *oxide.* There is, of course, no phosphorus in iron oxide and no iron in phosphorus oxide. But both oxides do contain ...., which is present in every oxide.

Jot down the substance and then go to [41] .

32 Your answer is wrong. *Physical* processes are those in which the substance remains unchanged and only a change of its state and associated properties occurs.

During electrolysis water is decomposed by means of direct current into two substances (oxygen and hydrogen) having quite different properties from those of water. Thus, electrolysis produces a *transformation of substance*, and not just a change of state.

Processes in which the substances involved are transformed into different substances are *chemical* processes.

3

Electrolysis is therefore a .... process.

Jot down the missing word and then go to 3.

---

33 Your answer is quite right.

Pure oxygen is *not* a catalyst.

Any combustion process will certainly proceed slower in air than in pure oxygen, but *only* because air contains a lower proportion of oxygen (20%). Oxygen is *used up* in all combustion processes, whether they are carried out in the air or in pure oxygen.

A catalyst is a substance which speeds up a chemical reaction but which at the end of the reaction is present in the same form as at the beginning. The process itself is described as catalysis.

Take good note of this important definition!

If one holds a sugar cube in a gas flame, the sugar turns brown and begins to melt. The sugar, however, does not actually catch fire. But, if one tips a small amount of cigarette ash on to the sugar, it will ignite and burn away slowly.

Would you describe this as catalysis?

- Yes → 43
- No → 51

---

34 *chemical elements*

Elements can be combined together to form *chemical compounds.* For instance, the element *phosphorus* combines with the element *oxygen* to form a *phosphorus oxide.*

The element *sulfur* combines with the element *iron* to form the compound *iron sulfide.*

The element *oxygen* and the element *hydrogen* combine to form the compound *water* (which may be decomposed into the elements again by means of electrolysis). Thus, in the chemical sense, water is an *oxide* of hydrogen.

In which of the following groups of substances are *only elements* listed?

- Oxygen, air, hydrogen, iron → 5

- Phosphorus, hydrogen, sulfur, iron sulfide → 23
- Iron sulfide, phosphorus oxide, water → 15
- Iron, sulfur, hydrogen → 42

---

35 Your answer is wrong.

You need to concentrate more on your work, and re-read 72 .

---

36 Your answer is wrong. We have already discussed the separation of a variety of mixtures. Think, for instance, of:

the separation of liquid air or
the separation of iron filings and sulfur powder.

The means utilized to separate these mixtures always involved .... processes.

Jot down the missing word and go to 19 .

---

37 Your answer is quite right.

The combustion product is indeed *heavier*. This must be so as oxygen has been *added* to the phosphorus in this particular reaction.

In the combustion of coal, wood or heating oil it may seem as though these substances "disappear". In fact, they combine with oxygen to form gaseous combustion products which could be collected and weighed in a suitable apparatus. One could then demonstrate that the combustion products and the ash taken together weigh more than the starting material (before any oxygen has been added to it).

The same applies in the case of the rusting of iron: the iron becomes *heavier* during the rusting process owing to the uptake of oxygen.

No substance can be destroyed during any combustion process; all that can happen is that one substance is transformed into another. In such a process, therefore, the total "mass" remains constant.

If sulfur and iron are heated together and react to form iron sulfide, does the original mixture become

- heavier? → 67

- lighter? → 4
- neither heavier nor lighter? → 13

---

38 Yes, potassium chlorate gives off *oxygen* when heated.

Upon heating, potassium chlorate first melts to form a colorless liquid. Further heating produces gas bubbles which rise to the surface of the liquid. If a glowing wood taper is then thrust into the test tube, it starts to burn very vigorously. The gas being formed must therefore be oxygen.

This procedure represents a third method for the production of oxygen. Do you still remember the other two methods for producing oxygen?

Outline briefly the three methods, and then continue reading at 30 .

---

39 Your answer is wrong. You have allowed yourself to be misled by the conflagration which occurs when sulfur and iron combine.

Fire *may* be observed in a combustion process, though there are such processes which take place *without* any conflagration. In an extended use of the term, these latter processes are also referred to as "combustions" because they are closely related to the processes involving conflagrations (iron + oxygen → rust).

What then is the essential feature of any combustion?

It is the combination of a substance with ....

Jot down the missing word, and go on to 47 .

---

40 It's just as well that you didn't guess here. Consider for a moment the following points:

Method 1 (Liquefaction of air and separation of the nitrogen) requires large and unwieldy apparatus.

Method 2 (Electrolysis of water) is suitable for small amounts. Large scale production would entail the uneconomical use of too much electricity.

Method 3 (Heating of potassium chlorate) supplies small amounts of oxygen rapidly and conveniently.

Answer the question posed in 30 once again.

---

41 Your answer is quite right.

Oxides are compounds of *oxygen*.
Oxygen also combines with the element carbon. Now coal (which consists very largely of carbon) combines with oxygen (from the air for instance) in a process which is well known to all of us. Basically, this process involves the *combustion* of carbon.

However, oxygen is involved not only in the combustion of carbon, for *oxygen* must be present *at each and every combustion process. Without oxygen it is not possible for any combustion to take place!*

Apart from carbon, we have already come across two elements which catch fire when they burn. Jot down the names of these two elements. → 49

---

42 Your answer is right: iron, sulfur and hydrogen are elements.

If two (or more) *elements* combine together a compound is formed.

The compound *water* consists of the elements .... and ....

The compound *phosphorus oxide* consists of the elements .... and ....

Jot down the missing elements, and go on to 50 .

---

43 Your answer is quite right.

In this catalytic process the cigarette ash is indeed acting as the *catalyst.* In fact, the catalyst speeds up the chemical process of combustion of the sugar so much that, even if the gas flame is removed, the sugar will still go on burning. The cigarette ash itself is not changed in any way during this process.

In this experiment the cigarette ash has a rather surprising effect. Using laboratory jargon, the chemist would term this a so-called "impurity effect". Chemists are extremely careful to eliminate such effects in the course of their everyday activities. Accordingly, they take great pains to use only clean apparatus and pure chemicals.

Write down now the completed version of the following sentences:

A catalyst is a substance which .... a chemical reaction but which at the end of the reaction is present in the same form as at the beginning. The process itself is described as .... → 52

3

44 We have already discussed the separation of a variety of mixtures. Think, for instance, of:

the separation of liquid air or
the separation of iron filings and sulfur powder.

The means utilized to separate these mixtures always involved .... processes.

Jot down the missing word and then go to [19] .

---

45 Your answer is wrong.

You should concentrate more on your work, and read [72] once again.

---

46 Your answer is correct. *Oxygen* is produced when potassium chlorate is heated together with manganese dioxide as the catalyst. In the process the *compound* potassium chlorate is decomposed. We can thus speak here of a *chemical* process.

We have already come across a second chemical process in which oxygen is produced from a compound. Which of the following processes is the one referred to?

- Production from liquid air → [71]
- The electrolysis of water → [55]
- The burning of sugar in the presence of cigarette ash → [7]
- The combustion of phosphorus → [63]

---

47 *oxygen*

However, in the formation of iron sulfide from iron and sulfur, *oxygen* is *not* involved.

There is therefore no question of a combustion here. → [48] .

---

48 Your answer is correct. There is no combustion process involved when sulfur and iron combine together to form iron sulfide because *no oxygen* participates in this reaction. *Oxygen* must be present in every combustion process. (For the chemist a conflagration is *not* the essential feature in a combustion process.)

In general, the oxygen required for a combustion process comes from the air (candles, stoves). Any fire which is not burning too well may be kindled simply by supplying additional fresh air to it.

It should now be evident that a combustion process will proceed fastest in pure oxygen. For instance, a wooden taper which only glows in air will immediately begin to burn with a bright flame if thrust into a glass vessel full of oxygen. (In fact, this experiment with a taper may be used as a *test* for *oxygen.)*

Even iron will burn away quite fast in pure oxygen and with vigorous scintillation.

In summary then, all substances burn faster and more vigorously in pure oxygen than in air or in diluted oxygen.

The oxygen required for a given combustion may be taken from a compound, instead of taking it from the air or from a gas bottle. Such a compound must contain oxygen in a form which can be fairly easily released. One example of a compound of this type is *potassium chlorate.*

Let's suppose now that the following experiment is performed:

Some *potassium chlorate* is heated in a test tube. After it has melted, small pieces of wood charcoal are dropped in. An immediate and spontaneous conflagration is observed and the charcoal is quickly burnt up. When heated potassium chlorate gives off ...., which is responsible for the very vigorous combustion of the charcoal.

Continue your reading at [38] .

---

[49] *sulfur, phosphorus*

A combustion is not necessarily associated with fire and flames. In fact, the term "combustion" is used in an extended sense to describe a number of rather slow processes which proceed without any fire, flames or even a detectable amount of heat being produced. One instance of this type of slow process is the rusting of iron; in this combustion iron combines with oxygen in the air to form iron oxide (rust).

The process by which phosphorus and oxygen form the compound .... is a .... reaction.

Jot down the missing words, and then check them at [57] .

---

[50] Water consists of *hydrogen* and *oxygen.*

Phosphorus oxide consists of *phosphorus* and *oxygen.*

3

The element oxygen displays a great tendency to combine with other elements. We have already learnt that oxygen in the air combines with the element phosphorus to form a phosphorus oxide, i.e. phosphorus burns in air.

Oxygen also combines very readily with the element iron. We have all seen how easily iron rusts, e.g. how nails, railroad tracks or the iron parts of a bicycle rust. In this process, i.e. in this combustion, iron combines with oxygen and the compound iron oxide is formed (usually referred to as rust).

What element is to be found in *all* oxides?

- Phosphorus → 58
- Oxygen → 41
- Iron → 31

---

51 Why not? The process is a catalytic one. The cigarette ash acts as the catalyst for the burning of the sugar. Without the cigarette ash the sugar would not burn. In fact, only when the ash is added does the sugar begin to burn slowly. The catalyst remains completely unaltered at the end of the experiment. Accordingly, all of the conditions for catalysis are fulfilled.

Continue your reading at 43 .

---

52 Now check your answer:

A catalyst is a substance which *speeds up* a chemical reaction but which at the end of the reaction is present in the same form as at the beginning. The process itself is described as *catalysis*.

We have come across two substances which can catalyze chemical processes.

Now write out the completed versions of the sentences below.

1. The production of oxygen from .... is catalyzed by ....
2. The burning of .... is catalyzed by .... → 60

---

53 No, a *chemical* process is involved. You would have come to the right conclusion if you had taken the following facts into consideration:

1. The *compound* potassium chlorate is decomposed to produce oxygen.

2. A catalyst is a substance which speeds up a *chemical* reaction.

Continue your reading at [46] .

---

54 Your answer is quite correct.

Combustible and flammable substances are frequently encountered in the laboratory. For this reason we feel it advisable to mention here certain measures which should be adopted when extinguishing fires.

It is not always correct to fight fires with water. This is especially true in the case of fires in the laboratory, but is also applicable in a number of other instances, e.g. for a fire in an automobile.

Whenever liquids having a density less than that of water, i.e. liquids such as gasoline or oil which float on water, catch fire, there is little point in using water to extinguish them.

One fights a fire of this type with extinguishers containing a substance which will effectively isolate the fire from the air (and thus from oxygen). In this way the fire can be put out.

An extremely hazardous type of fire is that involving burning clothes. In such a situation the individual affected must:

Under no circumstances run about

Maintain a calm presence of mind

Lie down and put out the fire with extinguishing blankets.

One should never run about as this would have the effect of increasing the supply of oxygen to the fire and so make the burning more intense.

Although discussed here within a laboratory context, the above rules apply to any outbreak of fire.

Having said this, let's return to the subject of the combustion process itself which we want to consider now in somewhat greater detail.

Let's go back to the combustion of phosphorus in air.

Will the combustion product formed, namely the phosphorus oxide, weigh

- More than the original amount of phosphorus? → [37]

3

- Less than the original amount of phosphorus? → 66
- Exactly the same as the original amount of phosphorus? → 74

---

55 Your answer is quite correct.

Two of the three methods of producing oxygen are *chemical* processes, namely:

1. The electrolysis of water.
2. The heating of potassium chlorate with manganese dioxide.

The third method is a *physical* one, namely:

3. The separation of liquid air.

In the case of the chemical processes we start out from compounds, namely from .... or from .... → 64

---

56 Your answer is wrong.

You should concentrate more on your work, and now re-read 72 .

---

57 The process by which phosphorus and oxygen form the compound *phosphorus oxide* is a *combustion* reaction.

The combination of a substance with oxygen (with or without a conflagration) is a combustion in the chemical sense. The chemist terms this process *oxidation.*

Combustion takes place whenever:

1. A combustible substance is present;
2. Oxygen is present;
3. The ignition temperature is attained (for combustions involving conflagrations).

We are all familiar with a whole range of combustible substances: gasoline, oil, wood, coal, phosphorus and iron.

Most of the combustible substances mentioned here will burn only above a certain temperature. In other words, we have to ignite them. After this, the combustion often proceeds very rapidly and heat and flames appear.

Phosphorus and iron combine with the oxygen in the air even at room temperature, though the process is very slow. At higher temperatures they burn much more rapidly and then there is a conflagration.

The following question may not be too easy to answer:

When heated together sulfur and iron combine to form iron sulfide and there is a conflagration. Is this process a combustion?

- Yes → 65
- No → 48
- I'm not sure → 39

---

58 Your answer is wrong. There are many different *oxides*. For instance, we have come across phosphorus *oxide*. We also know that iron oxide contains no phosphorus and that phosphorus oxide contains no iron. However, both of these oxides do contain .... which is always present in every oxide.

Jot down the missing substance, and then go to 41 .

---

59 In the chemical sense a combustion is any combination of a substance with oxygen, with or without a conflagration. For all combustion processes we find that the Law of Conservation of Mass is obeyed.

How did the law read?

Write down your definition, and then check it at 73 .

---

60 The production of oxygen from *potassium chlorate* is catalyzed by *manganese dioxide*.
The burning of *sugar* is catalyzed by *cigarette ash*.

Since we are able to produce oxygen from potassium chlorate merely by heating it, this implies that the potassium chlorate must contain oxygen.

Alternatively, we could also say that:

Potassium chlorate is an *oxygen*-containing *compound*.

Oxygen is present in potassium chlorate in a combined form. If we wish to prepare oxygen, we have to *liberate* it *from the compound* potassium chlorate. This is what actually happens when we heat potassium chlorate with manganese dioxide as the catalyst.

3

Is a chemical or a physical process involved here?

- A physical process → [53]
- A chemical process → [46]
- I'm not sure → [68]

---

61 In *industry* oxygen is produced from liquid air. The other two methods are especially suited for *the laboratory.*

Oxygen is absolutely essential to sustain life. In fact, without it breathing would not be possible. Now write down the following sentence in completed form:

People and animals breathe in .... and breathe out .... → [69]

---

62 Your answer is quite right.

In the breathing process oxygen is inhaled and carbon dioxide is exhaled.

One particular combustion process, which is an everyday occurrence in the chemical laboratory, takes place every time we use the *Bunsen burner*. We refer to the combustion of natural gas in the burner. A special device at the base of the burner enables air to be supplied to the flame. If the air supply is reduced, one obtains a yellow, luminous and relatively cool flame. The combustion of the natural gas will in this case be incomplete. On the other hand, when there is a good supply of air, the combustion process will be complete. The flame will then be very hot and have a pale blue appearance.

From this specific example, we are reminded again that every combustion involves:

1. *A combustible substance* and
2. The presence of *oxygen.*
   Moreover, whenever there is a conflagration,
3. The *ignition temperature* must be attained.

How can one produce a hot flame with a Bunsen burner?

- By reducing the air supply → [29]
- By increasing the air supply → [54]
- By increasing the gas supply → [70]

---

63 Your answer is wrong. A combustion is not a method for the production of oxygen. On the contrary, oxygen is *used up* in all combustion processes.

Read through 46 again.

64 In the chemical methods, we start out from *water* or from *potassium chlorate.* Both of these substances are compounds. Accordingly, it is *only by means of a chemical process* that these *oxygen-containing compounds may be decomposed* and thereby yield up oxygen.

In order to obtain oxygen from liquid air, we separate the air into its components by means of a physical process. This is possible only because air is a *mixture.*

In a mixture the components are *not bonded* to one another chemically. For instance, air is a mixture of *free oxygen* and *free nitrogen.*

Compounds, e.g. water, are decomposed by means of .... processes.

Mixtures, e.g. air, are separated by means of .... processes.

Write out completed versions of both sentences. → 72

65 Your answer is wrong. You have allowed yourself to be misled by the conflagration which occurs when sulfur and iron combine together.

A conflagration *may* occur during a combustion process, though there are processes in which this does *not* occur (iron + oxygen → rust). In the chemical sense these latter processes are still regarded as combustions.

What is the essential feature of any combustion process from the chemical standpoint?

A combustion is the combination of a substance with ....

Jot down the missing word, and go on to 47 .

66 Your answer is wrong. During the combustion process nothing is lost from the original phosphorus. On the contrary, something is added to it, namely a quantity of oxygen. The reaction product will therefore be .... (lighter/heavier).

Continue your reading at 37 .

67 Your answer is wrong. The mixture does not get heavier as nothing is added to it. What happens is that the two substances contained in the mixture, sulfur and iron, form the compound iron sulfide (without any material being added or taken away).

Read through 37 again.

68 In answering this question you should bear in mind that:

1. The *compound* potassium chlorate is decomposed in order to produce oxygen.

2. A catalyst is a substance which speeds up a *chemical* reaction.

Go back now to 60 .

69 People and animals breathe in *oxygen* and breathe out *carbon dioxide.*

Without oxygen neither breathing nor any combustion process would be possible. Explain briefly what combustion is, and then continue your reading at 59 .

70 Your answer is wrong. By increasing the gas supply (and keeping the amount of air constant) one will not obtain a hotter flame.

In fact, if the amount of gas is increased and the air supply is kept constant, there will be diminished combustion of the gas. Read through 62 again.

71 Your answer is wrong. Air is *not* a compound but a *mixture.* The separation of a mixture is not a *chemical* process but rather a *physical* one.

Read 46 once again.

72 Compounds are decomposed by means of *chemical* processes. Mixtures are separated by means of *physical* processes.

In the air oxygen is found as the free element whereas in water it is present in a chemically bound form. As there is a very abundant supply of both of these substances on earth, it is perhaps not surprising that oxygen is the most common element known. Even many rocks contain oxygen in combined form.

Oxygen is vital to living creatures. Every time we breathe in air we are inhaling oxygen and then exhaling carbon dioxide. Under conditions where there is no oxygen available for breathing, e.g. in outer space, man has to take his oxygen with him in order to survive.

Oxygen is used for welding, since combustion processes carried out in the presence of pure oxygen produce very hot flames (more on this follows later).

What is actually taking place when one breathes?

- Oxygen is inhaled and nitrogen is exhaled → 35
- Oxygen is inhaled and carbon dioxide is exhaled → 62
- Carbon dioxide is inhaled and oxygen is exhaled → 56
- Carbon dioxide is inhaled and nitrogen is exhaled → 45

---

73 In chemical processes mass (matter) is neither destroyed nor created.

Did you make any mistakes in the material you repeated? If so, you should work your way through this Program again as soon as possible.

End of Program 3.

---

74 Your answer is wrong. In the combustion of phosphorus none of the original phosphorus is lost, but rather something is added to it, namely a quantity of oxygen. The reaction product is therefore .... (lighter/heavier).

Continue your reading at 37 .

Program 4

Water

1 Water is an extremely abundant chemical compound. In fact, the greater part of the earth's surface is covered with water in the form of oceans. In addition to this, there is the water which makes up seas, lakes and rivers.

The misting over of windows on a cold day provides proof that a considerable amount of water vapor is always present in the atmosphere, even though this is normally invisible.

Water is also an essential component of all living things.

The water on the earth is in constant circulation. Water comes into the atmosphere by means of evaporation. There it forms mist, fog or clouds. The water in clouds can return to earth as dew, rain, hail or snow according to the weather conditions. It then collects as groundwater within the bowels of the earth and comes up to the surface again in the form of springs. The springs develop into streams and rivers which eventually find their way into lakes and seas. The whole circulation process commences from the beginning again when this water starts to evaporate into the atmosphere.

How does the water cycle described here come about?

- By means of a series of chemical processes → 26
- By means of a series of physical processes → 8
- By means of a series of physical and chemical processes → 18

2 Your answer is wrong. Direct electric current is indeed used in electrolysis. The production and utilization of direct electric current belongs essentially to the realm of physics. However, when direct current is passed through water, it brings about a *chemical* change.

Electrolysis is a chemical process in which substances are decomposed with the aid of direct electric current.

Write down this definition, and then continue your reading at 19

3 No. Please note that when chlorine is led into a gas jar full of air, the jar fills up with the chlorine *from the bottom up.*

It may be useful to remember here that objects having a density greater than that of water sink to the bottom when placed in water.

Read 17 once again.

---

4 The *Celsius* temperature scale.

Water ist frequently used as a solvent. Water can dissolve solid substances (salt, sugar, soda), liquid substances (sulfuric acid, alcohol) and gaseous substances (carbon dioxide in mineral water).

Domestic water and mineral water contain small amounts of *calcium* compounds which are dissolved out of the ground. This type of water is referred to as *hard* water. On the other hand, rain water contains no dissolved *calcium* compounds; it is described as *soft* water.

(Calcium is an element; its most important compounds will be discussed later.)

Hard water, e.g. mineral water, contains ....

Soft water, e.g. ...., contains no ....

Jot down the missing words.
→ 12

---

5 Your answer is wrong. If all you do is to hold your thermometer in melting ice, you will be able to determine only the 0 °C point.

To set up a scale, however, you need *two* fixed points. Accordingly, the 100° point must also be determined by immersing the thermometer in boiling water. Division of the distance between the 0° point and the 100° point into .... equal parts will give the Celsius thermometer scale.

Jot down the missing number. → 13

---

6 No. Please complete the following word equation:

Chlorine detonating gas = a mixture of hydrogen and ....

Detonating gas = a mixture of .... and oxygen.

Continue your reading at 15 .

4

---

7 You must, at some time, have made the following observation. If one fills a vessel to its brim, e.g. a bucket with sand, it is possible to get the sand to settle more by means of shaking and tapping the bucket. The bucket is then no longer quite full.

The same amount of sand which had previously filled the bucket occupies less space (has a smaller volume) after it has been shaken and tapped. This means that the sand is now more densely packed.

The density of water may also be increased, though not by shaking and tapping; the method here is to cool the water down.

Read through 16 again.

---

8 Your answer is quite right: the water cycle in nature comes about by means of a series of *physical* processes.

As a chemical compound, the water is not changed in any way by these processes. All that happens is that it is converted from the liquid state to the vapor state by evaporation and then from the vapor state back to the liquid state by condensation.

Let's take a look now at a diagram of the water cycle:

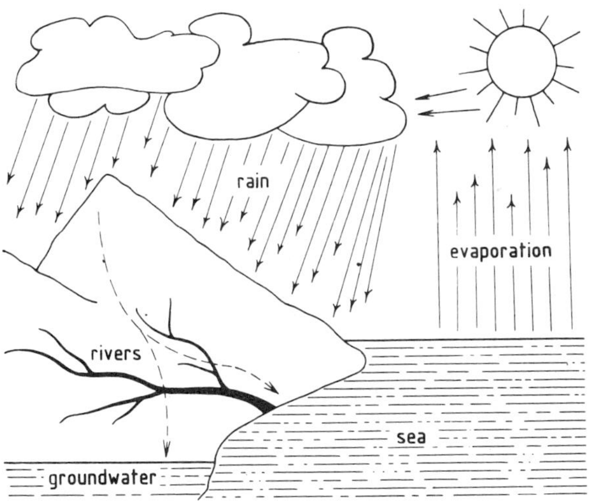

Is this diagram of the water cycle correct as it stands?

4

- Yes → 16
- No → 21

---

9 No, you have mixed up something here.
Take another look at the diagram in 19 . Think very carefully about which gas is formed at the cathode and which at the anode.

---

10 It's just as well that you didn't guess. Jot down the following definition:

Electrolysis is a chemical process in which substances are decomposed with the aid of direct electric current.

Continue your reading at 19 .

---

11 Your answer is partly correct. Hydrogen is indeed formed during the electrolysis of water. In industry, however, hydrogen is usually produced by the electrolysis of brine. In this process one also obtains the important products .... and ....

Jot down the two products obtained. → 42

---

12 Now check your answer:

Hard water, e.g. mineral water, contains *calcium compounds.* Soft water, e.g. *rain water*, contains no *calcium compounds.*

Hard and soft water may be distinguished by their taste. Hard water tastes refreshing whereas pure, soft rain water tastes flat and insipid.

There is also another way of distinguishing between hard and soft water.

When soap solution is added to domestic water, flocculation is observed in the water. If the experiment is repeated with rain water, however, the solution remains clear and a foam is easily produced.

Write down the properties of hard and soft water (taste, behavior with soap solution) using only key words.

Hard water: ....
Soft water: .... → 20

---

13 Your answer is correct.

After determining the 0° and 100° points the distance between them is divided into 100 equal parts.

In this way one obtains the .... temperature scale.

Jot down the name. → 4

14 When water freezes it expands. Accordingly, 1 kg of ice takes up more space than 1 kg of water. Ice is thus not so 'densely packed' as water. The density of ice is thus less than the density of water.

What happens when a lump of ice is put into water? This question is easily answered if it is remembered that less dense substances float on more dense ones.

- The ice will float on the water → 22
- The ice will sink in the water → 30

15 Your answer is quite right.

Chlorine detonating gas = a mixture of hydrogen and chlorine.

Detonating gas = a mixture of hydrogen and oxygen.

The new chemical compound formed from the hydrogen and chlorine of chlorine detonating gas is called *hydrogen chloride.*

Hydrogen chloride is a pungent gas which is very soluble in water. The aqueous solution of hydrogen chloride has an acid reaction; the actual acid formed is referred to as hydrochloric acid.

Chlorine is one of the most active chemical elements known. It combines readily with many other elements. For this reason chlorine is not found in the free state in nature but occurs only in the form of compounds. The most important naturally occurring chlorine compound is common salt. This is designated in chemical terms as sodium chloride, since it consists of the elements sodium and chlorine.

Write down the names and the properties of the two chlorine compounds mentioned above.

Then continue your reading at 34 .

4

16 Your answer is quite right.

Let's now examine in some detail one very important property of water.

Water has a density of 1 g/cm^3, which means that 1 cm^3 of water weighs 1 g at +4 °C. If water at +20 °C is poured into a long narrow tube, its level marked off, and the water is then cooled down, one can observe that the level falls.

In which of the tubes shown below has water the greatest density?

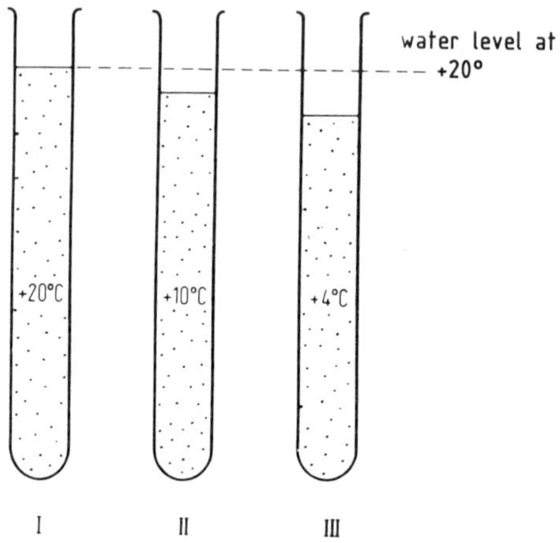

- In tube number I → 32
- In tube number II → 40
- In tube number III → 24
- I don't know → 7

17 *Detonating gas* is formed.

Detonating gas contains hydrogen and oxygen in the same ratio by volume as that of the two gases formed during the electrolysis of water, namely in the ratio:

$$\text{Hydrogen} : \text{Oxygen} = 2 : 1$$

Hydrogen too is produced by means of an electrolytic process in industry, though in this case the electrolysis is carried out on brine rather than on water. In addition to hydrogen, the products sodium hydroxide and chlorine are also formed. Although for the present only *chlorine* will be discussed here, we shall return to the subject of sodium hydroxide at a later stage.

Chlorine is a gas having a greenish-yellow appearance. **Chlorine is poisonous.**

When chlorine is led into a gas jar (in a fume hood!), the air will be expelled from the jar. The jar will gradually be filled with the chlorine *from the bottom up*.

Which of the three statements below is correct?

The density of chlorine is

- Greater than that of air → 25
- Less than that of air → 45
- The same as that of air → 3

---

18 Your answer is wrong. The water cycle depends only upon physical processes. The water itself is not transformed *as a substance* in any of these processes. Only its state changes when it is converted from liquid to vapor or from vapor to liquid. The processes which make up the water cycle are exclusively physical ones.

Read through 1 again.

---

19 Your answer is quite correct.

Electrolysis is a chemical process in which substances are decomposed with the aid of direct electric current.

If direct electric current is passed through acidified water, the water will be decomposed into its components. Oxygen will be formed at the anode (positive electrode) and hydrogen at the cathode (negative electrode). We may therefore conclude that water consists of hydrogen and oxygen. Incidentally, this experiment also gives us information on the *relative proportions* of these two components of water.

4

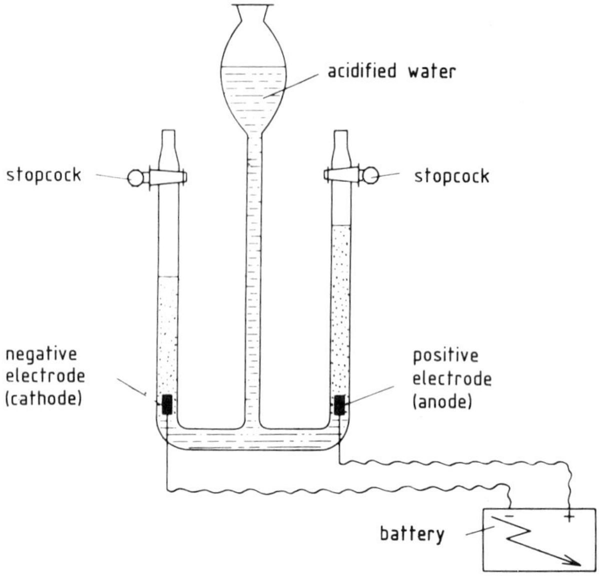

Hofmann's apparatus for the decomposition of water.

At the beginning of the experiment both of the side arms in Hofmann's apparatus are filled with water right up to the stopcocks. When the apparatus is connected to the battery, gas bubbles start to rise up from the electrodes and collect beneath the stopcocks. After a while, the situation depicted in Fig. 6 will be reached.

What are you able to deduce from this figure?

- Twice as much hydrogen as oxygen is formed → 29
- Twice as much oxygen as hydrogen is formed → 9
- Equal volumes of oxygen and hydrogen are formed → 39

---

20 Now check your answer:

Hard water: taste is good and refreshing; flocculates with soap solution; difficulty in forming foam.

Soft water: taste is flat and insipid; dissolves soap and forms a clear solution; foam is easily produced.

Soft water may be obtained from hard water by means of distillation. An apparatus for the distillation of water is illustrated below:

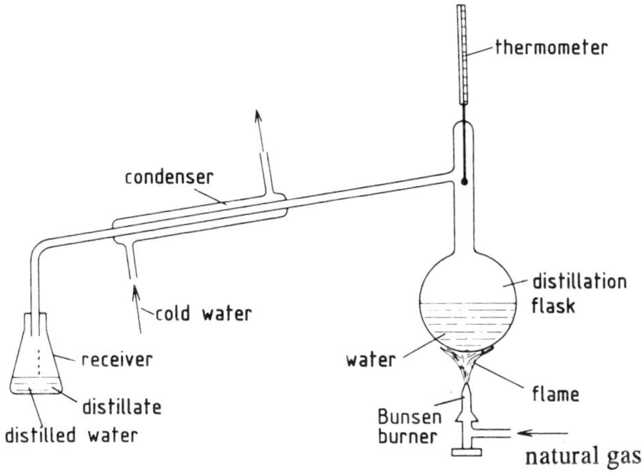

By studying this diagram carefully, you should be able to complete the following sentences. Jot down the missing words in their correct sequence. Water is evaporated in a .... The water vapor formed moves up past the .... and is then condensed back into water again in a .... For cooling purposes .... is used. The water now purified by distillation drips into the .... The distillate produced in this way is called ....

Continue reading at 28 .

---

21 Your answer is wrong. The diagram is *completely correct* in that it shows how water from the seas enters the atmosphere by means of evaporation, condenses there into clouds, and then returns to earth again as precipitation. The latter forms the groundwater which is the source of streams and rivers. These in turn eventually flow into the sea, thereby completing the water cycle.

Have another good look at the diagram; you will find it in 8 .

---

22 Your answer is quite right.

Water freezes at 0 °C and boils at 100 °C under a pressure of 1 atmosphere (= the pressure of a column of mercury 760 mm high). Both of these temperatures, **0 °C** and **100 °C**, are termed fixed points and the Celsius thermometer is defined with respect to them. This thermometer is constructed by dividing the distance between the two fixed points on the stem of a mercury thermometer into 100 equal parts. Any such division of a thermometer stem is referred to as a thermometer scale or a temperature scale.

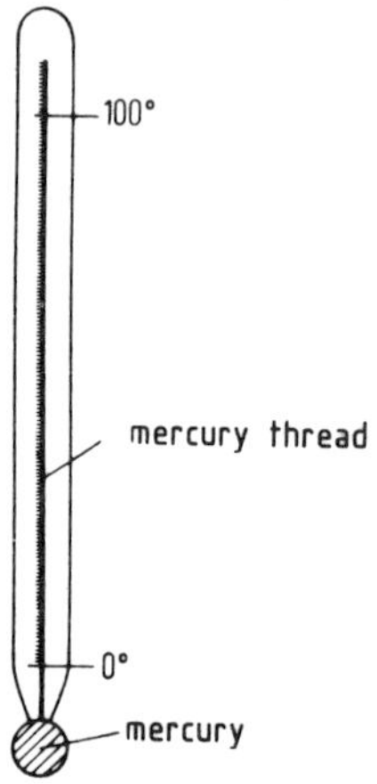

What is the simplest way of determining the two fixed points on a thermometer?

- By copying the corresponding points from other thermometers → 31
- By immersing the thermometer in boiling water → 43
- By immersing the thermometer in boiling water and then in melting ice → 13
- By immersing the thermometer in melting ice → 5

23 You have recently learnt that:

$$\text{Water} \xrightarrow{\text{electrolysis}} 2 \text{ volumes of hydrogen} + 1 \text{ volume of oxygen.}$$

The volumes are thus in the ratio of .... : .... → 37

24 Your answer is quite correct.

You have undoubtedly had the following experience.
After filling a suitable vessel, e.g. a bucket, to the brim with sand and then shaking and tapping it, the sand becomes compacted. The bucket will then no longer be completely full.

The same volume of sand which had filled the bucket when loosely packed occupies a smaller volume after it has been compacted. In fact, the sand is now packed more densely. The density of water can also be increased, though not in this case by shaking and tapping but rather by cooling it down.

At +4 °C water has a greater density than it has at (say) +10 °C or +20 °C. This is not surprising, for all substances contract when they are cooled down. However, if water is cooled to below +4 °C, something completely unexpected happens. At this point the water actually stops contracting and starts to expand again!

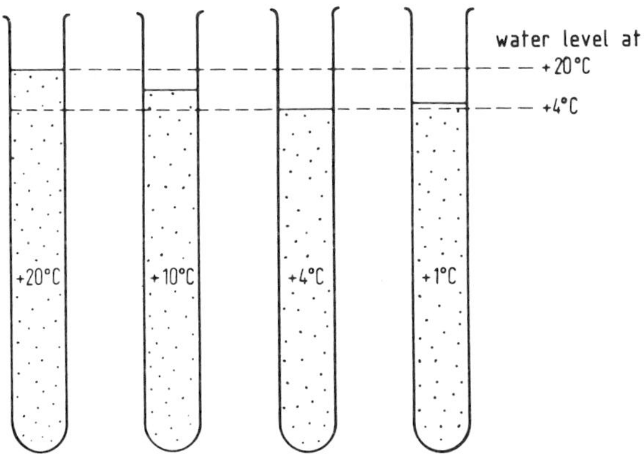

Write out the following sentences with either the word *increases* or *decreases* inserted at the appropriate positions:

1. When water is cooled from +10 °C to +4 °C, its density .... and its volume ....

2. When water is cooled from +4 °C to +1 °C, its density .... and its volume .... → 33

---

25 Your answer is quite right.

The density of chlorine is greater than that of air. Chlorine is a pungent and poisonous gas. In small quantities chlorine may be used as a bleaching and disinfecting agent. Mention should also be made of the fact that the water in swimming pools, and in many areas the domestic water supply, is sterilized by the addition of small amounts of chlorine. Chlorine is a very reactive element. It combines for instance with hydrogen.

Caution! A gas jar filled half with chlorine and half with hydrogen will **explode** with a loud bang if a flame is brought near to the mouth of the jar or if the jar is exposed to bright light (e.g. the sun, a flash light or an ultraviolet lamp).

For this reason such a mixture is termed *chlorine detonating gas.*

What does chlorine detonating gas consist of?

- Hydrogen and oxygen → 46
- Hydrogen and chlorine → 15
- Chlorine and oxygen → 6

---

26 Your answer is wrong. During the various stages of the cycle the water *as a substance* remains unchanged. It is merely changed from the liquid to the vapor state and then back again from the vapor to the liquid state. There are therefore no chemical processes involved in the water cycle, for in all chemical processes substances are transformed into different substances.

Read through 1 again.

---

27 Your answer is right! Distilled water is *soft* water.

It is even purer than rain water which, although free from calcium compounds, is usually contaminated with other impurities.

Water is essential for human beings, animals and plants. Moreover, the important uses of water in everyday life are manifold. Water is used for instance in cooking and washing. In industry too it is absolutely indispensible as for example, a cleaning, scrubbing and dissolving agent. In the form of steam it is employed exensively for heating purposes, and in the form of ice for cooling purposes.

In order to determine the composition of water, one can subject it to the process of electrolysis.

What do you understand by the term electrolysis?

- A physical process in which direct electric current is used → 2
- A chemical process in which substances are decomposed with the aid of direct electric current → 19
- A chemical transformation of substance which is speeded up by the addition of small amounts of other substances → 35
- I don't know → 10

28 Check your answers now:

distillation flask

thermometer

condenser

cold water

receiver

distilled water

Give a detailed outline of the distillation of water, and make sure that you use all six of the above terms in your description. → 36

If you encounter any difficulties here, read through 20 again.

29 Yes. Twice as much hydrogen as oxygen is formed. In addition to the 2 volumes of hydrogen there is also 1 volume of oxygen formed during the electrolysis.

Write out the completed statement below:

Water $\xrightarrow{\text{electrolysis}}$ .... volumes of .... + ... volume of ....

→ 38

30 Your answer does *not* agree with observations that you yourself must have made in winter time! → 33

31 Your answer is wrong. How could you know that the standard thermometer is identical in every detail to the one you wish to calibrate?

Read through 22 once again.

32 Your answer is wrong.

You will find an explanation of this at 7 .

4

33 Now check your answers:

1. When water is cooled from +10 °C to +4 °C, its density *increases* and its volume *decreases.*

2. When water is cooled from +4 °C to +1 °C, its density *decreases* and its volume *increases.*

If the water is cooled down further it will freeze and form ice at 0 °C. In the process there will be a further expansion, i.e. another increase in the volume.

For this reason it is necessary to empty unprotected water pipes whenever there is a danger of frost. Otherwise the expansion of the water as it starts to freeze could well burst the pipes.

Is the density of ice less than that of water?

● Yes → 22

● No → 41

● The densities are equal → 49

● I don't know → 14

34 Check your answers now:

Hydrogen chloride: gas, pungent, readily soluble in water, the solution has an acid reaction and is called hydrochloric acid.

Common salt (= Sodium chloride): compound of sodium and chlorine, solid, readily soluble in water, tastes salty.

For revision purposes:

Water is a chemical compound which is in abundant supply all over the earth. It freezes at 0 °C and boils at 100 °C. The Celsius temperature scale is based on these two fixed points. The density of ice is less than that of water.

Water consists of hydrogen and oxygen. The electrolysis of water yields two volumes of hydrogen and one volume of oxygen.

How is hydrogen produced industrially?

● From the electrolysis of water → 11

- From the electrolysis of brine → [42]

35 Your answer is wrong. You have mixed up two quite different concepts.

A chemical process which is speeded up by the addition of small amounts of other substances is *catalysis.*

But a chemical process in which substances are decomposed with the aid of direct current is termed *electrolysis.*

Jot down the definition of electrolysis, and then continue reading at [19] .

36 Your outline of the distillation of water should have included the following points:

Water is evaporated in a *distillation flask.* The temperature of the vapor is measured by a *thermometer.* The vapor is condensed back into water in a *condenser.* For cooling purposes *cold water* is used. The *distilled water* drips into a *receiver* as the distillate.

Do you still remember the concepts of hard and soft water?

Which sort of water is distilled water?

- Soft water → [27]
- Hard water → [44]
- I don't know → [48]

37 Your answer is quite right.

In the electrolysis of water hydrogen and oxygen are formed in a volume ratio of 2 : 1.

Water is a compound of the elements hydrogen and oxygen. When hydrogen and oxygen combine, i.e. when hydrogen is burnt, water is formed.

Caution! The reaction of hydrogen with oxygen proceeds so vigorously that mixtures of hydrogen and oxygen **explode** when they are ignited. A particularly violent explosion occurs when the mixture consists of two parts of hydrogen and one part of oxygen by volume. For this reason such a mixture is referred to as *detonating gas.*

What is formed when the gases produced during the electrolysis of water are mixed together?

The mixture known as .... will be formed.

4

Jot down the name, and then go on to 17 .

---

38 Water $\xrightarrow{\text{electrolysis}}$ 2 volumes of *hydrogen* + 1 volume of *oxygen.*

In the electrolysis of water .... is formed at the anode and .... at the cathode.

Jot down the missing words. → 47

If you are doubtful here about what to write, reat through 19 again.

---

39 Your answer is wrong. The quantities of gas which have collected beneath the stopcocks are certainly not equal! Under the stopcock on the left twice as much gas has collected as that under the stopcock on the right. Go back to 19 .

---

40 Your answer is wrong. You will find an explanation of this at 7 .

---

41 Your answer is wrong. You will find an explanation of this at 14 .

---

42 Your answer is quite right. Industrially, hydrogen is produced from the electrolysis of brine. The by-products of this process are *sodium hydroxide* and *chlorine.*

Chlorine is a greenish yellow, pungent and **poisonous** gas having a density greater than that of air. A mixture of chlorine and hydrogen is known as .... Chlorine combines with hydrogen to form .... The aqueous solution of hydrogen chloride is called .... The most important compound of chlorine found in nature is ....

Jot down the missing words, and then go to 50 .

---

43 Your answer is wrong. If you immerse your thermometer in boiling water, you will certainly be able to determine where the 100 °C fixed point lies on the scale. To set up the whole scale, however, one must have *two* fixed points. The 0 °C fixed point also has to be determined by immersing the thermometer in melting ice. Then division of the distance between the 0 °C mark and the 100 °C mark into .... equal parts will produce the Celsius thermometer scale.

Jot down the missing number, and then go to 13 .

---

44 No. As a result of distillation the water becomes purified. Impurities such as calcium compounds are left behind in the distillation flask.

Accordingly, distilled water is .... (hard/soft) water. → 27

---

45 No. Please note that when chlorine is led into a gas jar full of air, the jar fills up with the chlorine *from the bottom up.*

It may be useful to remember here that objects having a density greater than that of water sink to the bottom when placed in water.

Read through 17 once again.

---

46 No, please note here that:

Chlorine detonating gas = a mixture of hydrogen and ....

Detonating gas = a mixture of .... and oxygen.

Continue your reading at 15 .

---

47 Now check your answer:

During the electrolysis of water *oxygen* is formed at the anode and *hydrogen* at the cathode. Water thus consists of *oxygen* and *hydrogen.*

What is the ratio of the volumes of oxygen and hydrogen produced from the electrolysis of water?

- Hydrogen : Oxygen = 2 : 1 → 37
- Hydrogen : Oxygen = 1 : 1 → 39
- Hydrogen : Oxygen = 1 : 2 → 9
- I don't know → 23

---

48 As a result of distillation the water becomes purified. Impurities such as calcium compounds are left behind in the distillation flask.

Accordingly, distilled water is .... (hard/soft) water. → 27

49

Your answer is wrong. You will find an explanation of this at 14 .

50

Check your answers now:

A mixture of chlorine and hydrogen is known as *chlorine detonating gas.* Chlorine combines with hydrogen to form *hydrogen chloride.* The aqueous solution of hydrogen chloride is called *hydrochloric acid.* The most important compound of chlorine found in nature is *sodium chloride (common salt).*

End of Program 4.

Program 5

Hydrogen

5

1 Hydrogen is a chemical element we have already learnt quite a lot about in the two preceding Programs.

Hydrogen is a colorless, odorless and tasteless gas.

Can hydrogen burn?

- I don't know → 26
- Yes → 12
- No → 16

2 You should have included the following points in your answer:

Before carrying out any chemical reactions with hydrogen, one must ensure that the apparatus is completely **free from oxygen.** In this way serious explosions can be avoided. To check on this we can make use of the **detonating gas test.**

We shall discuss now three methods for the production of hydrogen. *One* method you have already come across. This is the method based on:

- The evaporation of water → 10
- The electrolysis of water → 28
- I don't know → 35

3 *detonating gas, water*

Gases have a fixed density, just as liquids and solids do. Some gases have a density greater than that of air whilst others have a density less than that of air. Put simply, one could say that there are heavy and light gases.

To determine whether hydrogen is a heavy or light gas, we could perform the following experiment:

Two gas jars which are full of hydrogen are hung up. One is hung in such a way that its opening is at the bottom, whilst the other is hung so that its opening is at the top. After a short time (ca. 1 minute) a flame is held close to the opening of each jar.

5

No reaction is observed for the jar with its opening at the top: the hydrogen has escaped by moving upwards out of the jar.

In the case of the jar with its opening at the bottom, there is a mild explosion: this jar still contains hydrogen.

These observations should enable you to answer the following question:

Is hydrogen lighter or heavier than air?

- Lighter → 11
- Heavier → 27
- I don't know → 20

---

4 Your answer is quite right.

Initially, the copper combined with oxygen from the air to form copper oxide, i.e. there was *oxidation* of the copper to copper oxide.

In the second part of the experiment hydrogen was passed over the copper oxide. This hydrogen combined with the oxygen in the copper oxide to form water, i.e. there was a *reduction* of the copper oxide to copper.
Thus, hydrogen may be used for carrying out reductions.

Write out now the following summary of the experiment (completing where necessary):

1. *Oxidation*

Copper + ... → ....

2. *Reduction*

Copper Oxide + Hydrogen → .... + ... → 15

---

5 Your answer is wrong. The method of producing hydrogen by passing steam over red-hot coke is an *industrial process* which is not suited to the laboratory.

Go back now to 23 .

---

6 Your answer is wrong.

You have just learnt that hydrogen is the *lightest* of all gases, i.e. hydrogen has the *lowest* density of all gases. *Because of the low density* of hydrogen, balloons filled with hydrogen gas are impelled upwards, i.e. they rise very readily in the air.

Go back now to [23] .

7 *hydrogen*

detonating gas

Jot down the three methods used for the production of hydrogen. → [19]

8 Your answer should have covered the following points:

When the stopcock H is opened the gas in volume B can escape. The hydrochloric acid can then flow from A via C to B. In the globe B there is a pile of zinc metal resting on a sieve. As soon as the hydrochloric acid comes into contact with the zinc, hydrogen gas is generated.

What happens when the stopcock H is closed again?

Please write out your answer in full. → [17]

9

1. Hofmann's apparatus for the decomposition of water
2. Kipp's gas generating apparatus
3. Water gas

What is a) an oxidation and
b) a reduction?

a) An oxidation is a .... process in which ....

b) A reduction is a .... process in which ....

Write out the complete definitions. → [29]

5

10 No. The water will not be changed chemically in any way when it evaporates. Water vapor is after all still water. The process of evaporating water is a purely physical one.

Hydrogen, of course, is a quite different substance from water. Accordingly, hydrogen may be obtained from water only by means of a chemical process.

Think back to our discussion on water. There we said that water could be decomposed with the aid of direct electric current into hydrogen and oxygen. Hydrogen is given off at the cathode (at the negative electrode) and oxygen at the anode (at the positive electrode).

One method for the production of hydrogen thus involves the .... of water.

Jot down the missing word. → 28

11 Your answer is quite right.

Hydrogen is even much lighter than air. Hydrogen is in fact the lightest of all gases. One liter of it weighs around 1/10 gram.

Special **precautions** are necessary when working with hydrogen. **Safety spectacles** must be worn and the apparatus should be mounted behind a **protective screen.** The apparatus must be **free from oxygen.** Hydrogen and oxygen form a mixture which **explodes violently** when ignited (detonating gas).

Before carrying out any chemical reaction involving hydrogen, the **detonating gas test** must first be performed: **Hydrogen** is passed through the apparatus and, after a reasonable length of time, a sample of the out-flowing gas is collected in a test tube (with its mouth pointing downwards). A flame is then held at the mouth of the test tube. If there is a minor **explosion** the apparatus still contains quite a lot of **oxygen**; enough oxygen in fact to form **detonating gas!**

If this happens, one must continue to flush hydrogen through the apparatus until such time as a further gas sample burns quietly and without any explosion. Only then can one be sure that the apparatus is free of oxygen.

Why is the detonating gas test performed? Write out a brief answer, and then check it at 2 .

12 Your answer is quite correct.

Hydrogen can indeed burn. In fact it combines very violently with oxygen.

Caution! A mixture of oxygen and hydrogen **explodes** with a loud bang when ignited. For this reason the mixture is called .... In the combustion of hydrogen .... is formed. Write down the missing words, and then continue your reading at [3] .

13

The possible uses of hydrogen include:

1. The inflation of airships and balloons
2. The welding of metal objects

Vast amounts of hydrogen are used by large chemical companies, e.g. in the production of *ammonia* and *methyl alcohol.* You will hear more on this later.

To conclude, we would like to mention one last use of hydrogen. This will involve us in two new definitions:

An oxidation is a chemical process in which a substance combines with oxygen.

A reduction is a chemical process in which oxygen is removed from a compound.

An example of an *oxidation* is any *combustion* process.

An example of a *reduction* will now be given.

Suppose that the following two-part experiment is carried out:

1. A copper wire is heated in the air. The reddish, metallic copper turns black. This is because copper oxide has been formed, since copper combines with oxygen in the air. Copper oxide is a compound of copper and oxygen.

2. The now black copper wire which we heated in the air is inserted into a glass tube and hydrogen is passed over it. This gas is flushed through until a sample of the out-flowing gas burns quietly when ignited (only in this way can we be sure that the apparatus is free of oxygen: **detonating gas test**). The outside of the glass tube is now heated at the point where the copper wire is resting. After a short while the metallic luster of the copper is restored. In addition, one observes that water droplets condense out on the colder parts of the apparatus.

Which of the following statements is correct?

- The first part of this experiment involves oxidation whereas the second part involves reduction of the copper → [4]

- The first part of this experiment involves reduction whereas the second part involves oxidation of the copper → 25
- Both parts of this experiment involve reduction of the copper → 30
- Both parts of this experiment involve oxidation of the copper → 34
- I need some help → 21

---

14 Your answer is wrong.

You have just learnt that hydrogen is the *lightest* of all gases, i.e. hydrogen has the *lowest* density of all gases. *Because of the low density* of hydrogen, balloons filled with hydrogen gas are impelled upwards, i.e. they rise very readily in the air.

Go back now to 23 .

---

15 Check your answers now:

1. *Oxidation:*

Copper + *Oxygen* → *Copper Oxide*

2. *Reduction:*

Copper Oxide + Hydrogen → *Copper* + *Water*

We conclude with two rather familiar questions.

Which is the lightest of all gases?
What is the mixture of hydrogen and oxygen called?

Jot down your answers and then go on to 7 .

---

16 Your answer is wrong.

Hydrogen can indeed burn. In fact, it combines very violently with oxygen.

Caution! A mixture of oxygen and hydrogen **explodes** with a loud bang when ignited.

For this reason the mixture is called ....
In the combustion of hydrogen .... is formed.

Jot down the missing words and then go on to [3] .

---

17 Your answer should have covered the following points:

When the stopcock H is closed, the pressure which builds up from the hydrogen still being generated forces the hydrochloric acid from B via C into A. As the hydrochloric acid will then no longer be in contact with the zinc, the production of hydrogen gas will cease.

Write out the following completed sentences:

Hydrogen is obtained when hydrochloric acid reacts with ....
A particularly convenient gas generator is that known as .... apparatus. → [24]

---

18 With the aid of your diagram write out an answer to the following question:

What happens when the stopcock H is opened? → [8]

19

1. The electrolysis of water or brine.

2. The action of acids on metals.

3. The action of steam on red-hot coke.

Methods 1 and 2 are carried out using special apparatus. What is the name of this apparatus in each case?
In method 3 a gas mixture of hydrogen and carbon monoxide is produced. What is this mixture called?

Jot down your answers and go on to [9] .

---

20 Imagine that the following experiment is performed:

Two bottles are filled with oil. Both are sealed and then placed in a water-bath so that they are completely immersed in water. The bottles are opened, one with its mouth pointing upwards and the other with its mouth pointing downwards. From the first of the bottles oil will immediately begin to flow out and rise to the surface, for oil is lighter than water. The oil in the second bottle, having its mouth pointing downwards, will not flow out.

Read through [3] again.

5

21 In the first part of the experiment:

We heat up some copper in the air. As a result the copper *combines with oxygen.* Copper oxide, a compound of copper and oxygen, is formed.

In the second part of the experiment:

We pass hydrogen over the copper oxide. The *oxygen is removed* from the copper oxide, and so copper is formed again.

Compare these outlines with the definitions given in [13] .

22 Your answer is quite correct.

When hydrogen burns a very hot flame is produced, especially if oxygen is used instead of air. Hot flames of this type are utilized in *welding.*

Special **precautions** have to be observed when hydrogen-oxygen mixtures are prepared for the purpose of welding. Accordingly, a specially constructed burner is used as an oxy-hydrogen blow torch as follows:

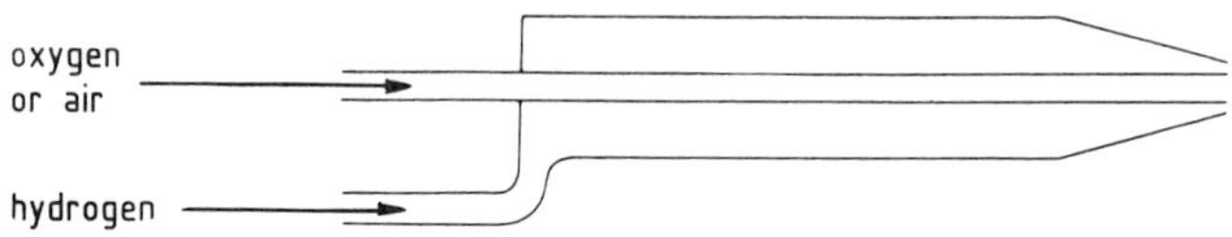

In this burner, known as a *Daniell torch*, hydrogen flows through the outer housing whilst oxygen is fed in along the pipe located in the middle. In this way it is possible to keep the two gases separate until the moment of ignition. Such a device enables one to avoid any build up of the dangerous hydrogen-oxygen mixture (detonating gas).

When actually using such a torch or burner one opens first the hydrogen inlet, ignites the gas, and then opens the oxygen or air valve. The reverse procedure is adopted when extinguishing the flame: first the oxygen or air supply is turned off, and then the hydrogen supply.

Copy down the diagram, write out two possible uses of hydrogen, and then continue reading at 13 .

5

23 The hydrogen in water is bound to *oxygen*, as we have learnt from electrolysis experiments.

What is the most important method of preparing hydrogen in the laboratory?

- The electrolysis of water → 31
- The action of acids on metals → 33
- The action of steam on red-hot coke → 5

24 *zinc*

Kipp's

So far we have discussed only two methods of producing hydrogen, namely:

1. The electrolysis of water or brine.
2. The action of acids on metals.

We now mention a third method which represents an important way of producing hydrogen industrially. When steam is passed over red-hot coke a mixture of hydrogen and carbon monoxide is formed. Because of its origin, this gas mixture is referred to as water gas.

If the carbon monoxide is removed from water gas, one obtains *hydrogen*.

We shall describe later how this may be accomplished.

Now write out completed versions of the sentences:

Water gas is a mixture of .... and ....
It is formed when .... is passed over .... → 32

25 No, please read through 21 , and analyze the experiment described there again.

26 Hydrogen can indeed burn. In fact, it combines very violently with oxygen.

Caution! A mixture of oxygen and hydrogen **explodes** with a loud bang when ignited.

5

For this reason the mixture is called ....
In the combustion of hydrogen .... is formed.

Jot down the missing words and then go on to 3 .

---

27 Your answer is wrong. Just imagine that the following experiment is performed:

Two bottles are filled with oil. Both are sealed and then placed in a water-bath so that they are completely immersed in water. The bottles are opened, one with its mouth pointing upwards and the other with its mouth pointing downwards. From the first of the bottles oil will immediately begin to flow out and rise to the surface, for oil is lighter than water. The oil in the second bottle having its mouth pointing downwards, however, will not flow out.

Read through 3 again.

---

28 Your answer is right. Hydrogen can be produced by the *electrolysis* of water.

In industry hydrogen is produced in a similar way. However, it is not water which is electrolyzed but rather brine, i.e. an aqueous solution of common salt.

Hydrogen is also formed by the action of acids on metals. The most convenient method of producing hydrogen in the *laboratory* makes use of the action of hydrochloric acid on zinc. Special gas generating equipment is used for this purpose, e.g. *Kipp's apparatus* (the inventor of this apparatus was called Kipp).

Now read the following paragraph several times until you have understood it fully; compare what is written with the information in the diagram.

When the stopcock H is opened the gas in volume B can escape. The hydrochloric acid can then flow from A via C into the globe B. In globe B there is a pile of zinc metal resting on a sieve. As soon as the hydrochloric acid comes into contact with the zinc, hydrogen gas is generated which flows out through the open stopcock H. When the stopcock H is closed, the pressure which builds up from the hydrogen still being generated forces the hydrochloric acid from B via C into A. As the hydrochloric acid will then no longer be in contact with the zinc, the production of hydrogen gas will cease.

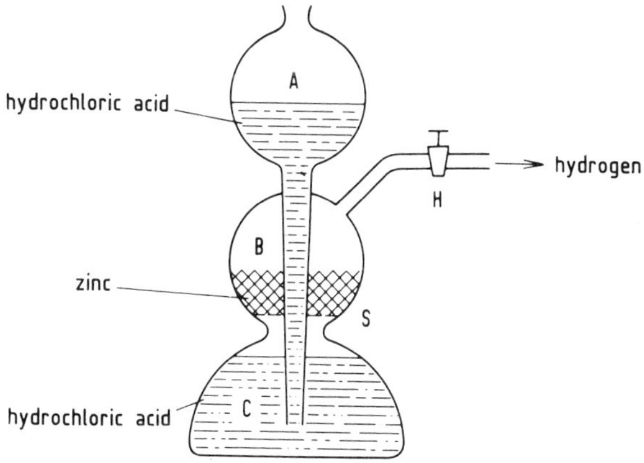

Copy down the above diagram of Kipp's apparatus (with all the labels!), and then continue reading at [18] .

29

a) An oxidation is a *chemical* process in which *a substance combines with oxygen.*

b) A reduction is a *chemical* process in which *oxygen is removed from a compound.*

End of Program 5.

30 No, please read through [21] , and analyze the experiment described there again.

31 Your answer is only partially correct.

Hydrogen can be produced in the laboratory from the electrolysis of water. This method is, however, rather inconvenient and far too slow. Such a method is used only occasionally for the preparation of hydrogen in the laboratory. Electrolysis is used though in industry. In the process adopted other valuable by-products are also produced. From the electrolysis of a brine solution the products formed are sodium hydroxide, chlorine and hydrogen.

Go back now to [23] .

32 Water gas is a mixture of *hydrogen* and *carbon monoxide.* It is formed when *steam* is passed over *red-hot coke.*

5

Hydrogen is available commercially in steel bottles which are painted red. Whilst oxygen steel bottles have a *right-handed* thread, those used for hydrogen have a *left-handed* thread. As a consequence of this, it is not possible to screw an oxygen valve into a hydrogen gas bottle and vice versa.

Unlike oxygen, which occurs naturally in the form of the element (e.g. in the air), hydrogen is seldom found in the free state in our environment. Hydrogen is far too reactive for that. Only in a few isolated instances is hydrogen found in the ground, e.g. together with natural gas or oil deposits. Thus, hydrogen usually occurs not as the element but in a combined state in various compounds. The most important of these compounds is water.
(In the upper reaches of the atmosphere – and especially in outer space – there is an abundance of hydrogen.)

With which element is hydrogen combined in water?

Jot down the name of the element, and then go on to 23 .

---

33 Your answer is quite right.

The most important method of preparing hydrogen in the laboratory involves the action of acids on metals, e.g. the action of hydrochloric acid on zinc.

Hydrogen is the lightest of all gases and as such is suitable for filling airships and balloons. Its drawback for this type of usage is, however, its

- Low density → 6
- High flammability → 22
- High density → 14

---

34 No, please read through 21 , and analyze the experiment described there again.

---

35 Think back to our discussion on water. There we said that water could be decomposed with the aid of direct electric current into hydrogen and oxygen. Hydrogen is given off at the cathode (at the negative electrode) and oxygen at the anode (at the positive electrode).

One method of producing hydrogen thus involves the .... of water.

Jot down the missing word, and then go on to 28 .

Program 6

Theoretical Foundations (I)

6

1 In the first five Programs a wide range of differing substances has been introduced. The substances that we have studied include:

iron, sulfur, iron sulfide, sulfur dioxide, nitrogen, oxygen, hydrogen, water, hydrogen chloride and chlorine.

In our discussion we have indicated how one may prepare

iron sulfide from iron and sulfur and
hydrogen chloride from chlorine and hydrogen.

In the present Program we shall be probing somewhat deeper into the topics covered this far, and in particular we shall be taking a look at the theoretical foundations of our subject.

Do you still remember the experiment in which iron sulfide was prepared from sulfur and iron?

First we made up a mixture of sulfur and iron by grinding these two elements together in a mortar and this was put into a test tube. After a short initial heating of the lower end of the test tube with a small flame, the mixture began to glow at the spot where the heat was applied. Now, although the test tube was then immediately removed from the flame, the glowing became increasingly intense until finally it enveloped all of the mixture. After cooling, the test tube was found to contain a uniform gray mass, i.e. the compound iron sulfide. It proved to be impossible to extract from this mass either the iron with a magnet or the sulfur with carbon disulfide.

What kind of process is involved in this preparation of iron sulfide from iron and sulfur?

- A physical process → 39
- A chemical process → 15
- I don't know exactly → 28

2

Solid:	Sulfur Phosphorus
Liquid:	Bromine

Gaseous: Oxygen
Nitrogen
Hydrogen
Chlorine

6

Up to the present we have always written out in full the names of the elements we have discussed in the various experiments described so far.

It is, however, much more practical when writing down chemical reactions to use abbreviated symbols rather than the full names of the elements. In fact, there is a standard abbreviation for each element which is known as its chemical symbol.

As examples, we now list the following symbols:

Hydrogen	H	Phosphorus	P
Nitrogen	N	Sulfur	S
Oxygen	O	Chlorine	Cl
Sodium	Na	Iron	Fe
Magnesium	Mg	Copper	Cu

Please memorize these symbols as we shall be using them later on instead of the chemical names.

What are the chemical symbols for the following elements?

Nitrogen
Magnesium
Chlorine

Jot down the relevant symbols, and then check your answers at 13.

---

3 Your answer is wrong. Phosphorus oxide is *not an element* but rather a compound of the elements phosphorus and oxygen.

Hydrogen chloride is *not an element* but rather a compound of the elements chlorine and hydrogen.

Go back now to 12 .

---

4 The missing words are:

electron shells

Write out the following sentence and insert, where appropriate, the missing words: negatively, positively or neutral:

Viewed externally an atom as a whole is electrically ...., because every atom has the same number of .... charged protons in its nucleus as .... charged electrons in its electron shells.

Check your answer at [14] .

5 Your answer is wrong.

Go now to [2] , write out the list of elements with their symbols given there, and then work your way through this Program again starting from [2] .

6 Your answer is wrong. As the name suggests, electrons are found in the electron sheath and not in the nucleus.

Continue your reading at [36] .

7 *chlorine, hydrogen*

The product which is formed, i.e. which is present at the end of a reaction, is the *end* product or *reaction* product.

From the .... materials sulfur and iron the end product of the reaction is ....

Jot down the missing terms, and then go on to [24] .

8 Your answer is wrong. You must have just guessed here, for all you need do is to count up the electrons in the diagram to arrive at the correct answer.

Read through [46] once again.

9 Your answer is wrong. You have just copied down the diagram of the hydrogen atom and written in the symbols of its component particles. Go back to [48] , compare your drawing with the figure there, and check to see if you have mixed up the three particles and their symbols.

10 Your answer is wrong.

Repeat now from [38] .

6

11 Your answer is correct if you wrote down the following metals:

iron, copper, gold, silver, aluminum, lead.

All metals are solid (under normal conditions).
All non-metals are solid or gaseous (under normal conditions).

There are, however, two exceptions:

Mercury is the only *liquid* metal (under normal conditions).
Bromine is the only *liquid* non-metal (under normal conditions).

(Normal conditions are: 1 bar pressure and 0 °C.)

About 75% of the 106 known elements are metals and the remaining 25% are non-metals.

(Incidentally, there are elements which may be classified as both metals and non-metals on the basis of their properties. The following elements however can be uniquely classified.)

Metals: iron, magnesium, copper, lead, gold, silver, platinum, zinc, aluminum and mercury.

Non-metals: sulfur, oxygen, nitrogen, hydrogen, chlorine, phosphorus and bromine.

State which of the *non-metals* listed here are solid, liquid or gaseous (under normal conditions).

Solid: ......
Liquid: ......
Gaseous: ...... → 2

12 Elements are substances which cannot be further decomposed by chemical means.

In which of the following groups are *only elements* listed?

- Copper oxide, nitrogen water → 31
- Oxygen, hydrogen, chlorine, iron → 21
- Phosphorus oxide, hydrogen chloride, sulfur → 3
- Hydrogen, detonating gas, water gas → 41

13

Nitrogen	N
Magnesium	Mg
Chlorine	Cl

Jot down now the symbols for the following elements:

Sodium
Sulfur
Copper

Check your answers at 25 .

14

The correct sequence of missing words is:

neutral
positively
negatively

Jot down the missing words in the sentences:

The smallest unit of an element capable of existence is an .... Every element consists of untold millions of identical ....

Check your answer at 30

15 Your answer is quite correct. The preparation of iron sulfide from iron and sulfur is a *chemical* process.

Let's now use this particular reaction to introduce some new nomenclature.

Since chemists refer to a chemical process as a *reaction,* we should also speak of the *reaction* of sulfur with iron. The sulfur *reacts* with the iron and iron sulfide is formed.

Write out the missing words in the following sentences:

Hydrogen chloride is formed when chlorine .... with hydrogen. This is referred to as a .... process or a .... → 33

16 Your answer is quite correct.

The nucleus of an atom consists of protons and neutrons, whilst the electrons make up the electron sheath.

Complete the following sentence:

The electrons are located in the electron sheath and move in fixed orbits, i.e. they are found in the ....

Check your answer at [4] .

6

---

[17] Your answer is wrong.

The nucleus consists of two types of particle.

You will find an explanation of this at [36] .

---

[18] Your answer is wrong. Repeat now from [38] .

---

[19] Your answer is wrong.

Go now to [2] , copy down the list of elements with their symbols given there, and then work your way through the Program once again from [2] .

---

[20] The missing words are:

conversion
compound

Jot down the missing words in the sentences:

In a chemical reaction the .... materials react with each other. In the process the .... is formed.

Check your answer at [40] .

---

[21] Your answer is quite right.

The chemical elements may be classified into two large groups, namely:

1. The metals
2. The non-metals

This classification may be made on the basis of certain easily recognizable properties:

Metals have a characteristic metallic *luster* and they conduct heat and electric current well.

Non-metals do not conduct heat and electric current well.

Which of the following elements are metals?

Oxygen, iron, copper, chlorine, gold, nitrogen, hydrogen, silver, aluminum, lead.

The metals listed here will probably be familiar to you as useful working materials. It should therefore not be too difficult to identify the metals and to jot down their names. → 11 .

---

22 Your answer is wrong. Read the explanation given at 74 .

---

23 Your answer is quite correct.

The fact that water may be decomposed into hydrogen and oxygen by means of electrolysis, demonstrates that water is a compound of the elements hydrogen and oxygen.

We are now going to discuss the *elements* in some detail.

To start with, you are asked now to write out again the definition of an element.
→ 12

---

24 *starting* materials
iron sulfide

Instead of saying "iron *reacts* with sulfur" one could equally well say "iron and sulfur are *converted* into iron sulfide". Similarly, the words *reaction* and *conversion* may be used interchangeably.

This is because we have two concepts here with essentially the same meaning.

Write down an equivalent word for each of the two expressions below:

reaction ≡ ....

reaction product ≡ .... → 37

25

Sodium Na
Sulfur S
Copper Cu

6

What elements are represented by the following symbols?

O
P
Cl

Jot down the names of the elements in question, and then check your answer at 35 .

26

The constituent particles of atoms are:

protons
electrons
neutrons

Which of these particles form the nucleus of an atom?

- Neutrons and electrons → 18
- Protons and electrons → 67
- Neutrons and protons → 16
- Neutrons, protons and electrons → 10

27

Your answer is quite right.

The hydrogen atom consists of two particles: a proton (p) which forms the nucleus and an electron (e) which forms the electron sheath around the nucleus. The hydrogen atom, the simplest possible atom, has *no* neutron in its nucleus.

The other elements, which are all heavier than hydrogen, consist of atoms having a larger number of component particles.

The next heaviest element is the inert gas helium (He). We have not discussed this up to the present time.

The helium atom has the following structure:

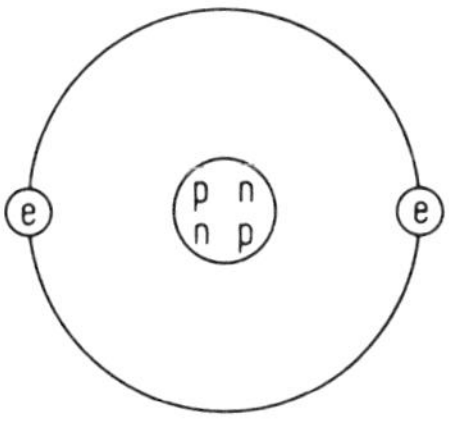

Copy down this diagram of the helium atom, and then write out the completed version of the following sentence:

The helium atom (He) has .... protons and two .... in its nucleus. It also has .... electrons in its ....

Check your answer at [46] .

---

[28] We have already learnt that:

Physical processes are those in which the substance remains unchanged and only a *change of state and associated properties* occurs.

Chemical processes are those in which the substances involved are *transformed into different substances.*

In the preparation of iron sulfide from sulfur and iron a *transformation of substance* occurs. The properties of iron sulfide are quite different from those of iron or sulfur.

Iron is attracted by a magnet.
Sulfur is soluble in carbon disulfide.
Iron sulfide is neither attracted by a magnet nor is it soluble in carbon disulfide.

The preparation of iron sulfide from sulfur and iron is thus a .... process. Continue your reading at [15] .

---

[29] The missing words are:

atomic nucleus

electron sheath

Whilst *electrons* make up the electron sheath, the nucleus of the atom is constructed from *neutrons* and *protons*. (The hydrogen atom is an exception to this rule in that its nucleus consists of only one proton.)

We thus know now where the protons, neutrons and electrons in an atom are located.

What are the constituent particles of the nucleus?

- Electrons, neutrons and protons → 17
- Electrons and protons → 58
- Neutrons and protons → 48
- Neutrons and electrons → 6

---

30 The missing words are:

atom
atoms

Write down the completed version of the sentence:

When two elements combine together in a chemical reaction, a .... takes place and a .... is formed.

Check your answer at 20 .

---

31 Your answer is wrong. Copper oxide is *not an element* but rather a *compound* of the elements copper and oxygen.

Water is *not an element* but rather a *compound* of the elements hydrogen and oxygen.

Go back now to 12 .

---

32 Yes, the preparation of iron sulfide from sulfur and iron is a *synthesis*.

The opposite of synthesis is *analysis*.

In chemistry the term *analysis* is used to signify not only the *decomposition of a substance*, but also the *investigation* of that substance, particularly in regard to its *composition* or purity.

In the past it was necessary to *decompose* any substance whose composition was to be determined. The word *analysis* thus came to be used for the investigation of a substance. Nowadays, there are methods available for determining the composition of a substance *without decomposing it.* Such procedures are, however, still described as *analytical* methods.

The investigation of air described in Program 2, which demonstrated that air is a mixture consisting primarily of 20% oxygen and 80% nitrogen, was in fact an *analysis* of the air.

Which of the procedures listed below is an analytical one?

- The combustion of hydrogen → 51
- The electrolysis of water → 23
- The oxidation of phosphorus → 43

---

33 The missing words are:

reacts, chemical, reaction.

As we *start out* from sulfur and iron when preparing iron sulfide, these two substances are referred to as the *starting materials.*

What are the starting materials in the following reaction?

Chlorine + Hydrogen → Hydrogen Chloride

Jot down the names of the two starting materials, and then go on to 7 .

---

34 Your answer is wrong.

You have ignored the rule which states that:

In every atom the number of protons is equal to the number of electrons.

Continue reading now at 53 .

---

35 The names of the elements are:

O Oxygen
P Phosphorus
Cl Chlorine

Now jot down the symbols for the following elements:

Hydrogen
Iron
Magnesium

Check your answers at 45 .

---

36 In atoms there are three constituent particles, namely:

Protons
Neutrons
Electrons

When considering the structure of the atom, we distinguish between the atomic *nucleus* and its surrounding *electron sheath.* The nucleus is constructed from protons and neutrons.

.... }	→	Nucleus	} Atom
.... }			
Electrons	→	Electron sheath	

Copy down this scheme, insert the missing words, and then go on to 48 .

---

37 Now check your answers:

reaction ≡ *conversion*
reaction product ≡ *end product*

Jot down the missing words in the following sentences. Where there is more than one suitable word, write down both of them.

The .... materials become glowing hot in the .... (....) of sulfur and iron. When the reaction is over, the reaction mixture cools off and we obtain iron sulfide as the .... (....) product.

Check your answers at 55 .

---

38 The missing words are:

atom
protons, neutrons, electrons

Now exactly how is an atom constructed?

The *nucleus* is located at the center of the atom. The electrons are located around the nucleus in the form of *electron shells*. Because the nucleus is surrounded by the electron shells, one frequently speaks of an *electron cloud* around the nucleus. (In the sketch below the relative sizes of the nucleus and the electron are not to scale.)

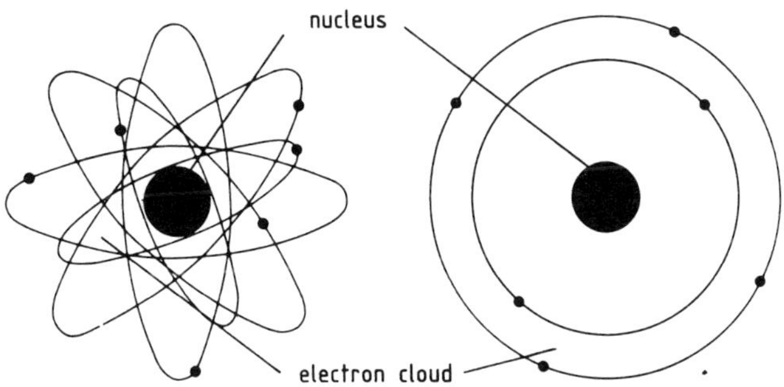

Atomic structure

Simplified representation of atomic structure

Now jot down the missing words in the following sentences:

Atoms have at their center an .... This is surrrounded by an ....

Check your answers at [29] .

---

[39] No; an explanation of this is given at [28] .

---

[40] The missing words are:

starting materials
end product or *reaction product*

Write out the completed version of the sentence:

The 106 elements known today may be classified into two groups: 25% are .... and 75% are .... .

Check your answers at [49] .

6

41 Your answer is wrong. Detonating gas is *not an element*, but rather a *mixture* of the elements hydrogen and oxygen.

Water gas is *not an element*, but rather a *mixture of* hydrogen and carbon monoxide.

Go back now to 12 .

42 Yes, hydrogen and chlorine are the *starting materials* for the preparation of hydrogen chloride.

The preparation of hydrogen chloride from hydrogen and chlorine may also be described as the *synthesis* of hydrogen chloride.

Whenever two or more substances combine together in a chemical reaction, we speak of a *synthesis*. The words preparation, production and synthesis are often used interchangeably in the present context.

Which of the procedures listed below is a synthesis?

- The formation of hydrogen and oxygen by electrolyzing water → 50
- The formation of a sugar solution by dissolving sugar in water → 60
- The formation of iron sulfide by reacting iron an sulfur → 32

43 Your answer is wrong. In the oxidation of phosphorus a phosphorus oxide is formed from phosphorus and oxygen. This is a *synthesis*, and *not* an analytic reaction!

Repeat now from 42 .

44 Your answer is quite right.

In every atom the number of protons is exactly equal to the number of electrons. Thus, because the phosphorus atom has 15 protons in its nucleus, it must also have 15 electrons in its electron cloud.

Both protons and electrons bear electric charges.

The proton is *positively* charged whereas the electron is *negatively* charged.

Neutrons bear no electric charge and so they are electrically *neutral*.

What will be the net charge on an atom viewed as a whole? Will it be positively or negatively charged, or will it be electrically neutral?

- An atom is overall positively charged as it contains positively charged protons in its nucleus → 76

- An atom is overall negatively charged as it contains negatively charged electrons in its electron cloud → 66

- An atom is overall electrically neutral. Because the number of protons equals the number of electrons, the positive and negative charges will cancel each other → 57

45

Hydrogen	H
Iron	Fe
Magnesium	Mg

Which elements do the following symbols represent?

Na
Fe
Cu

Jot down their names, and then check your answers at 54 .

---

46 The helium atom (He) contains *two* protons and two *neutrons* in its nucleus. It also has *two* electrons in its *electron cloud.*

The next heaviest element is the metal lithium (Li); so far we have not even mentioned this element.

The lithium atom has the following structure:

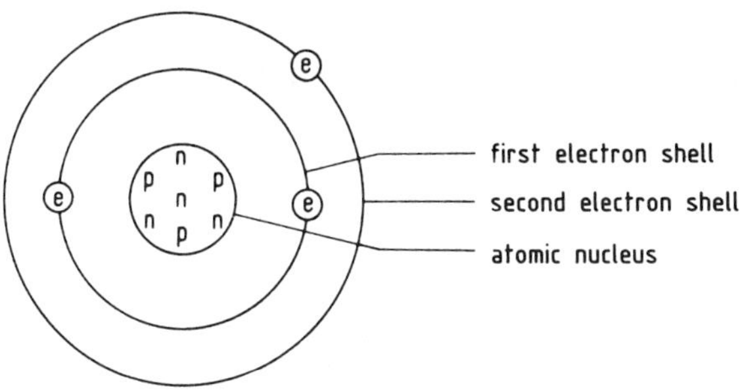

One notable feature of this structure is the fact that the three electrons which make up the electron cloud are actually in two different orbits. These orbits are described as *electron shells*.

Please copy down this diagram of the lithium atom.

6

How many electrons are located in the first shell and how many in the second shell of the lithium atom?

	First shell	*Second shell*		
●	1	2	→	70
●	2	1	→	56
●	3	0	→	8

47

The missing words are:

proton, positive
electron, negative
neutral

Let's now briefly recapitulate:

Write down the three component particles from which all atoms are constructed:

....
....
....

Check your answers at 26 .

48

Your answer is correct if you wrote:

Protons and neutrons together make up the atomic nucleus. Let's now take a closer look at the structure of atoms. Different types of elements are possible only because the three constituent particles of the atom can be arranged together in many differing combinations.

In the diagrams which follow we shall be using the following new abbreviations (symbols) to represent particles:

A neutron	n
A proton	p
An electron	e

Hydrogen (H), the lightest known element, also has the lightest atoms.

A hydrogen atom has the following structure:

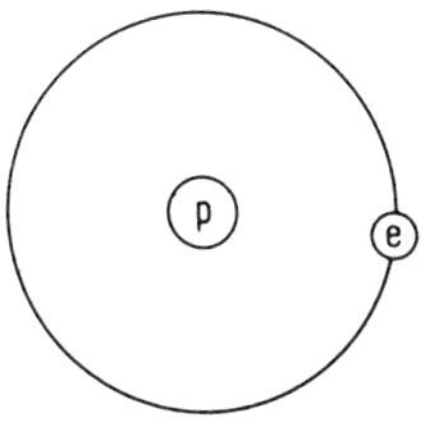

Copy down this diagram of the hydrogen atom, and then answer the question below.

What are the constituent particles in the hydrogen atom?

- 1 neutron, 1 proton and 1 electron → 75
- 1 neutron and 1 proton → 9
- 1 neutron and 1 electron → 64
- 1 proton and 1 electron → 27

---

49 The missing words are:

non-metals
metals

Jot down the missing word in the sentence:

In order to make the writing out of elements more convenient, an abbreviated representation or .... has been introduced for each.

Check your answer at 61 .

---

50 Your answer is wrong. The electrolysis of water is *not* a synthesis.

In a synthesis substances are *combined* together whereas in electrolysis water is *decomposed.*

Go back now to 42 .

---

51 Your answer is wrong. When hydrogen burns water is formed from the hydrogen and oxygen. This process is thus a synthesis, namely the *synthesis* of water, and not an analytic process.

Repeat now from 42 .

52 The missing words are:

atom
nitrogen atom
phosphorus atom

The smallest unit of the element copper capable of existence is the *copper atom.*

The smallest unit of the element chlorine capable of existence is the *chlorine atom.*

It should now be evident that each element is made up from a differing type of atom.

Atoms are in fact *exceedingly* minute. They cannot be seen under any microscope. This should be immediately obvious to you when we mention that 1 gram of hydrogen contains 602000000000000000000000 hydrogen atoms. This incredible number (602 followed by 21 zeros) is not worth committing to memory; it is given only to indicate how *tiny* hydrogen atoms are. The mass of a hydrogen atom is thus equal to 1 gram divided by this unbelievably large number. A similar situation prevails in the case of the other elements. Yet, in spite of their minute size, atoms are actually constructed from even smaller particles.

The investigations of atomic physics have revealed that all atoms of the various elements consist basically of three particles. These constituent particles are:

The proton
The neutron and
The electron.

Now jot down the missing words in the sentences below:

The smallest unit of an element capable of existence is an ....

The constituent particles of an atom are called ....

Check your answers at 38 .

53 The nucleus of the phosphorus atom consists of 15 protons and 16 neutrons. A general rule holds that in every atom the number of protons is equal to the number of electrons.

The phosphorus atom has 15 protons in its nucleus.
The phosphorus atom will thus have .... electrons in its electron cloud.

Jot down the missing number, and then go to 44 .

---

54 The names of the elements are:

Na sodium
Fe iron
Cu copper

Which elements have the symbols N, O and S?

- Sodium, phosphorus and nitrogen → 19
- Hydrogen, chlorine and oxygen → 5
- Nitrogen, oxygen and sulfur → 63

---

55 The missing words are:

starting materials
reaction (conversion)
end (reaction) product

If you made more than two mistakes here, it would be in your own interest to go back to 33 . If you got everything right, you are coming along well.

Hydrogen reacts with chlorine to form hydrogen chloride.

How would you describe the hydrogen and chlorine in this instance?

- As end products → 68
- As reaction products → 22
- As starting materials → 42
- I don't know → 74

6

56 Your answer is correct.

In the lithium atom there are two electrons in the first electron shell and one electron in the second shell.

We may continue in the same sort of way with the other atoms. As more protons and neutrons are successively added to the nucleus, further electrons are fed into the electron shells which make up the electron cloud.

A general rule which applies here is that:

In every atom the number of protons in the nucleus is always equal to the number of electrons in the electron cloud.

The hydrogen atom	has 1 proton and 1 electron.
The helium atom	has 2 protons and 2 electrons.
The lithium atom	has 3 protons and 3 electrons.

The nucleus of the phosphorus atom consists of 15 protons and 16 neutrons.

How many electrons will there be in the electron cloud of a phosphorus atom?

- 31 electrons → 34
- 16 electrons → 73
- 15 electrons → 44
- 10 electrons → 65
- I don't know → 53

57 Your answer is quite right.

In every atom the number of protons equals the number of electrons. The charges thus mutually cancel each other and the atom as a whole is electrically *neutral.*

Write out now completed versions of the sentences:

The nucleus of the hydrogen atom consists of one .... (whose charge is ....). The electron cloud of the hydrogen atom consists of one .... (whose charge is ....).

Overall, the hydrogen atom is electrically .... → 47

58 Your answer is wrong. As the name implies, the electrons are found in the electron cloud and not in the nucleus.

Continue your reading at 36 .

59 The missing words are:

poorly
bromine

Write down completed versions of the sentences:

Metals conduct electric current .... Under normal conditions there is only one liquid metal, namely ....

Check your answers at 69 .

60 Your answer is wrong. The dissolving of sugar in water is *not* a synthesis.

A synthesis is a *chemical* process whereas the dissolving of sugar in water is a *physical* process.

Go back now to 42 .

61 *symbol*

Jot down the symbol for each of the following elements:

Hydrogen
Nitrogen
Sulfur
Oxygen
Iron

Check your answers at 71 .

62 Now check your answer:

Elements are substances which cannot be further decomposed by chemical means.

Now you may be wondering:

What do elements actually consist of?

The smallest unit of an element capable of existence is an atom.

6

The smallest unit of the element oxygen capable of existence is an oxygen atom.

The smallest unit of the element sulfur capable of existence is a sulfur atom.

The smallest unit of the element iron capable of existence is an iron atom.

Jot down now the missing words below:

The smallest unit of an element capable of existence is an ....

The smallest unit of the element nitrogen capable of existence is a ....

The smallest unit of the element phosphorus capable of existence is a ....

Check your answers at 52 .

---

63 Your answer is quite right.

N is the symbol for nitrogen, O the symbol for oxygen, and S the symbol for sulfur.

With the aid of chemical symbols we can very simply and conveniently formulate the starting materials in a reaction as well as the end products which are produced.

The first chemical reaction we came across was the reaction of iron with sulfur in which iron sulfide was formed.

In words, the reaction may be written out as:

Iron + Sulfur → Iron Sulfide

If symbols now replace the words the reaction becomes:

$$Fe + S \rightarrow FeS$$

When oxygen is passed over heated copper, black copper oxide is tormed.

In words, we may write out the reaction as:

Copper + Oxygen → Copper Oxide

Write this reaction out now in terms of symbols, and then check your answer at [72].

64 Your answer is wrong. You have sketched the hydrogen atom yourself and have entered in the symbols for its constituent particles.

Go back now to [48], and check your sketch with the figure shown there, to see if you have mixed up the symbols for the three constituent particles of atoms.

65 Your answer is wrong.

You have ignored the rule which states that:

In every atom the number of protons is always equal to the number of electrons.

Continue your reading at [53].

66 No, you should note that a positive and a negative electric charge mutually cancel each other.

As the number of protons is always equal to the number of electrons, the number of positive charges is equal to the number of negative charges in an atom.

Accordingly, an atom as a whole is electrically .... [57].

67 Your answer is wrong. Repeat now from [38].

68 Your answer is wrong. Read the explanation given at [74].

69 The missing words are:

well
mercury

Jot down the name and symbol of each of the non-metals in the following list of elements:

Hydrogen, sulfur, copper, chlorine, iron, nitrogen, phosphorus, oxygen.

Check your answers at 77 .

6

70 Your answer is wrong. You must have guessed here, for all that you need do here is to count up the electrons in the figure to arrive at the correct answer.

Read through 46 once again.

71

Hydrogen	H
Nitrogen	N
Sulfur	S
Oxygen	O
Iron	Fe

If you got some of these symbols wrong, you are advised to work your way through this Program again fairly soon.

Write out now completed versions of the sentences:

Non-metals conduct electric current ....

Under normal conditions there is only one liquid non-metal, namely ....

Check your answer at 59 .

72 $Cu + O \rightarrow CuO$

Before we start to write down all of the reactions discussed above in terms of symbols, it will be necessary first to take a closer look at the elements.

We already know that:

Elements are substances which ....

Write out this sentence, insert the missing words, and then check your answer at 62 .

73 Your answer is wrong.

You have ignored the rule which states that:

In every atom the number of protons is always equal to the number of electrons.

Continue your reading at 53 .

---

74 *Starting* materials are those substances which one *starts* with in a chemical reaction.

End products or reaction products are those substances which are present at the *end* of the reaction. They are in fact substances formed during the reaction.

When hydrogen reacts with chlorine to form hydrogen chloride, the *hydrogen and chlorine* are the ....

Continue your reading at 42 .

---

75 Your answer is wrong. You have sketched the hydrogen atom yourself and should therefore know that it consists of *two* particles.

Go back now to 48 .

---

76 No, you should be aware that a positive and a negative electric charge mutually cancel each other.

As the number of protons is always equal to the number of electrons, the number of positive charges is equal to the number of negative charges in an atom.

Accordingly, an atom as a whole is electrically .... → 57

---

77 The non-metals in the list are:

Hydrogen	H
Sulfur	S
Chlorine	Cl
Nitrogen	N
Phosphorus	P
Oxygen	O

End of Program 6.

Program 7

Theoretical Foundations (II)

1 In this Program you will learn how the composition of a chemical compound may be expressed in terms of a *chemical formula*. In addition, we shall also be discussing how a chemical reaction may be represented by means of an *equation*. We shall, of course, assume that you are fully conversant with the material covered in the previous Program 5.

Now, do you remember what the symbols Fe, P, O, H, S, and K represent?

Jot down the names of the elements represented here thus:

Fe :
P :
etc.

Then check your answers at [10] .

---

2 *atoms, atoms*

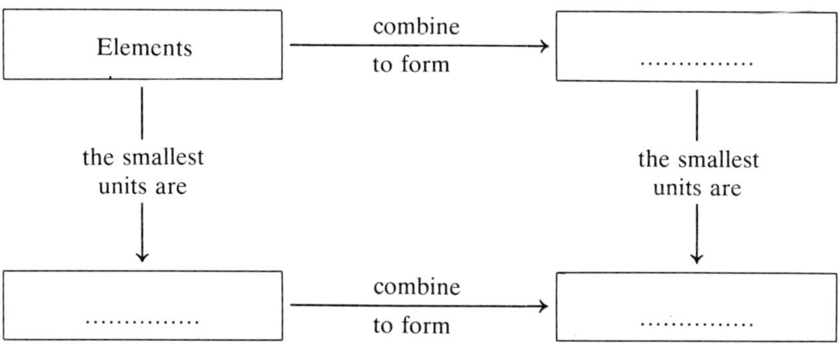

Complete this flow chart and then check your answers at [11] .

---

3 Phosphorus is an *element* whereas phosphorus oxide is a *compound*. Your answer therefore cannot be right.

Go back to [20] , and take a careful look at the terms used to describe the smallest units of *elements* and the smallest units of *compounds*.

4 No. Consider for a moment the fact that the electrically neutral chlorine atom accepts one electron, which bears a *negative electric* charge. The number of protons is then no longer equal to the number of electrons. In fact, the number of electrons will exceed the number of protons by *one*.

A chlorine atom which has accepted an electron thus bears a .... charge. → 32

---

5 Your answer is wrong.

Earlier on we learnt that water is composed of hydrogen and oxygen. Water is therefore a *compound.*

Write out the completed version of the sentence:

The smallest unit of a chemical compound capable of existence is a .... → 12

---

6 No, water actually consists of two *elements,* namely hydrogen and oxygen. In the water molecule, however, three *atoms* are united together, namely two hydrogen atoms and one oxygen atom. The fact that there are two hydrogen atoms is indicated in the formula H_2O by the subscript 2 on the right of the H atom.

How many atoms are there in a molecule of the compound H_2S?

- Two atoms → 15
- Three atoms → 29

---

7 Your answer is wrong.

You will find an explanation of this at 13 .

---

8 No, the 3 in front of the H_3PO_4 indicates that the *whole* molecule appears three times.

The number of atoms present is therefore:

3 x 3 H atoms =	9 H atoms
3 x 1 P atoms =	3 P atoms
3 x 4 O atoms =	12 O atoms
Total number	24 atoms

How many atoms in all are there in $4H_2SO_3$? → 30

---

9 The gist of your answer should be that:

An ion is an electrically charged atom (or an electrically charged group of atoms).

Sodium ions ($Na^{\oplus}$) and chlorine ions ($Cl^{\ominus}$) bear opposite electric charges.

It is well known that opposite charges attract each other.

This electrical attraction is responsible for the formation of the compound sodium chloride (NaCl).

Sodium chloride consists of sodium ions and chlorine ions.

Sodium chloride (NaCl) and the hydrogen molecule (H_2) provide convenient examples of two quite different types of chemical bonding.

In diatomic molecules we make a distinction between:

1. Compounds in which the bonding electron pair is located on one of the two atoms, this resulting in the formation of ions, e.g. in NaCl.

2. Compounds in which the bonding electron pair is located midway between the two bonded atoms, e.g. in H_2.

In polyatomic molecules corresponding distinctions are made for each of the individual bonds present.

Which of the following compounds consists of ions?

- H_2 → 25
- NaCl → 19
- H_2 and NaCl → 72

10

Fe:	iron	P:	phosphorus	O:	oxygen
H :	hydrogen	S:	sulfur	K:	potassium

7

Write out appropriate definitions as answers to the following questions:

1. What is a chemical element?
2. What is a chemical compound?
3. What is an atom?

Check your answers at [20] .

---

[11] Elements

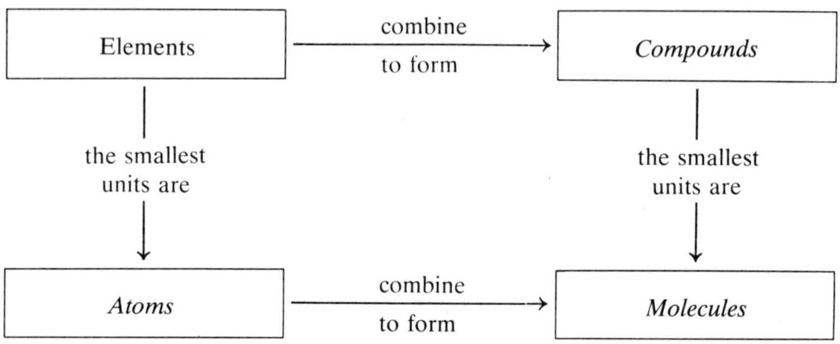

What are the smallest possible units of

phosphorus and
phosphorus oxide

capable of existence?

Phosphorus	Phosphorus oxide	
● atoms	molecules	→ [21]
● atoms	atoms	→ [3]
● molecules	molecules	→ [33]
● molecules	atoms	→ [41]

---

[12] Your answer is quite right.

The smallest unit of water capable of existence is a *molecule* (a water molecule).

Let's turn now to the following question:

If we consider the smallest possible units of iron sulfide, water, magnesium, hydrogen chloride and iron, what types of different units will be represented?

In all there will be a total of:

- 5 molecules → 7
- 4 molecules and 1 atom → 36
- 3 molecules and 2 atoms → 28
- 2 molecules and 3 atoms → 44
- 1 molecule and 4 atoms → 52
- 5 atoms → 61

13 In order to fix in your mind the difference between atoms and molecules, you are advised to learn the following sentences by heart:

The smallest unit of a chemical *compound* capable of existence is a *molecule.*

The smallest unit of a chemical *element* capable of existence is an *atom.*

Or are you now uncertain as to which of the substances listed are elements and which are compounds?

Iron sulfide was obtained by reacting sulfur powder with iron filings. Water is formed during the combustion of hydrogen. Magnesium is an element. Hydrogen chloride consists of hydrogen and chlorine. Iron is an element.

Go back now to 12 .

14 When setting up equations one must ensure that the same number of *atoms* appears on both the left and the *right* of the arrow (or equality sign).

However, in setting up equations one must ensure not only that the number of atoms on each side is equal but also that for *every element* the *same* number of atoms is present, on both sides.

Thus, if *one* iron atom appears on the left hand side of an equation, then *one* iron atom must also appear on the right hand side. In this context it does not matter whether the iron is present as the element or in a compound.

We shall now consider molecules in which a given element appears more than once.

7

H_2O is water, or more precisely: H_2O is the *formula* for water. How many atoms is the water molecule composed of?

- Two atoms → 6
- Three atoms → 29

---

15 You are not taking sufficient care! Your answer is wrong.

The small subscript 2 in the formulas H_2O and H_2S tells us that two hydrogen atoms are present in each of these molecules.

H_2S consists of	2 hydrogen atoms and 1 sulfur atom
making a total of	3 atoms
H_2O thus consists of	2 hydrogen atoms and 1 oxygen atom
making total of	...... atoms

Continue your reading at 29 .

---

16 $Cl^{\ominus}$

What is an ion? Write down your answer in full, and then go on to 9 .

---

17 Your answer is quite correct.

The molecule having the formula H_3PO_4 consists of 8 atoms, namely of 3 hydrogen atoms, one P atom and 4 O atoms.

The small subscripts in formulas refer *only* to the atom *immediately preceding* them.

If, on the other hand, a number is written in *front* of a formula, e.g. 2FeS, this indicates that the corresponding number of molecules is present. In our case this would mean that two molecules of FeS were present.

How many atoms in all are present in $3H_3PO_4$?

- 11 atoms → 40
- 14 atoms → 8
- 24 atoms → 30

7

---

18 The correct equation is: $2Na + Cl_2 \rightarrow 2NaCl$

If you remembered here that chlorine is *di*atomic, you surely got this equation right. Do not forget when setting up equations that certain elements always occur in *di*atomic form. Jot down now the names and symbols for these elements:

Bromine	Br_2
....	......
....	......
....	......
....	......

Check your answers at 26 .

---

19 Your answer is quite right.

Sodium chloride (NaCl) consists of ions. The bond is formed as a consequence of the attractive force existing between the oppositely charged ions ($Na^{\oplus}$ and $Cl^{\ominus}$). This type of bonding is referred to in general as:

Ionic Bonding

In the hydrogen molecule the bond comes about as a result of the formation of an electron pair located midway between the two hydrogen atoms. This electron pair actually belongs to the electron clouds of *both* the hydrogen atoms. The type of bonding here is referred to in general as:

Atomic Bonding

For the present we shall therefore distinguish between two different types of bonding, namely:

1. Ionic Bonding
2. Atomic Bonding

What type of bonding is present in the chlorine molecule (Cl–Cl)?

- Ionic bonding → [66]
- Atomic bonding → [31]
- I don't know → [71]

---

20 The gist of your answer should be that:

1. *Elements* are substances which cannot be further decomposed by chemical means.

2. If two or more substances combine together to form a new substance having different properties, the new substance is a *chemical compound.*

3. The smallest unit of an element capable of existence is an *atom.*

We shall now proceed to a new definition:

The smallest unit of a compound capable of existence is a *molecule.*

What does a molecule consist of? Remember that:

Compounds are formed during a chemical reaction by the *combination of elements.* Molecules (the smallest units in a compound) are thus formed by the *combination of* ...., the smallest units of elements.

Conversely, *compounds* can be *decomposed* into *elements* by means of a chemical reaction. The molecules of the compound are then split up into ....

Jot down the missing words, and then check your answers at [2] .

---

21 Your answer is quite right.

Phosphorus is an *element*; accordingly the smallest unit of phosphorus capable of existence is an *atom* (a phosphorus atom).

Phosphorus oxide is a *compound*; accordingly the smallest unit of phosphorus oxide capable of existence is a *molecule* (a phosphorus oxide molecule).

What is the smallest unit of water capable of existence?

- An atom → 5
- A molecule → 12

---

22 Your answer is quite correct.

The sodium has donated one electron which, together with one electron from the chlorine, forms an electron pair that is located on the chlorine atom:

Na :Cl

As a result of this, one *electron* is now *missing* from the electron cloud of the sodium atom and, correspondingly, there is *one new electron* in the electron cloud of the chlorine atom.

Now we know that electrons bear an electric charge.

What electric charge will a chlorine atom have after accepting an electron?

- It will be negatively charged → 32
- It will be positively charged → 4
- It will be electrically neutral → 74
- I don't know → 63

---

23 No, protons and neutrons are the constituent particles of the *nucleus* of an atom.

The .... are located in the surrounding charge cloud. → 69

---

24 Your answer is quite right.

A sodium atom which has donated one electron is electrically *positively* charged. As the number of protons now *exceeds* the number of electrons by *one,* the sodium atom as a whole will now bear *one positive charge.*

Electrically charged atoms (or groups of atoms) are called *ions*. An ion is either positively or negatively charged.

A positive charge is indicated by means of a plus sign written as a superscript whereas a negative charge is indicated by means of a minus superscript sign.

Accordingly, the sodium ion is written as $Na^{\oplus}$.

7

Now write down the symbol for a chlorine ion. → 16

---

25 No. We have already made it clear that ions are formed when the bonding electron pair is located on one of the two bonded atoms. This is certainly not the case for the hydrogen molecule.

Repeat now from 49 .

---

26 The answers are:

Chlorine Cl_2, Oxygen O_2, Nitrogen N_2, Hydrogen H_2, Iodine I_2, Fluorine F_2.

What factors are missing from the following equation?

$$H_2 + N_2 \rightarrow NH_3$$

Check your answer at 35 .

---

27 The correct equation is: $2H_2 + O_2 \rightarrow 2H_2O$

Write down now the completed version of the equation below, i.e. insert the missing starting material and the missing factors.

$$Na + \ldots\ldots \rightarrow NaCl$$

Check your equation at 18 .

---

28 Your answer is quite right.

At this point we shall indicate how chemical processes may be described by means of *equations*.

If we wish to describe the reaction of sulfur with iron to form iron sulfide, we make use of the appropriate symbols of the elements involved. In this particular instance we would write:

$$Fe + S \rightarrow FeS$$

We have now set up an *equation*. This kind of representation is referred to as an equation because in front of and after the arrow (or, equivalently, on the left and right of the arrow) the *same* number of atoms (of the same type) must always be present.

Fe here means not just iron but specifically 1 atom of iron.
S means specifically 1 atom of sulfur.
FeS means specifically 1 molecule of iron sulfide.

As the FeS molecule consists of two atoms, two atoms must appear on both the left and right hand sides of the equation. The two atoms on the right, of course, together form one molecule.

Fe	+	S	→	FeS
1 atom of iron		1 atom of sulfur		1 molecule of iron sulfide (consisting of one atom each of iron and sulfur)

Instead of an arrow (→) one often uses the equality sign (=) in chemical equations, thus:

$$Fe + S = FeS$$

Write down now the completed version of the following sentence:

When setting up equations one must ensure that both on the left and .... of the arrow (or equality sign) the same number of .... is present.

Check your answer at [14] .

---

[29] Your answer is quite correct.

H_2O consists of three atoms. The small subscript 2 on the right of the hydrogen in the formula indicates that there are two hydrogen atoms present.

How many atoms are there in a molecule of H_3PO_4?

- 8 atoms → [17]
- 3 atoms → [48]
- I don't know → [39]

---

30 Your answer is quite right.

There are in all 24 atoms present.

If a *number* is placed *in front of* the symbol of an element, e.g. 2Fe, this indicates that there are as many atoms present as the number specifies. In this case it would mean that there are two iron atoms present.

8S thus means eight sulfur atoms, 5P means five phosphorus atoms, and so on.

In the formulas of *compounds,* however, a small subscript number placed after the symbol of an element indicates the number of atoms of that type present in the molecule. An example of this was provided by the formula

H_3PO_4

which we discussed above.

Now write down using appropriate symbols and numbers each of the following:

Four sodium atoms ....
One iron atom ....
The formula for water ....
The formula for iron sulfide ....
The formula for sulfuric acid ....
(The sulfuric acid molecule consists of two hydrogen atoms, one sulfur atom and four oxygen atoms.) → 38

31 Yes, an atomic bond is involved here.

An atomic bond is always present when the bonding electron pair is located midway between the two bonded atoms. (In this case there can be *no* ionic bond present. No ions are formed, for the two *atoms* in question are bonded together. This fact gives rise to the name: *atomic* bonding.)

It is necessary to distinguish between ions and atoms. What is the principal difference between an ion and an atom?

Write out your answer, and then go on to 42 .

32 Your answer is correct: the chlorine atom becomes *negatively* charged after accepting an electron.
The number of its electrons will then exceed the number of its protons by *one*.

In other words, we could say that:
The number of negative charges now exceeds the number of positive charges by *one* in the chlorine atom.
Thus, after the acceptance of one electron, the chlorine atom as a whole bears *one negative charge.*

What will be the charge on a sodium atom after it has *donated* one of its electrons?

- It will be negatively charged → 57
- It will be positively charged → 24
- It will be electrically neutral → 73
- I don't know → 45

---

33 Phosphorus is an *element* whereas phosphorus oxide is a *compound.* Your answer cannot therefore be right.

Go back now to 20, and take good note of what the smallest units of *elements* and *compounds* are called.

34

1. In the hydrogen molecule the electron pair is located *midway* between the two hydrogen atoms.

2. In sodium chloride the electron pair is located *on* the chlorine atom.

Thus, in sodium chloride the electron pair belongs to the *chlorine atom.*

Where does this electron pair come from?

- From two of the sodium electrons → 67
- From two of the chlorine electrons → 58
- From one of the sodium electrons and one of the chlorine electrons → 22

---

35 The correct equation is: $3\ H_2 + N_2 \rightarrow 2\ NH_3$

The difficulty in balancing this equation arises from the fact that there are three hydrogen atoms on the right, i.e. an *uneven* number. On the left, however, there can only ever be an even number of hydrogen atoms: **2** in H_2 which upon doubling gives **4** (2 H_2) and upon trebling **6** (3 H_2). Now doubling of the right hand side (2 NH_3) also results in an *even* number of hydrogen atoms, namely 6.

Write out the completed version of the equation:

$$C + ... \rightarrow CO$$

Check your equation at 43 .

---

36 Your answer is wrong.

You will find an explanation of this at 13 .

---

37 $4\ P + 5\ O_2 \rightarrow 2\ P_2O_5$

Two factors are also missing in the following equation:

$$H_2 + O_2 \rightarrow H_2O$$

Remembering that oxygen is diatomic, we are *obliged* to start on the left hand side with two oxygen atoms (written as O_2). You are now asked to write down the correct equation, and then check your answer at 27 .

---

38 Check your answers now:

Four sodium atoms	4 Na
One iron atom	Fe (without a 1!)
The formula for water	H_2O
The formula for iron sulfide	FeS
The formula for sulfuric acid	H_2SO_4

In both the case of the *gaseous* non-metals (under normal conditions) and the non-metals fluorine (F), bromine (Br) and iodine (I), allowance must be made for the fact that:

All of these non-metals (apart from the inert gases) normally occur in the form of *molecules* consisting of *two* atoms.

Accordingly, they should be written down as:

$$H_2, N_2, O_2, F_2, Cl_2, Br_2, \text{and } I_2$$

since these particular non-metals are *diatomic.*
We are thus dealing here with molecules formed not from the *combination* of atoms *from two or more elements* but rather exceptionally from the combination of atoms from one *single* element.

Note down now which of the following elements are diatomic, giving the symbol in each case:

> Potassium, Oxygen, Hydrogen, Iron, Copper, Sodium, Bromine, Magnesium, Chlorine, Nitrogen, Phosphorus, Sulfur.

The diatomic elements are: O_2, ... → 47

7

---

39 H_3PO_4 consists of three different *elements,* namely hydrogen, phosphorus and oxygen.

The small subscript 3 after the H tells us that three H atoms are present in each molecule.

Furthermore, every molecule of the compound H_3PO_4 also contains 1 P atom and 4 O atoms.

Go back now to 29 .

---

40 No, the 3 in front of the H_3PO_4 indicates that the *whole molecule* appears three times.

The atoms present may thus be added up as follows:

3 x 3 H atoms =	9 H atoms
3 x 1 P atom =	3 P atoms
3 x 4 O atoms =	12 O atoms
In total	24 atoms

How many atoms in all are there in 4 H_2SO_3? → 30

---

41 Phosphorus is an *element* whereas phosphorus oxide is a *compound.* Your answer cannot therefore be right.

Go back now to 20 , and take good note of what the smallest units of *elements* and *compounds* are called.

7

42 Ions bear an electric charge.
Atoms are as a whole electrically neutral.

What types of electric charge are found on ions and how are they symbolized?

Write your answer down, and then go to 54 .

---

43 The correct equation is: $2\ C + O_2 \rightarrow 2\ CO$

Here too it is necessary to double the right hand side to attain an even number of oxygen atoms.

How is it that atoms of a given element appear only *once* in some compounds whereas they appear *several times* in other compounds? For instance, the *hydrogen* in hydrogen chloride appears only once (HCl) but twice in water (H_2O).

The answer is that each element has a certain *valence*. This term specifies the *number of bonds* which the atom concerned can form with other atoms.

Hydrogen is *monovalent*, for it can form *one* bond.
Oxygen is *bi*valent, for it can form *two* bonds.

If one symbolizes the bond between two atoms by means of a line, the hydrogen molecule can be represented as:

H–O–H

By counting up the lines representing bonds, decide now which elements are monovalent, which bivalent and which tervalent in the following formulas:

$$\text{H–S–H} \quad \text{H–Cl} \quad \text{Fe=S} \quad \underset{\displaystyle \text{H}}{\text{H–}\underset{|}{\text{N}}\text{–H}} \quad \text{Fe=O}$$

The monovalent elements are: ....
The bivalent elements are: ....
The tervalent elements are: ....

Check your answers at 53 .

---

44 Your answer is wrong.

You will find an explanation of this at 13 .

45 No. You must not forget that the electrically neutral sodium atom donates one electron (and with it goes one negative charge).

After donating this electron, the number of protons in the sodium atom will *exceed* the number of electrons by *one*.

Thus, a sodium atom which has donated one electron is .... charged.

→ 24

7

46 $H_2 + Cl_2 \rightarrow 2\ HCl$

No doubt you realized that a 2 should be placed in front of the HCl. Such numbers placed *in front of* atomic symbols in an equation are called *factors*. Thus, a factor of two was required in front of the HCl.

In the following equation two factors are missing.

$$P + \quad O_2 \rightarrow 2\ P_2O_5$$

Write out the completed version of this equation, and then check your answer at 37 .

47 The diatomic elements are: O_2, H_2, Br_2, Cl_2, N_2

Whenever these non-metals react chemically, they always appear in the equation with an *even* number of atoms. Thus, because they are *di*atomic, they must be present as 2, 4, 6, etc. atoms.

2 oxygen atoms are written as: O_2
4 oxygen atoms are written as: $2\ O_2$
6 oxygen atoms are written as: $3\ O_2$

In compounds, however, these elements can appear with any number of atoms, e.g.

H_2O
SO_2
O_3
H_2SO_4
P_2O_5

(Look particularly at the oxygen in these examples.)

Write down now each of the following chemical entities using symbols for the elements in question:

2 oxygen atoms, 8 nitrogen atoms, 4 chlorine atoms, 12 bromine atoms. → 56

7

---

48 Your answer is wrong.

H_3PO_4 clearly consists of three different *elements* (hydrogen, phosphorus and oxygen). Both the hydrogen and the oxygen are *multiply* present in the molecule. The small subscript numbers (3 and 4) tell us respectively the number of hydrogen and oxygen atoms present.

Go back now to 29 .

---

49 *electron pair*

We shall continue using the method of writing out structural formulas in which a line symbolizes a bond between two atoms. The line actually represents the *electron pair* responsible for the formation of the bond.

In many instances the electron pair is located (more or less) *midway* between the two atoms bonded together. Examples of this are found in the molecules

H–H Cl–Cl

In other instances, however, the electron pair is (more or less) *displaced* over towards one of the two atoms involved. This is especially evident in the case of sodium chloride (NaCl):

Na –Cl or, alternatively, Na :Cl

Write down answers now to the following questions:

1. Where is the electron pair located in the hydrogen molecule?

2. Where is the electron pair located in sodium chloride? → 34

---

50 No, protons and neutrons are the constituent particles of the atomic *nucleus*.

The .... are located in the surrounding charge cloud. → 69

51 **No, all you need do here is to count the lines radiating from the nitrogen atom in the structural formula**

```
H–N–H
  |
  H
```

to determine the valence of the nitrogen.

Repeat now from 43 .

---

52 Your answer is wrong.

You will find an explanation of this at 13 .

53

The monovalent elements are:	hydrogen, chlorine
The bivalent elements are:	iron, oxygen, sulfur
The tervalent element is:	nitrogen

As chlorine is thus monovalent, it can form a single bond:

H–Cl

This explains why two hydrogen atoms appear in the water molecule whereas only on appears in the hydrogen chloride molecule. It is due quite simply the different valencies of oxygen and chlorine.

H–O–H H–Cl

Such formulas enable us to form a mental image of the structure of these molecules. In fact it is for this reason that formulas of this sort are referred to as *structural formulas.*

The structural formula of the water molecule is drawn with an .... atom in the middle and a .... atom on both its left and right.

What are the missing words here? Check your answer at 62 .

---

54 Ions may be either positively (+) or negatively (−) charged.

What kinds of chemical bonding are you now familiar with?

....
.... → 64

---

55 Yes, the equation
$2\ H_2O \rightarrow 2\ H_2 + O_2$

is in fact correct. On both its left and right hand sides there is a total of four H atoms and two O atoms. (By the way, the equation represents the formation of hydrogen and oxygen during the electrolysis of water.)

In the equation below (in which hydrogen reacts with chlorine to form hydrogen chloride) a factor is missing in front of the HCl, i.e. the equation is incorrect as it stands.

$H_2 + Cl_2 \rightarrow \quad HCl$

Write down the correct version of this equation, and then check your answer at 46 .

---

56 Now check your answer:

O_2, 4 N_2, 2 Cl_2, 6 Br_2

All other elements (with the exception of the inert gases) normally occur in the form of large clusters of atoms. As the exact size of these clusters is not usually known, such elements are regarded as being *monatomic* for purposes of writing down chemical equations. Thus, they may appear in equations with either an *even* or an *uneven* number of atoms, e.g. Fe, 2 Fe, 3 Fe, 4 Fe, etc.

Write down now each of the following chemical entities using symbols for the elements in question:

4 iron atoms, 6 chlorine atoms, 5 phosphorus atoms, 2 bromine atoms, 1 sulfur atom, 3 sodium atoms, 4 nitrogen atoms, 8 hydrogen atoms.

Check your answers at 65 .

---

57 No. You must not forget that the electrically neutral sodium atom donates one electron (and with it goes one negative charge).

After donating this electron, the number of protons in the sodium atom will *exceed* the number of electrons by *one*.

Thus, a sodium atom which has donated one electron is ....
charged. → 24

---

58 No, the electron pair is made up of one electron from the sodium atom and one from the chlorine atom.

Repeat now from 77

---

59 $H^{\bullet} + {}^{\bullet}H \rightarrow H : H$

By forming this electron pair, the two hydrogen atoms have become *bonded* together to form a hydrogen molecule. Thus a *chemical bond* results from this formation of an electron pair.

In the past we have symbolized the bond between two atoms by means of a line, e.g. in the structural formulas:

H–O–H H–Cl H–H

We now know, of course, what this line represents. The line represents the .... which is responsible for the chemical bond.

Jot down the missing words, and then go on to 49 .

---

60 Your answer is correct. The nitrogen (N) in NH_3 is tervalent.

The valence specifies the number of bonds which a given atom is capable of forming with other atoms.

What is responsible for the formation of bonds between atoms?

In order to make the following explanations clearer, we shall start by referring back to the structure of atoms.

You will recall that atoms consist of a nucleus with a surrounding charge cloud.

What are the constituent particles of this cloud called?

- Protons → 50
- Electrons → 69
- Neutrons → 23

---

61 Your answer is wrong.

You will find an explanation of this at 13 .

---

62 In the water molecule (H–O–H) there is an *oxygen* atom in the middle. On both its left and right hand sides there is a *hydrogen* atom.

In addition to a structural formula, every compound also has a *constitutional formula.* Water, for instance, has the constitutional formula H_2O.

This formula tells us nothing about the structure of a molecule; it merely specifies the number and type of each atom in the molecule. Lines representing bonds are *not* drawn in such formulas.

Copy down the formulas below and then decide which are structural formulas and which constitutional formulas.

H–Cl, NH_3, H–N–H (with H bonded below N), HCl, H_3PO_4, Fe=O, Fe=S, FeS, H_2SO_4, Na_2O, NaCl

→ 70

---

63 No. You must not forget that the electrically neutral chlorine atom accepts one electron, which is *negatively* charged. The number of protons is thus no longer equal to the number of electrons. In fact, the number of electrons now exceeds the number of protons by *one.*

A chlorine atom which has accepted an electron will therefore be .... charged. → 32

---

64 *Ionic Bonding*
Atomic Bonding

Jot down now one example each of a compound having ionic bonding and one having atomic bonding. → 76

---

65 4 Fe, 3 Cl_2, 5 P, Br_2, S, 3 Na, 2 N_2, 4 N_2

Did you get all of them right? If so, you have worked well. If not, then you should in your own interest repeat from 30 .

Let's get back now to chemical equations. We have already learnt that: Both on the left and right hand sides of an *equation* the *same* number of each type of atom must appear.

Is this condition satisfied in the following equation, i.e. is the equation correct as it stands?

$$2\ H_2O \rightarrow 2\ H_2 + O_2$$

- Yes → 55
- No → 75

---

66 No. You have not taken into account the structural formula of the chlorine molecule

Cl–Cl.

The bonding electron pair is located midway between the two chlorine atoms.

Thus, this molecule contains an .... bond. → 31

---

67 No, the electron pair is made up of one electron from the sodium atom and one from the chlorine atom.

Repeat now from 77 .

---

68 The two dots represent an *electron pair.*

The two hydrogen *atoms* which make up the hydrogen *molecule* contribute *one electron* each.

Copy down the equation below, and insert dots to represent the missing electrons and electron pair.

$H + H \rightarrow H \quad H$ → 59

---

69 Yes, the electrons are found in the surrounding charge cloud.

For this reason one speaks of the *electron cloud* around the nucleus of an atom. In the electron cloud of the hydrogen atom there is just one electron. We shall formally represent this electron by means of a dot placed next to the symbol for hydrogen thus:

$$H^{\bullet}$$

What is the significance of this dot next to the hydrogen? Jot down your answer, and then go on to [77] .

---

[70] Now check your answers:

Structural formulas: H–Cl, $\begin{matrix} H{-}N{-}H \\ | \\ H \end{matrix}$, Fe=O, Fe=S

Constitutional formulas: NH_3, HCl, H_3PO_4, FeS, H_2SO_4. Na_2O, NaCl

NH_3 is the constitutional formula of a compound we have not yet come across. However, you already have sufficient knowledge to ascertain the valence of nitrogen in this compound simply by looking at its structural formula above.

What is the valence of the nitrogen in NH_3?

- Monovalent → [78]
- Bivalent → [51]
- Tervalent → [60]

---

[71] You have not taken into account the structural formula of the chlorine molecule

Cl–Cl.

The bonding electron pair is located midway between the two chlorine atoms.

Thus, this molecule contains an .... bond. → [31]

---

[72] Your answer is only partially correct.

Sodium chloride (NaCl) consists of ions.

In the hydrogen molecule (H_2), however, the bonding electron pair is located midway between the two hydrogen atoms.

Repeat now from 49 .

---

73 No. You must not forget that the electrically neutral sodium atom donates one electron (and with it goes one negative charge).

After donating this electron, the number of protons in the sodium atom will *exceed* the number of electrons by *one*.

Thus, a sodium atom which has donated one electron will be .... charged. → 24

---

74 No. You must not forget that the electrically neutral chlorine atom accepts one electron, which is *negatively* charged. The number of protons is thus no longer equal to the number of electrons. In fact, the number of electrons exceeds the number of protons by *one*.

A chlorine atom which has accepted an electron will therefore be .... charged. → 32

---

75 You have unfortunately not allowed for the fact that the 2 in front of the H_2O relates to the *whole* H_2O molecule, which implies that there are two O atoms on the left hand side of the equation.

$$2\,H_2O \rightarrow 2\,H_2 + O_2$$

The total numbers of atoms on each side are thus:

4 H atoms	4 H atoms
2 O atoms	2 O atoms

The equation is therefore balanced.

Continue your reading at 55 .

---

76 Ionic bonding: $Na^{\oplus}Cl^{\ominus}$ or NaCl
Atomic bonding: H–H, Cl–Cl

End of Program 7.

7

---

77 The dot represents the *electron* forming the electron cloud of the hydrogen atom.

If two hydrogen atoms combine together to form a hydrogen molecule thus:

$$H + H \rightarrow H_2$$

the two electrons actually form an *electron pair*:

$$H^{\bullet} + {}^{\bullet}H \rightarrow H \,{:}\, H$$

What is the significance of the two dots between the two hydrogen atoms on the right hand side of the equation?

Jot down your answer, and then go to 68 .

---

78 No, all you need do here is to count the lines radiating from the nitrogen atom in the structural formula

```
H–N–H
  |
  H
```

to determine the valence of the nitrogen.

Repeat now from 43 .

Program 8

Sodium and Potassium

1 We shall devote this particular Program to a detailed discussion of the metals sodium (Na) and potassium (K) and their most important compounds. After carefully working your way through the Program, you should be able to describe the reactions of these two metals with both water and oxygen. It should also be possible for you to write out chemical equations for the preparation of sodium, potassium, sodium hydroxide and potassium hydroxide. Furthermore, we shall also be saying something about the occurrence and detection of these metals.

We have already explained that all elements may be classified either as metals or as non-metals. The metals can be further sub-divided into the two categories of

heavy metals and light metals.

You should now be able to complete the sentences below.

Elements may be classified as metals and ....
Metals may be classified as .... and ....

Check your answers at [7] .

2 It is just as well not to guess. You should, however, be careful not to take it too easy!

We have already learnt that there are both *metals* and *non-metals.* Their oxides are respectively referred to as *metal* oxides and *non-metal* oxides.

Metal oxides form *metal hydroxides* with water. Aqueous solutions of metal hydroxides are known as *alkalis. Non-metal* oxides on the other hand form *acids* with water.

Go back now to [35] .

3 No. You are not sufficiently involved! We have already described the following sodium compounds:

Sodium chloride (table salt, rock salt)	NaCl
Sodium hydroxide	NaOH

Sodium carbonate (soda)	Na_2CO_3
Sodium nitrate	$NaNO_3$

Jot down now the names and chemical formulas of the corresponding potassium compounds. → 39

---

4 Your answer is wrong. Chlorine is indeed formed, though *no* oxygen is.

8

Think carefully about this question again, and then you should be able to decide on the right answer at 49

---

5 No. You have not read 35 carefully enough!

Alkalis are formed when the reaction product of a *metal oxide* and water is dissolved in water.

However, phosphorus and sulfur are *non-metals*, and so phosphorus oxide and sulfur dioxide are thus *non-metal oxides.* When phosphorus oxide or sulfur dioxide reacts with water an .... is formed. This will turn blue litmus paper red. → 48

---

6 No. You are not sufficiently involved! We have already described the following sodium compounds:

Sodium chloride (table salt, rock salt)	$NaCl$
Sodium hydroxide	$NaOH$
Sodium carbonate (soda)	Na_2CO_3
Sodium nitrate	$NaNO_3$

Jot down now the names and chemical formulas of the corresponding potassium compounds. → 39

---

7 Elements may be classified as metals and *non-metals.* Metals may be classified as *heavy metals* and *light metals.*

One distinguishes between light and heavy metals on the basis of their density. The density of a substance is the mass (in g) of one cubic centimeter (cm^3) of it.

Light metals: have a density less than 5 g/cm^3
Heavy metals: have a density greater than 5 g/cm^3

Here are some examples:

Metal	Density (g/cm^3)	Metal	Density (g/cm^3)
Aluminum	2.7	Copper	8.9
Lead	11.3	Sodium	0.97
Iron	7.8	Platinum	21.5
Gold	19.3	Silver	10.5
Potassium	0.86	Zinc	7.2

Now make separate lists of the light metals and the heavy metals. → 15

---

8 That's right! Sodium and potassium are *highly reactive* elements. For instance, they react with oxygen very rapidly.

Both of these metals are always monovalent in their compounds.

If we also mention here that oxygen (O) is bivalent, you should be able to write down the chemical formulas for sodium oxide and potassium oxide.

See if you really can write down both of these formulas, and check your answers at 28 .

If this presents problems, start by reading through 36 .

---

9 No. You are not sufficiently involved! We have already described the following sodium compounds:

Sodium chloride (table salt, rock salt)	NaCl
Sodium hydroxide	NaOH
Sodium carbonate (soda)	Na_2CO_3
Sodium nitrate	$NaNO_3$

Jot down now the names and chemical formulas of the corresponding potassium compounds. → 39

---

10 Your answer is wrong. You will have to take a rather more careful look at the most important sodium and potassium compounds listed in 42 .

---

11 Your equation is correct if you have written:

$$4K + O_2 \rightarrow 2K_2O$$

Sodium and potassium are always stored under *naphtha.*

Can you think why this is done? Write out an answer to this before going on to 25 .

---

12 $NaCl, Na_2CO_3, NaNO_3, K_2CO_3, KNO_3, NaOH, KOH$

What simple analytical test may be performed to determine whether a compound contains sodium or potassium?

If sodium or a sodium compound is placed in the non-luminous flame of a Bunsen burner, the flame will immediately become colored with an *intense yellow.* In the case of potassium or potassium compounds, the flame is colored *violet.*

These different colorations may be used as a test for both sodium and potassium compounds. The coloration is exceptionally intense for sodium and its compounds. Even small traces of sodium compounds produce a yellow coloration of the flame.

Now complete the following sentences:

Soda will color a Bunsen flame ....
Potash will color a Bunsen flame ....
Table salt will color a Bunsen flame .... → 44

---

13 The correct version of the equation is:

$$K_2O + H_2O \rightarrow 2KOH$$

Potassium hydroxide is formed in this reaction.

Potassium hydroxide and sodium hydroxide are both soluble in water. The aqueous solution of sodium hydroxide is sometimes referred to as:

caustic soda

and that of potassium hydroxide as:

caustic potash

We have previously described the use of litmus as an indicator. The blue color of the litmus dye turns red when the dye is added to the solution of an acid in water.

What color change will there be in a strip of paper soaked in litmus dye when it is put into an aqueous solution of sodium or potassium hydroxide?

The red color of the litmus paper turns................ → 23

---

14 Sodium and potassium are stored under *naphtha.*

Sodium hydroxide is also formed when sodium oxide comes into contact with water.

Sodium oxide + Water → Sodium hydroxide.

Write out the equation for this reaction, and then check your answer at 26 .

---

15 Light metals: Aluminum, Potassium, Sodium
Heavy metals: Lead, Iron, Gold, Copper, Platinum, Silver, Zinc

The two metals *sodium* and *potassium* discussed in this Program are thus *light metals.*

Explain in a short sentence why sodium (Na) and potassium (K) are light metals. → 22

---

16 Your answer is wrong. You will have to take a rather more careful look at the most important sodium and potassium compounds listed in 42 .

---

17 Your answer is wrong. Hydrogen is certainly produced during the electrolysis of brine, but *no* sodium is.

You are not really paying attention. How could the highly reactive sodium exist in water?

Just think about it again, and then you should be able to come up with the right answer at 49 .

---

18 Your answer is wrong. You will have to take a rather more careful look at the most important sodium and potassium compounds listed in 42 .

19

$4Na + O_2 \rightarrow 2Na_2O$

Now work out the equation for the formation of potassium oxide (potassium = K)! Check your equation at 11 .

20

Your answer is wrong. We have already said that the freshly exposed surfaces become covered with a thin layer of oxide *after a few seconds.* Thus, both sodium and potassium react *immediately* with the oxygen in the air.

If, on the other hand, a freshly exposed metal surface stays shiny for a long time, it is because it remains unchanged, i.e. no reaction with oxygen occurs. An example of such a metal ist gold.

An element which readily reacts with another element is said to be *reactive.*

An element which does not react (or only slowly reacts) with another elements is said to be *unreactive* or *inert.*

Sodium and potassium are both .... elements.

Jot down the missing word, and then continue reading at 8 .

21

Na_2CO_3 is sodium carbonate
KNO_3 is potassium nitrate

In everyday life you have doubtless come across:

Table salt, Soda and Potash.

What are the chemical terms for these three substances?

- Sodium hydroxide, sodium carbonate, potassium nitrate → 3
- Sodium chloride, sodium carbonate, potassium nitrate → 6
- Sodium oxide, potassium oxide, potassium chloride → 9
- Sodium chloride, sodium carbonate, potassium carbonate → 38

22

Sodium and potassium are light metals *because their density is less than 5 g/cm*^{3}.

Both of these metals are as soft as wax and can easily be cut with a knife. The surfaces exposed in this way have a silvery-white metallic luster. The luster remains for only a few seconds in the air and, after that, the exposed surface of the metal takes on a bluish-gray shimmer. This coloration is caused by the formation of a thin layer of oxide. The oxygen required for the formation of this oxide layer is simply taken from the air.

Just as sodium immediately gets coated with a layer of sodium oxide, potassium correspondingly gets coated with a layer of .... → [30]

---

[23] The red color of the litmus paper turns *blue.*

The aqueous solutions of sodium hydroxide and of potassium hydroxide are alkalis. The indicator litmus changes from red to *blue* in *alkalis.*

We have described two different ways for making sodium hydroxide (both of which can be used for making potassium hydroxide), namely:

1. The reaction of sodium with water.
2. The reaction of sodium oxide with water.

Write out the equations for reactions 1 and 2. *It is important to get as much practice as possible in writing out equations!*

The stepwise procedure to be followed in setting up equations is again fully described in [24]. It would probably be best to read through [24] before attempting to write out the equations for 1 and 2. Check your equations at [35].

---

[24] The procedure for setting up chemical equations is as follows:

1. Write out the chemical process involved in words, e.g.

   Sodium + chlorine → sodium chloride

   (This step can be omitted when you get more experienced.)

2. Replace the chemical names of the elements and compounds with the corresponding symbols and formulas.

   $Na + Cl \rightarrow NaCl$

3. Watch out for the diatomic gases!

   $Na + Cl_2 \rightarrow NaCl$

4. Check whether the number of atoms for each element which appears in the equation is the same on both the left and right hand sides of the equation.

   In our example this is *not* the case with the chlorine, as on the left there are two Cl atoms but only one on the right.

5. If the numbers of atoms are not equal at stage 4, it will be necessary to add factors to the equation.

   In our example we shall have to add a 2 thus:

   $Na + Cl_2 \rightarrow 2NaCl$

   The number of chlorine atoms is now the same on the left and right hand sides of the equation. However, this is no longer true for the sodium atoms. An additional factor must therefore be added to balance up the Na thus:

   $2Na + Cl_2 \rightarrow 2NaCl$

   The above represents the final and correct version of the equation. → 23

---

25 As both of these metals are extremely sensitive to oxygen, they have to be stored under *naphtha*. In this way the oxygen in the air is prevented from coming into contact with the sodium or potassium.

For practice you should now write down the equations for the reaction of sodium and potassium with oxygen. Check your equations at 34 .

---

26 The correct version of the equation is:

$Na_2O + H_2O \rightarrow 2NaOH$

Now write out the equation for the reaction of potassium oxide with water, and check your answer at 13 .

27

Rock salt is sodium chloride	$NaCl$
Sodium saltpeter is sodium nitrate	$NaNO_3$
Potassium saltpeter ist potassium nitrate	KNO_3

What are the chemical names for rock salt, soda and potash? → 38

28

Sodium oxide Na_2O
Potassium oxide K_2O

(If you found this difficult, you should now read 36 to clarify matters.)

Na_2O is formed during the reaction of sodium with oxygen. Let's represent this process

Sodium + Oxygen → Sodium oxide

by means of an equation. The starting materials (sodium and oxygen) will appear on the left hand side:

$Na + O_2 \rightarrow$ (oxygen is diatomic!)

The end product (sodium oxide) will appear on the right hand side:

$Na + O_2 \rightarrow Na_2O$

This equation, however, is not correct as it stands. In any equation, the number of atoms of a given element must be the *same* on both the left and right hand sides.

We have here:

On the left hand side 1 Na; on the right hand side 2 Na
On the left hand side 2 O; on the right hand side 1 O

It is evident that the appropriate *factors* are missing!

Try to find these factors yourself, and then check your answer at 19 .

29

The process by which sodium is produced is called *molten electrolysis.*

Potassium is produced by the electrolysis of molten KOH.

To sum up now what we have learnt so far:

Sodium and potassium are chemically very active elements. Sodium reacts with water to form sodium hydroxide and hydrogen:

$2Na + 2H_2O \rightarrow 2NaOH + H_2$

Potassium reacts with water to form potassium hydroxide and hydrogen; its reaction is much more violent than that of sodium. Both metals have to be stored under naphtha in view of their great sensitivity to oxygen and water vapor.

Neither of these elements is found in the free state in nature.

Sodium and potassium are light metals.

They are prepared by the electrolysis of molten sodium chloride or potassium chloride (molten electrolysis):

$$2NaCl \rightarrow 2Na + Cl_2$$

Sodium and potassium hydroxides may be obtained from sodium or potassium chloride respectively by electrolysis of their aqueous solutions:

$$2NaCl + 2H_2O \rightarrow 2NaOH + Cl_2 + H_2$$

Sodium and potassium hydroxides are very corrosive solid substances. Even their aqueous solutions are corrosive.

Take great care when handling these compounds!

Take good care also when working with metallic sodium on account of its sensitivity to water! Never touch sodium with your fingers!

The most important sodium and potassium compounds are:

Sodium chloride (table salt, rock salt), sodium carbonate (soda), sodium nitrate (sodium saltpeter), potassium carbonate (potash), potassium nitrate (potassium saltpeter), sodium hydroxide (caustic soda) and potassium hydroxide (caustic potash).

Write out the chemical formulas of these compounds. → 12

---

30 Just as a freshly exposed sodium surface *immediately* becomes coated with a layer of sodium oxide in air, one of potassium similarly becomes coated with a layer of *potassium oxide.*

These processes, which occur on freshly exposed sodium or potassium surfaces, reveal just how reactive these metals are.

Which of the two sentences regarding sodium and potassium is correct?

- Both metals are highly *inert* → 20

- Both metals are highly *reactive* → 8

---

31 Your answer is wrong. During the electrolysis of brine hydrogen is indeed formed, but *no* oxygen is (Oxygen and hydrogen are formed during the electrolysis of water!).

Think about this question again, and then see if you can come up with the right answer at 49.

---

32 Sodium hydroxide (NaOH) and potassium hydroxide (KOH) are both highly corrosive.

What are the following compounds called?

Na_2CO_3 and KNO_3

- Sodium chloride and potassium carbonate → 10
- Sodium nitrate and potassium chloride → 16
- Sodium carbonate and potassium nitrate → 21
- Sodium carbonate and potassium chloride → 18

---

33 Now check your equation:

$$2K + 2H_2O \rightarrow 2KOH + H_2$$

KOH is *potassium hydroxide.*

In view of the great reactivity of sodium and potassium toward water, it is absolutely vital that both of these light metals are stored under the liquid mentioned. Otherwise they would *react* not only *with* the *oxygen* in the air but also with the *water vapor* it contains.

Jot down the name of the liquid under which sodium and potassium are stored, and then check your answer at 14.

34

$$4Na + O_2 \rightarrow 2Na_2O$$

$$4K + O_2 \rightarrow 2K_2O$$

Both of these metals also display a high degree of reactivity toward *water.*

Imagine now the following experiment is performed:

A small chunk of sodium is thrown into some water. At once a violent reaction starts accompanied by a hissing noise. The *heat* generated during the reaction causes the temperature of the sodium to rise so much that it melts (sodium actually melts at 98°C). The molten ball of sodium fizzes and moves around on the surface of the water. Hydrogen is liberated in the course of the reaction. The sodium ball gets smaller and smaller until it finally vanishes.

Two observations are particularly significant here, viz.

1. The molten sodium floats on the surface of the water.
   Explanation: The density of sodium (0.97 g/cm^3) is less than that of water (1.00 g/cm^3)

2. The sodium ball disappears. Why is this?

- The sodium ball evaporates → [45]
- The sodium reacts with water to form a new compound soluble in water → [46]
- I don't know → [47]

---

[35]

1. $2Na + 2H_2O \rightarrow 2NaOH + H_2$
2. $Na_2O + H_2O \rightarrow 2NaOH$

If you made mistakes in either the first or both equations, you should now repeat from [46] .

If you made mistakes only in the second equation, you should repeat from [14] .

In general it may be said that:

When *metal oxides* react with water *metal hydroxides* are formed. Their aqueous solutions are usually *alkalis,* i.e. they turn red litmus paper blue.

Earlier on we discussed the reaction of other types of oxide with water, e.g. that of phosphorus oxide with water. What is formed when either phosphorus oxide or sulfur dioxide reacts with water?

(Think out your answer carefully as the question is not as easy as it looks!)

- Alkalis → 5
- Acids → 48
- I don't know → 2

---

36 You should now write down the chemical formulas for sodium oxide and potassium oxide. Sodium oxide consists of sodium (Na) and oxygen (O).

Sodium is monovalent.
Oxygen is bivalent.
One oxygen atom can combine with two sodium atoms.
The chemical formula is thus Na_2O.

You should now be in a position to write down the formula for potassium oxide. After doing this continue reading at 28 .

---

37 The formation of solid sodium hydroxide by evaporation of the solution is thus represented by the equation:

$$Na^{\oplus} + OH^{\ominus} \rightarrow NaOH$$

Potassium hydroxide may be prepared by the same procedure. One starts out from potassium chloride (KCl). Work out the equation for this reaction, and then check your answer at 60 .

If this proves to be rather difficult for you, some assistance is offered at 68 .

---

38 Table salt is sodium chloride.

Soda is sodium carbonate.
Potash is potassium carbonate.

How is sodium produced?

Sodium is prepared by means of molten electrolysis. We have already mentioned that sodium chloride is a solid. When heated strongly it turns into a liquid, i.e. it melts. The *molten mass* is electrolyzed – hence the name *molten electrolysis*.

If we start out from sodium chloride, the products of molten electrolysis will be sodium and chlorine.

See if you can write down the equation for this reaction. As you will not find this particularly easy to do, you may need some help. In this case you should turn to 66 .

If you are able to manage it on your own, check your answer at 57 .

39

Potassium chloride	KCl
Potassium hydroxide	KOH
Potassium carbonate	K_2CO_3
Potassium nitrate	KNO_3

What are the chemical formulas and names for rock salt, sodium saltpeter and potassium saltpeter? → 27

---

40 The equation takes the form:

$$H_2O \rightarrow H^{\oplus} + OH^{\ominus}$$

In an aqueous solution of sodium chloride there are four different types of ion. These arise from the ionization of the sodium chloride and from the ionization of a tiny fraction of the water molecules present.

What are the four ions in question? Jot down the names and symbols of these ions, and then check your answers at 53 .

41

a) NaCl without cobalt glass	yellow
b) NaCl with cobalt glass	no apparent coloration
c) NaCl + KCl without cobalt glass	yellow
c) NaCl + KCl with cobalt glass	violet
e) KNO_3 without cobalt glass	violet
f) KNO_3 with cobalt glass	violet

Now one last question:

What is the color of the cobalt glass? → 70

---

42 *caustic soda* and *caustic potash*

So far we have discussed the following compounds of sodium and potassium:

Sodium chloride (also known as rock salt or table salt)	$NaCl$
Potassium chloride	KCl
Sodium hydroxide	$NaOH$
Potassium hydroxide	KOH

Other important commercial products include:

Sodium carbonate (soda)	Na_2CO_3
Potassium carbonate (potash)	K_2CO_3
Sodium nitrate (sodium saltpeter)	$NaNO_3$
Potassium nitrate (potassium saltpeter)	KNO_3

It is worth making the effort to learn these names and formulas. We shall be going into more detail on these substances later on.

Which two of the above compounds are highly corrosive? Jot down their names and formulas! → 32

43 Your equation is correct if you have written:

$$2Na + 2H_2O \rightarrow 2NaOH + H_2$$

Potassium reacts with water in an analogous way, though in this case the reaction is *much more violent.* The potassium not only melts, but the greater amount of heat generated during the reaction is sufficient to *ignite* the hydrogen produced. The potassium thus rushes around on the surface of the water enveloped in small flames.

Write out the equation for the reaction of potassium with water, jot down the name of the potassium compound formed, and then check your answers at 33 .

44 Soda and rock salt color the flame *yellow* as they are sodium compounds. Potash colors the flame *violet* as it is a potassium compound.

If sodium and potassium compounds are together brought into the flame, one sees only the *yellow coloration* due to the sodium. The latter is so intense that it completely masks the violet of the potassium.

In order to detect the presence of both sodium and potassium from the coloration of the flame, one can make use of a cobalt glass. This is a dark blue glass which allows *no yellow* light to pass through, although it will let *violet* light through. Now, if the yellow coloration of a Bunsen flame due to sodium or sodium compounds is viewed through a cobalt glass, no coloration will be seen. However, if the violet coloration due to potassium or potassium compounds is viewed through the glass, one will see a violet coloration. It is therefore possible to establish the presence of potassium in the flame by means of a dark blue cobalt glass, even when its coloration is masked by the yellow color of sodium compounds which may also be present.

Suppose now that a mixture of sodium and potassium nitrates is placed in the flame. How will the flame appear

1. through a cobalt glass?

2. without a cobalt glass?

Jot down your answers, and then check what you have written at [56] .

---

[45] Your answer is wrong. The process taking place when a small lump of sodium is thrown into water is obviously a *chemical* process.

Heat is generated during the course of the reaction. The sodium does *not* evaporate, but rather reacts with the water. It is for this reason that it gets used up. → [46]

---

[46] Correct, Na reacts with water and gets used up in the process. What products are formed during this reaction?

First, we have *hydrogen* which comes off as a gas.

The second product is dissolved in the water. If this water is evaporated a white residue will remain behind. The residue is *sodium hydroxide* which has the formula NaOH.
Now describe this reaction *in words:*

Sodium + ... → .... + ... → [55]

---

[47] The process taking place when a small lump of sodium is thrown into water is obviously a *chemical* process.

Heat is generated during the course of the reaction. The sodium does *not* evaporate, but rather reacts with the water. It is for this reason that it gets used up. → 46

---

48 Yes. Phosphorus oxide and sulfur dioxide form *acids* with water. (If this answer was your first choice, you are coming along very nicely.) Both sulfur dioxide and phosphorus oxide are non-metal oxides.

We can now generalize and say that:

Metal oxides form metal hydroxides with water; their aqueous solutions are termed alkalis.
Non-metal oxides form acids with water.

Owing to the great reactivity of sodium and potassium, these two metals are never found in the free state in nature, i.e. as pure metals, but always in the form of compounds. The most important compounds of these metals are sodium chloride (table salt or rock salt, NaCl) and potassium chloride (KCl). We may picture the formation of sodium chloride from sodium and chlorine according to the equation:

Sodium + Chlorine → Sodium chloride

Write down the corresponding equation for this reaction, and then check your answer at 58.

---

49 Sodium chloride may be obtained as follows:

1. From the elements according to the equation:

   $$2\ Na + Cl_2 \rightarrow 2\ NaCl$$

   (It would be uneconomical to produce sodium this way in industry. The methods listed below are substantially cheaper.)

2. By mining existing deposits of rock salt.

3. By evaporation of brine solutions pumped out of salt mines.

4. By evaporation (or vaporization) of sea water.

Sodium chloride is the starting product for the preparation of sodium hydroxide, sodium and sodium compounds generally.

Sodium hydroxide is synthesized industrially from an aqueous brine solution. The solution is electrolyzed by means of direct electric current. (We have already discussed the electrolysis of water in *Hofmann*'s apparatus for the decomposition of water, during which hydrogen and oxygen are formed.)

When an aqueous solution of sodium chloride (brine) is electrolyzed, sodium hydroxide (NaOH dissolved in water) is formed.

What other products are formed during this electrolysis?

8

- Oxygen and hydrogen → 31
- Sodium and hydrogen → 17
- Chlorine and oxygen → 4
- Chlorine and hydrogen → 61

50

Starting material: NaCl (on the left hand side of the equation)

End products: Na and Cl_2 (on the right hand side of the equation)

$$NaCl \rightarrow Na + Cl_2$$

Only the factors are now missing from this equation.
On the right we have two Cl atoms. Hence, two Cl atoms must also appear on the left.

$$2\ NaCl \rightarrow Na + Cl_2$$

One factor is still missing though, which you are now asked to supply! → 57

51

The complete equation takes the form:

$$NaCl \xrightarrow{\text{dissolved in water}} Na^{\oplus} + Cl^{\ominus}$$

The negatively charged ($Cl^{\ominus}$) are referred to as *chloride* ions.

Even a small proportion of the water molecules (H_2O) is always present in the form of ions, i.e. as positively charged hydrogen ions ($H^{\oplus}$) and negatively charged *hydroxyl* ions ($OH^{\ominus}$).

Work out the equation for the decomposition of a water molecule into ions, and then check your answer at 40 .

52 At the anode, i.e. the positive electrode, the negatively charged ions are discharged, in this case the *chloride ions.*

At the cathode, i.e. the negative electrode, the positively charged ions are discharged, in this case the *sodium ions.*

$$Cl^{\ominus} \xrightarrow{\text{anode}} \ldots\ldots\ldots\ldots\ldots$$

$$Na^{\oplus} \xrightarrow{\text{cathode}} \ldots\ldots\ldots\ldots\ldots$$

Write out completed versions of both these equations. → 62

53 An aqueous solution of brine contains positively charged sodium and hydrogen ions ($Na^{\oplus}$ and $H^{\oplus}$) as well as negatively charged chloride and hydroxyl ions ($Cl^{\ominus}$ and $OH^{\ominus}$).

When such a solution is poured into an electrolysis apparatus, e.g. into Hofmann's apparatus for the decomposition of water, and the two electrodes are connected up to a battery, the *negatively* charged ions will migrate to the *positive* electrode, i.e. to the anode, and the *positively* charged ions will migrate to the *negative* electrode, i.e. to the cathode.

Note down which of the four ions present in aqueous brine will migrate to the anode and which to the cathode.

The .... migrate to the anode.
The .... migrate to the cathode.

Check your answers at 63 .

54 At the anode the chloride ions ($Cl^{\ominus}$) are discharged. Chlorine gas (Cl_2) is then formed.

$$2Cl^{\ominus} \xrightarrow{\text{anode}} Cl_2$$

At the cathode the hydrogen ions ($H^{\oplus}$) are discharged. Hydrogen gas (H_2) is then formed.

$$2H^{\oplus} \xrightarrow{\text{cathode}} H_2$$

Which of the four types of ion present in a brine solution are not discharged?

Jot down your answer, and then go to [67] .

---

55 Sodium + *Water* → *Sodium hydroxide* + *Hydrogen.*

Sodium hydroxide has the chemical formula NaOH.

8

Now set up the equation for this reaction by going through the following sequence of steps:

1. Put in first on the left hand side the symbol and the formula for the starting materials.

2. Put in second on the right hand side the symbol and the formula for the end products. (Do not forget that hydrogen is diatomic!)

3. Check wether the same number of atoms for each element appears on both the left and right hand sides of the equation.

4. If this is not so, the equation must then be balanced by the addition of appropriate factors.

Check your equation at [43] .

---

56 The flame coloration appears violet through the cobalt glass whereas without the glass it appears yellow.

Here are a few further examples. What coloration will be observed if the following compounds are placed in the flame?

a) NaCl	without cobalt glass
b) NaCl	with cobalt glass
c) NaCl + KCl	without cobalt glass
d) NaCl + KCl	with cobalt glass
e) KNO_3	without cobalt glass
f) KNO_3	with cobalt glass.

Jot down your answers and then go to [41] .

---

57 The equation for the electrolysis of molten sodium chloride is:

$$2NaCl \rightarrow 2Na + Cl_2$$

(Did you manage this one on your own? If so, you have learnt quite a lot!)

Even in the case of molten electrolysis, the ions of the starting material migrate to the electrodes and are discharged there.

Which ions are formed when the starting material decomposes?

Jot down your answer, if possible in the form of an equation, and then proceed to 65 .

8

58

The correct form of the equation is:

$$2Na + Cl_2 \rightarrow 2NaCl$$

There are enormous deposits in the earth of rock salt which may be extracted by mining. Sometimes it is also feasible to extract the salt by dissolving it in water. One then obtains an aqueous solution of sodium chloride, known as brine, from which solid sodium chloride (rock salt) can be obtained by evaporating off the water.

In some countries sodium chloride is produced by simply letting the water evaporate from sea water. Sea water contains around 2.7% sodium chloride.

Now list four methods by which sodium chloride may be obtained. → 49

59

1. Decomposition of the starting materials into ions:

$$NaCl \rightarrow Na^{\oplus} + Cl^{\ominus}$$

$$H_2O \rightarrow H^{\oplus} + OH^{\ominus}$$

2. Discharge of the hydrogen and chloride ions:

$$\ldots\ldots\ldots \xrightarrow{\text{cathode}} \ldots\ldots\ldots$$

$$\ldots\ldots\ldots \xrightarrow{\text{anode}} \ldots\ldots\ldots$$

Write out completed versions of these two equations, and then go to 69 .

60

1. Decomposition into ions:

$$KCl \rightarrow K^{\oplus} + Cl^{\ominus}$$

$$H_2O \rightarrow H^{\oplus} + OH^{\ominus}$$

2. Discharge of the hydrogen and chloride ions:

$$2\ H^{\oplus} \xrightarrow{\text{cathode}} H_2$$

$$2\ Cl^{\ominus} \xrightarrow{\text{anode}} Cl_2$$

3. Formation of solid potassium hydroxide by evaporation of the solution:

8

$$K^{\oplus} + OH^{\ominus} \rightarrow KOH$$

Sodium and potassium hydroxides are white, strongly hygroscopic substances. **Both are extremely corrosive. Great care must therefore be taken when working with sodium or potassium hydroxide (in particular, protect your eyes!) Even aqueous solutions of both compounds are highly corrosive.**

What are the aqueous solutions of sodium hydroxide and potassium hydroxide sometimes referred to as? → 42

---

61 Your answer is quite right. Apart from sodium hydroxide (dissolved in water), hydrogen and chlorine are also formed during the electrolysis. Chlorine is produced at the anode whereas hydrogen is formed at the cathode.

How can the process of electrolysis be explained?

We already know that the compound sodium chloride is made up of ions, that is of positively charged sodium ions ($Na^{\oplus}$) and negatively charged chloride ions ($Cl^{\ominus}$). When sodium chloride is dissolved in water it separates into these ions.

Now complete the following equation:

$$NaCl \xrightarrow{\text{dissolved in water}} \ldots\ldots\ldots\ldots\ldots + \ldots\ldots\ldots\ldots\ldots$$

Check your equation at 51 .

62

$$2Cl^{\ominus} \xrightarrow{\text{anode}} Cl_2$$

$$Na^{\oplus} \xrightarrow{\text{cathode}} Na$$

Did you manage to get both equations right? If so, you are working very well. Special care is necessary with the first equation as the chlorine must be written down as a diatomic molecule.

What is the name of the process by which sodium may be produced from sodium chloride?

Jot down the name, and then go on to 29 .

---

63 The $Cl^{\ominus}$ and $OH^{\ominus}$ ions migrate to the anode whereas the $Na^{\oplus}$ and $H^{\oplus}$ ions migrate to the cathode.

At each electrode one of the types of ion is discharged. The other two simply remain dissolved in the water. If you remember that hydrogen and chlorine gases are formed during the electrolysis of brine to produce caustic soda, you should now be able to answer the following question:

Which ion is discharged at the anode and which at the cathode?

Jot down your answers and go on to 54 .

---

64 The positive electrode is called the *anode*. The *negatively* charged ions migrate to it and are discharged there.

The negative electrode is called the *cathode*.
The *positively* charged ions migrate to it and are discharged there.

Go back now to 65 .

---

65 $NaCl \rightarrow Na^{\oplus} + Cl^{\ominus}$

Sodium and chloride ions are thus formed.

Which ions migrate to the anode and are discharged there?
Which ions migrate to the cathode and are discharged there?

The .... are discharged at the anode.
The .... are discharged at the cathode.

Jot down your answers, and then go to 52 .
Some assistance is available at 64 .

---

66 The starting material is molten sodium chloride.
The end products are sodium and chlorine.
Do not forget that chlorine is diatomic, i.e. written as Cl_2!
Can you manage it on your own now? If so, check to see if your answer is right at 57 .
A detailed explanation is given at 50 .

---

67 The $Na^{\oplus}$ and $OH^{\ominus}$ ions remain in solution.

If one goes on electrolyzing until all of the chloride ions have been discharged and then evaporates the solution, the $Na^{\oplus}$ and $OH^{\ominus}$ ions together will form solid sodium hydroxide (NaOH).

$$Na^{\oplus} + OH^{\ominus} \rightarrow NaOH$$

Thus, the following equations may be written for the production of sodium hydroxide from the electrolysis of brine:

1. Decomposition of the starting materials into ions:

$NaCl \rightarrow$ .... + ...
$H_2O \rightarrow$ .... + ...

Write out completed versions of these equations. → 59

---

68 We can set up the following equations for the production of potassium hydroxide (KOH) from the electrolysis of an aqueous solution of potassium chloride:

1. Decomposition of the starting materials into ions (2 equations).
2. Discharge of the ions at the anode and cathode (2 equations).
3. Formation of KOH by evaporation of the solution (1 equation).

Work out what these equations are now, and then check your answers at 60 .

---

69 2. Discharge of the hydrogen and chloride ions:

$$2\,H^{\oplus} \xrightarrow{\text{cathode}} H_2$$

$$2\,Cl^{\ominus} \xrightarrow{\text{anode}} Cl_2$$

3. Formation of solid sodium hydroxide from sodium and hydroxyl ions by evaporation of the solution.

Write out the relevant equation for this latter process. → 37

---

70 Cobalt glass is dark blue.

End of Program 8.

Program 9

Theoretical Foundations (III)

1 In every chemical laboratory several chemical *balances* are to be found, some of which are high precision instruments. They are there, amongst other things, for the purpose of *weighing out* exact quantities of the substances used in chemical experiments.

Now how is a chemist able to calculate precisely how much of this or the other substance he will be needing for a given experiment?

The present Program has been specifically designed to answer that question. After working your way carefully through it, you should be in a position to perform such calculations yourself.

You will no doubt still be able to recall the preparation of iron sulfide from sulfur and iron filings. The basis for all the calculations we shall be discussing here is the chemical equation for the reaction involved. We are thus going to start off by asking you to write out the equation for the reaction of sulfur and iron.

After doing this, you should check your equation at [10] .

2 The unit of atomic mass is 1/12th of the mass of the carbon atom.

> The atomic mass of an element specifies how many times greater its mass is than 1/12th the mass of the carbon atom.

This important statement should be committed to memory.

Now, elements combine together in the ratio of their atomic masses.

In iron sulfide we know that one atom of sulfur is combined with one atom of iron. This means that, upon heating, 32 g of sulfur will react with 56 g of iron to produce 88 g (32 + 56 = 88) of iron sulfide.

How many grams of iron would be required to react with 16 g of sulfur to form iron sulfide?

- 16 g. → [13]
- 28 g → [26]

- 14 g → 34
- I need some help → 24

---

3 Your answer is wrong. If iron and sulfur atoms were of the same mass, the same quantities by mass of iron and sulfur would react together. This, however, is *not* the case!

Repeat now from 20 .

9

---

4 No. You'll have to think again about your answer.

Continue reading now at 45 .

---

5 Your answer is wrong. You appear to have forgotten that oxygen occurs as a diatomic molecule, namely as O_2.

Furthermore, you must clearly differentiate between the molecular mass and the gram molecule!

How much, for instance, is one mole of oxygen? → 28

---

6 Your answer is wrong. You have obviously not set up the equation for the reaction properly. The equation for the combustion of hydrogen to form water is as follows:

$$2H_2 + O_2 \rightarrow 2H_2O$$

All you have to do now is to insert the atomic masses of hydrogen (1) and oxygen (16) to obtain the ratio by mass in which hydrogen reacts with oxygen. Reduce this ratio to its simplest possible form by cancelling where appropriate.

What now is the ratio by mass of H : O that you have calculated?

- 4 : 32 → 21
- 1 : 8 → 31

---

7 Your answer is wrong. Rework these three examples using the following scheme:

32 g of sulfur react with 56 g of iron.

1 g of sulfur reacts with $\frac{56}{32}$ g of iron.

16 g of sulfur react with $\frac{56}{32} \cdot 16$ g of iron.

Go back now to [30] .

8 Your answer is wrong. You should not forget that:

The molecule of iron sulfide is made up of one atom of sulfur and one atom of iron. This means that in any lump of iron sulfide the number of atoms of sulfur and of iron will be equal. Bearing in mind that the number of atoms of each element is the same and the fact that only 32 g of sulfur as against 56 g of iron are required in the preparation of iron sulfide, it is evident that the sulfur atom must be .... (lighter/heavier) than the iron atom.

Continue your reading at [19] .

9 The atomic mass of an element specifies how many times greater its mass is than 1/12th *the mass of the carbon atom.*

The atomic mass of an element in grams is the ....

Continue your reading at [37] .

10 $Fe + S \rightarrow FeS$

This equation actually tells us two things:

1. Iron reacts with sulfur to form iron sulfide.

2. One atom of iron reacts with one .... of sulfur to form one .... of iron sulfide.

Write out the completed version of the latter sentence, and then proceed to [20] .

11 You are not working carefully enough and have not properly understood the material we have covered so far.

Repeat now from 20 .

12 58.5 g of sodium chloride have to be dissolved.

9

In the case of *compounds* the amount is frequently specified in terms of the unit *"mol"*, whereas for elements the unit *"gram atom"* is used.

Example: *1 gram atom* of sulfur and *1 gram atom* of iron react together to form *1 mol* of iron sulfide.

This means that 32 g of sulfur react with 56 g of iron to form 88 g of iron sulfide

The term "mol" is often used instead of the term "gram atom", e.g. 1 mol of sulfur (= 32 g) or 2 mol of iron (= 112 g).

When calculating the number of moles, *especial care* must be taken *with the diatomic gases!*

Nitrogen has an atomic mass of 14. One gram atom of nitrogen is thus 14 g. However, 1 mol of nitrogen is 28 g because nitrogen always occurs as the diatomic molecule (N_2).

How much is 1 mol of oxygen?

- 16 → 5
- 16 g → 48
- 32 → 51
- 32 g → 28

13 No. You will need some help to get to the right answer.

Continue your reading at 24 .

14 Your answer is wrong. You are reminded again that:

One gram atom of an element means the atomic mass of that element in grams. Accordingly, we have:

16 g of oxygen equal to	1 gram atom of oxygen,
32 g of sulfur equal to	1 gram atom of sulfur, and
1 g of hydrogen equal to	1 gram atom of hydrogen.

Go back to 35 , and answer the question there again.

15 No. You should reconsider your answer again.

Continue your reading now at 45 .

16 The equation for the reaction is:

$$2H_2 + O_2 \rightarrow 2H_2O$$

Thus, 4g of hydrogen (atomic mass 1; H_2 has a molecular mass of 2; 2 x 2 = 4; we take 2x as there is a 2 in front of the H_2) react with 32 g of oxygen (atomic mass 16; O_2 has a molecular mass of 32). This means that they will react in the ratio of 4 : 32 or, when divided by 4, in the ratio of 1 : ....

Continue your reading at 31 .

17 It was a good idea to look up this section and so avoid any uncertainty or the making of mistakes.

The calculation of the molecular mass of sulfuric acid (H_2SO_4) is as follows:

2 atoms of hydrogen (atomic mass 1)	2 x 1 = 2
1 atom of sulfur (atomic mass 32)	1 x 32 = 32
4 atoms of oxygen (atomic mass 16)	4 x 16 = 64
Molecular mass of H_2SO_4	98

The calculation of the molecular mass of phosphorus pentoxide (P_2O_5) is as follows:

2 atoms of phosphorus (atomic mass 31)	2 x 31 =
5 atoms of oxygen (atomic mass 16)	5 x 16 =
Molecular mass of P_2O_5	

Insert the missing numbers, and then continue your reading at 27.

18 No. Additional help is given at 52.

19 Your answer is correct if you said that:
A sulfur atom is *lighter* than an iron atom.

Thus, atoms of *different* elements have *differing* masses!

9

Individual atoms cannot, of course, be weighed. They are far too tiny for that! However, we are certainly able to calculate the mass of an atom.

We know, for instance, that a sulfur atom has a mass of

0.000000000000000000000053 g.

Such numbers are not only extremely small but also rather unwieldy. For this reason, the concept of the *relative atomic mass* has been introduced. It is specified in terms of the carbon atom thus:

Carbon (C) has a relative atomic mass of 12.

From this definition it follows that:

Hydrogen (H) has a relative atomic mass of 1 and
Oxygen (O) has a relative atomic mass of 16.

What does this imply?

It means that the mass of a hydrogen atom is equal to 1/12th the mass of a carbon atom.

Similarly, the mass of an oxygen atom is 16 times greater than 1/12 the mass of a carbon atom.

Sulfur has a relative atomic mass of 32.

What does that imply? Please write out your answer, and then go on to 38.

20 Yes, the equation $Fe + S \rightarrow FeS$ tells us that:

One *atom* of iron reacts with one *atom* of sulfur to form one *molecule* of iron sulfide.

Thus, *one* atom of iron combines with *one* atom of sulfur. If we were to take more iron, some iron atoms would be left over at the end of the reaction. Similarly, if we were to take more sulfur, some sulfur atoms would be left over at the end of the reaction.

Experimentally, we can demonstrate that:

56 g of iron react with 32 g of sulfur to form 88 g of iron sulfide.

Thus, 56 g of iron combine with 32 g of sulfur.

Experiment 1. We start with 66 g of iron and 32 g of sulfur, and so at the end of the reaction we shall have something left over.

Experiment 2. We start with 56 g of iron and 40 g of sulfur, and so, once again, something will be left over.

	What is left over?	How much of it is left?
Experiment 1	. . . .	. . . . g
Experiment 2	. . . .	. . . . g

Check your answers at 29 .

---

21 Your answer is still not quite right.

The equation

$$2H_2 + O_2 \rightarrow 2H_2O$$

is certainly correct, as is the hydrogen to oxygen ratio of 4 : 32. However, as both these numbers can be divided by 4, the fixed ratio by mass in which hydrogen reacts with oxygen is actually ... : ...

Continue your reading at 31 .

---

22 Your answer is quite right.

Please take especial care not to mix up the notion of *atomic mass* with that of the *gram atom!*

Iron has an atomic mass of 56.
1 gram atom of iron equals 56 g.

Now copy down the completed version of each of the following sentences:

a) Sulfur has an atomic mass of ....
b) 3 gram atoms of iron equal..............
c) 2 gram atoms of sulfur equal..............
d) Hydrogen has an atomic mass of..............
e) 6 gram atoms of hydrogen equal..............

Check your answers at 33 .

---

23 It is the Law of *Fixed Proportions by Mass.*

In which of the following reactions do the starting materials combine together in a mass ratio of 7 : 3?

Work through each example below individually.

- $4P + 5O_2 \rightarrow 2P_2O_5$ → 44
- $S + O_2 \rightarrow SO_2$ → 18
- $C + O_2 \rightarrow CO_2$ → 55
- $4Fe + 3O_2 \rightarrow 2Fe_2O_3$ → 36
- I don't know → 52

---

24 If 32 g of sulfur are needed to react with 56 g of iron, then by taking simple ratios you should be able to work out how many grams of iron would be needed for 16 g of sulfur.

For 32 g of sulfur one needs 56 g of iron.

For 1 g of sulfur one needs $\frac{56}{32}$ g of iron.

For 16 g of sulfur one therefore needs $\frac{56}{32} \cdot 16$ g of iron.

Work out now the amount of iron needed, and then go on to 26 .

---

25 The mass of the iron atom is 56 times as great as 1/12th the mass of the carbon atom.

What is the unit of atomic mass?

Write out your answer, and then go on to [2] .

---

[26] Your answer is correct, for as 16 g of sulfur is just one half of 32 g we shall need on half of 56 g of iron, namely 56 g/2 – 28 g of iron.

32 g of sulfur, i.e. the atomic mass in grams, is referred to as one *gram atom* of sulfur.

One gram atom of an element is its atomic mass in grams

This statement should be memorized.

The atomic mass of oxygen is 16.
One gram atom of oxygen is 16 g.

Suppose that a vessel contains 48 g of oxygen. How many gram atoms of oxygen would it contain?

- 1 gram atom → [4]
- 2 gram atoms → [15]
- 3 gram atoms → [35]

---

[27] The molecular masses are: $H_2SO_4 = 98$
$P_2O_5 = 142$

If you came unstuck here or still feel somewhat uncertain, turn back to [17] , where the calculations are explained in detail.

Now, in the same way that

The atomic mass in grams is the *gram atom,*

we may also state that

The molecular mass in grams is the *gram molecule.*

How much is 1 gram molecule of sulfuric acid and one gram molecule of phosphorus pentoxide? → [39]

---

28 1 mol of oxygen (O_2) is 32 g.

Now perform the following calculations:

(You have already noted down the atomic masses needed to solve these problems.)

1. Work out the molecular masses of sodium carbonate (Na_2CO_3) and nitric acid (HNO_3).

2. How much are 4 mol of oxygen, 2 mol of nitric acid, 1 gram atom of iron, 2 mol of hydrogen, 3 grams of sulfur, and 3 mol of iron sulfide? → 54

9

29 We shall have left over in:

Experiment 1	Iron	10 g
Experiment 2	Sulfur	8 g

We have already established that: One atom of iron reacts with one atom of sulfur. Moreover: 56 g of iron react with 32 g of sulfur.

Thus, by weight less sulfur than iron is needed.

What may we conclude from this observation?

- One atom of sulfur is heavier than one atom of iron → 8
- One atom of sulfur is lighter than one atom of iron → 19
- One atom of sulfur has exactly the same weight as one atom of iron → 3

30

1. The molecular mass of a compound may be calculated by adding together the atomic masses of all the atoms constituting the compound as shown in the chemical formula.

2. One *gram molecule* or 1 *mol* of a compound is the molecular mass of that compound in grams.

The chemical equation

$$Fe + S \rightarrow FeS$$

informs us not only which elements and how many atoms of each type react together, but also makes a statement about the actual quantities of the individual elements involved in the reaction. As the atomic mass of iron (Fe) is 56 and that of sulfur (S) is 32, the above equation tells us that 56 g of iron will react with 32 g of sulfur to form 88 g of iron sulfide. In fact, iron and sulfur will *always* react together in the mass ratio of 56 : 32 whenever iron sulfide is being formed.

How many grams of iron will react with
1. 16 g of sulfur
2. 8 g of sulfur
3. 4 g of sulfur

to form iron sulfide?

	1.	2.	3.		
•	28 g of iron	8 g of iron	4 g of iron	→	7
•	The same amount of iron in each case			→	11
•	28 g of iron	14 g of iron	7 g of iron	→	40
•	I get quite different numbers			→	50
•	28 g of iron	14 g of iron	4 g of iron	→	47

31

Hydrogen and oxygen react together according to the equation:

$$2H_2 + O_2 \rightarrow 2H_2O$$

The ratio by mass of hydrogen to oxygen is 4 : 32. As each of these numbers is divisible by four, the simplified ratio becomes 1 : 8.

What is the Law of Fixed Proportions by Mass?

Write out the completed version of the sentence:

When two elements combine together, they do so in fixed .... determined by their ....

Check your answer at 42 .

32

The molecular mass in grams is the *gram molecule,* also referred to as the *mole.* (The abbreviation for this is mol.)

Iron sulfide has a .... of 88.
One .... of iron sulfide is 88 g.

Jot down the missing words, and then continue your reading at 48 .

---

33 a) Sulfur has an atomic mass of 32

b) 3 gram atoms of iron equal 168 g

c) 2 gram atoms of sulfur equal 64 g

d) Hydrogen has an atomic mass of 1

e) 6 gram atoms of iron equal 6 g

Hopefully, you did not forget the abbreviation "g" in the statements b), c) and e). Did you actually get everything right?

If some errors did creep in, you should now in your own interest repeat from 26.

It makes sense to go on to new material at this stage only if you feel you have really mastered what has been covered so far.

Can you still recall the two important statements introduced earlier on in this Program?

What ist the atomic mass? The atomic mass of an element specifies ....

What is a gram atom? One gram atom of an element is ....

Write out completed versions of these two statements, and then check your answers at 53.

34 No. You will need some help to reach the right answer.

Continue your reading at 24 .

35 Your answer is quite right.

3 gram atoms of oxygen are 48 g.

See if you can now answer this further question:

Suppose we have 32 g of oxygen, 32 g of sulfur and 32 g of hydrogen

How many gram atoms of oxygen, sulfur and hydrogen will we have?

- 2 gram atoms of oxygen, 1 gram atom of sulfur, and 32 gram atoms of hydrogen → 22
- 1 gram atom of oxygen, 1 gram atom of sulfur, and 1 gram atom of hydrogen → 14
- 1 gram atom of oxygen, 2 gram atoms of sulfur, and 16 gram atoms of hydrogen → 46

---

36 Your answer is quite correct.

In the reaction

$$4Fe + 3O_2 \rightarrow 2Fe_2O_3$$

the iron and oxygen react in the ratio by mass of 7 : 3.

At this point we should like to repeat the laws and important statements made in this Program.

What is the unit of atomic mass?

Write out your answer, and then check it at 43 .

---

37 The atomic mass of an element in grams is the *gram atom.*

Sulfur has an .... of 32.
32 g is one .... of sulfur.

Check your answers at 56 .

---

38 The mass of the sulfur atom is 32 times greater than 1/12th the mass of the carbon atom.

Thus, 1/12th the mass of the carbon atom is the

Unit of Atomic Mass.

(To be precise, the atomic mass unit is based on the atomic mass of the carbon isotope ^{12}C.)

In the table below we list the atomic masses of some important elements. (Instead of referring to "relative atomic mass", we shall in future be using the abbreviation *atomic mass*.)

The *atomic mass* is the mass of an atom in atomic mass units.

Copy down on a sheet of paper the following atomic masses and keep them handy as you will be needing them quite often as you work your way through the rest of this Program.

9

Hydrogen	1
Carbon	12
Oxygen	16
Nitrogen	14
Sodium	23
Phosphorus	31
Sulfur	32
Iron	56

Now complete the sentence:

The mass of the iron atom is .... times greater than .... the mass of the carbon atom. → 25

39

1 gram molecule of H_2SO_4 = 98 g
1 gram molecule of P_2O_5 = 142 g

Instead of "gram molecule" we shall be using the abbreviation "mole".

Now make sure that you commit to memory the following important statement:

> One gram molecule or one mole of a compound is the molecular mass of that compound in grams.

Amounts of chemicals used in experiments are frequently specified in terms of the *mole*. An example is provided by the instruction, "Dissolve one mole of sodium chloride in one liter of water".

How much sodium chloride would have to be dissolved in one liter in this instance? (The relevant atomic masses are Na = 23; Cl = 35.5.) → 12

40 Your calculations are correct.

32 g of sulfur react with 56 g of iron.

1 g of sulfur reacts with $\frac{56}{32}$ g of iron.

16 g of sulfur react with $\frac{56}{32} \cdot 16 = 28$ g of iron.

8 g of sulfur react with $\frac{56}{32} \cdot 8 = 14$ g of iron.

4 g of sulfur react with $\frac{56}{32} \cdot 4 = 7$ g of iron.

It should thus be clear that sulfur and iron react together only in a quite *definite* and *fixed ratio by mass.* This ratio is determined by the relative *atomic masses* of iron and sulfur.

Other elements also react together in a fixed ratio by mass.

If more of a given starting material in a reaction is taken than that required from the mass ratio, the *excess* amount of *material will not react.*

The Law of Fixed Proportions by Mass states that in general:

> When two elements combine together, they do so in fixed proportions by mass *determined by their relative atomic masses.*

Make sure that you learn this law by heart!

Now set up the equation for the combustion of hydrogen to form water, and then calculate:

In what ratio by mass will hydrogen and oxygen combine together when they form water? (The relevant atomic masses are H = 1; O = 16.)

- 1 : 16 → 6
- 1 : 8 → 31
- 4 : 32 → 21
- This question is too difficult for me → 16

41 The molecular mass of a compound is calculated by adding together the *atomic masses* of all the atoms which appear in the *formula* of that compound.

The molecular mass in grams is the ...., also known as the ....

What words are missing here? You will find these words given in 32.

42 When two elements combine together, they do so in fixed *proportions by mass* determined by their *relative atomic masses.*

What is this law known as?

It is the Law of .... → 23

43 The unit of atomic mass is 1/12th the mass of the carbon atom.

The atomic mass of an element specifies how many times greater its mass is than ....

The correct version is to be found at 9.

44 No. Additional help is given at 52.

45 We have said that:

One gram atom of an element is its atomic mass *in grams.*

Let's take an example:

The atomic mass of iron is **56.** The atomic mass of iron in grams is thus **56** g of iron. Whenever we have 56 *g of iron,* we refer to this as *one gram atom of iron.*

The question on how many gram atoms of oxygen there are in 48 g of oxygen may thus be answered as follows:

The atomic mass of oxygen is 16.
One gram atom of oxygen is 16 g of oxygen.
48 g/16 g thus gives us the number of gram atoms of oxygen.

How many gram atoms of oxygen are therefore in 48 g of oxygen? → 35

46 Your answer is wrong. It should not be forgotten that:

By one gram atom of an element we mean as many grams of this element as its atomic mass indicates. Accordingly, we have:

16 g of oxygen = 1 gram atom of oxygen,
32 g of sulfur = 1 gram atom of sulfur, and
1 g of hydrogen = 1 gram atom of hydrogen.

Go back to 35 and answer the question there again.

47 Your answer is wrong. Re-work the three problems using the following outline scheme:

32 g of sulfur react with 56 g of iron.

1 g of sulfur reacts with $\frac{56}{32}$ g of iron.

16 g of sulfur react with $\frac{56}{32} \cdot 16$ g of iron.

Now go back to 30 .

48 Iron sulfide has a *molecular mass* of 88.

One *mole* of iron sulfide is 88g.

What is the Law of Fixed Proportions by Mass?

Write out the statement of this law, and then check your answer at 57 .

49 Your answer is wrong. Although you have taken into account the difference between molecular mass and the gram molecule, you forgot that oxygen is diatomic, i.e. it has the formula O_2.

Now, how much in grams is one mole of oxygen? → 28

50 Re-work the three problems using the following outline scheme:

32 g sulfur react with 56 g of iron.

1 g of sulfur reacts with $\frac{56}{32}$g of iron.

16 g of sulfur react with $\frac{56}{32} \cdot 16$ g of iron.

Now go back to [30] .

9

[51] Your answer is wrong. Oxygen is indeed diatomic, i.e. it has the formula O_2, but you did not remember the difference between molecular mass and the gram molecule!

Now, how much in grams is one mole of oxygen? → [28]

[52] Let's work out the ratio for the first reaction:

$$4P + 5O_2 \rightarrow 2P_2O_5$$

Phosphorus has an atomic mass of 31. As there are four phosphorus atoms present, we have here

$$4 \times 31 = 124$$

Oxygen has an atomic mass of 16. As there are 5 x 2 = 10 oxygen atoms present, we have here

$$10 \times 16 = 160$$

Phosphorus and oxygen thus react in the mass ratio of

$$124 : 160$$

(divided by 2) or 62 : 80
(again divided by 2) or 31 : 40 (not further divisible!)

Go back to [23] and calculate the ratios for the other reactions likewise.

[53] The *atomic mass* of an element specifies how many times greater its mass is than 1/12th the mass of a carbon atom. One *gram atom* of an element is its atomic mass in grams.

We now want to say something about molecular mass. The *molecular mass* of a compound is calculated by adding together the atomic masses of all the atoms appearing in the formula of the compound.

Here are two worked examples:

Iron sulfide (FeS) is composed of one atom of iron and one atom of sulfur.

1 atom of iron (atomic mass of Fe = 56)	56
1 atom of sulfur (atomic mass of S = 32)	32
Molecular mass of iron sulfide	88

9

Water (H_2O) is composed of two atoms of hydrogen and one atom of oxygen.

2 atoms of hydrogen (atomic mass of H = 1)	2
1 atom of oxygen (atomic mass of O = 16)	16
Molecular mass of water	18

Calculate now the molecular masses of sulfuric acid (H_2SO_4) and phosphorus pentoxide (P_2O_5). → 27

54

1. The molecular mass of Na_2CO_3 = 106
   HNO_3 = 63

2. 4 mol O_2 = 128 g; 2 mol HNO_3 = 126 g; 1 gram atom Fe = 56 g; 2 mol H_2 = 4 g; 3 gram atoms S = 96 g; 3 mol FeS = 264 g.

Did you succeed in getting all of these right? If not, you should repeat from 19.

Now write out answers to the following questions:

1. How is the molecular mass of a compound calculated?

2. What is meant by a gram molecule or a mole? → 30

55

No. Additional help is to be found at 52.

56

Sulfur has an *atomic mass* of 32.

32g of sulfur represent one *gram atom* of sulfur.

The molecular mass of a compound is calculated by adding together the .... of all the atoms which appear in the .... of that compound.

Jot down the missing words, and then check your answers at 41 .

---

57 The Law of Fixed Proportions by Mass states that:

When two elements combine together, they do so in fixed proportions by mass determined by their relative atomic masses.

End of Program 9.

Program 10

Theoretical Foundations (IV)

1 In this Program we shall discuss the most important laws which describe the behavior of gases. It will be assumed that you are fully conversant with the material covered in the three previous theoretical Programs, i.e. in Programs 6, 7 and 9.

How many grams are there in 1 mol of

a) chlorine, b) hydrogen, c) oxygen?

- a) 35.5 g b) 1 g c) 14 g → [14]
- a) 71 g b) 2 g c) 28 g → [25]
- a) 106.5 g b) 3 g c) 42 g → [33]

10

2 Re-work this example again.

If a force of 18×10^5N is exerted on a surface of 6 m^2, then the pressure on each m^2 will be

$$\frac{18 \times 10^5 \text{ N}}{6 \text{ m}^2} = ? \text{ bar}$$

Go back now to [10] .

3 At 4 bar the enclosed gas will be compressed into a volume of 0.5 L.

This demonstrates that: When the pressure is *doubled* the volume will be *halved.*

In our particular example the gas had

at a pressure of 1 bar a volume of 2 L
at a pressure of 2 bar a volume of 1 L
at a pressure of 4 bar a volume of 0.5 L
at a pressure of 8 bar a volume of 0.25 L

It is thus clear that a well-defined *relationship exists between the pressure and volume* of a gas sealed within a container.

Now multiply together the pressure and the corresponding volume for each individual case cited above. The results obtained are thus as follows:

Pressure	x	Volume		
1	x	2	=	2
2	x	1	=	

and so on.

Compare the four products you obtain.

What strikes you about them?

- The numbers consistently decrease → 31
- The numbers consistently increase → 19 .
- The numbers are all identical → 43

---

4 In this particular example we have:

Pressure x Volume = 100.

We are now in a position to work out the volume corresponding to each of the pressures in the example, for we know that the product of pressure x volume will always equal 100.

What will the volume of the gas be at a pressure of 10 bar? → 21

---

5 1 cubic meter = 1000 L

Now calculate:

How many cm^3 there are in 1 L?
(Remember that 1 L = 1 dm^3 and that 1 dm = 10 cm)

- 10 → 60
- 100 → 18
- 1000 → 30

---

6 You appear to have made a mistake.

In the first case we have: 1 bar and 2 L

Thus, 1 x 2 = 2.

In the second case we have: 0.5 bar and 4 L

Thus, 0.5 x 4 = 2.

In the third case we have: 0.25 bar and 8 L

Thus, 0.25 x 8 = ...

In the fourth case we have: 0.125 bar and 16 L

Thus, 0.125 x 16 = ...

In this particular example, the product of pressure and its associated volume always yields the number ....

In fact, all the products have the same value.

Repeat now from 43 .

7 Your answer is wrong. We shall now solve this problem stepwise, starting at 77 .

8 You have not been sufficiently observant here, for otherwise you would have correctly noted what changes occur in the diagrams.

Read through 21 once again.

9 You appear to have made a mistake.

We shall now solve this problem stepwise, starting at 59 .

10 The volume will *decrease* as the pressure *increases.*

Expressions such as "will decrease" are not sufficiently precise for scientific work. In experiments it is necessary to know both the pressure and volume of gases exactly, i.e. in terms of actual figures.

The *pressure* is normally measured in terms of the *bar*.

1 bar is the pressure which a force of 10^5 N (N = newtons) exerts over a surface of 1 m^2. Thus, we have

$$1 \text{ bar} = 10^5 \text{ N/m}^2$$

(In the past pressure was frequently expressed in terms of the atmosphere (at or atm). 1 atmosphere is approximately equal to 1 bar.

Now calculate:

10 The pressure in bar when a force of 18 x 10^5 N is exerted over a surface of 6 m^2.

- The pressure is 18 bar → [2]
- The pressure is 9 bar → [20]
- The pressure is 6 bar → [28]
- The pressure is 3 bar → [39]
- I'm uncertain about the method of calculation → [47]

---

11 Your answer is wrong. Air or atmospheric pressure actually decreases with increasing height! Thus, in mountainous regions which are well above sea level the pressure will be .... than that at sea level. → [62]

---

12 You appear to have made a mistake.
We shall now solve this problem stepwise, starting at [59] .

---

13 Yes, in all cases the product of pressure and volume is equal to the same number, namely 2.

We are therefore in a position to calculate the increase in volume when the pressure is reduced.

Calculate now what the volume of a gas (100 L at 1 bar) will be if its pressure is reduced to 0.5 bar. → [24]

---

14 You have not allowed for the fact that the gases chlorine, hydrogen, nitrogen and oxygen occur *only as diatomic molecules,* i.e. as Cl_2, H_2, N_2 and O_2.

In your own interest you are advised to repeat Programs 7 and 9, preferably before starting again at 1 .

15 Your answer is wrong. We shall now solve this problem stepwise, starting at 77 .

16 The correct answer is: 1 m^3 = 1,000,000 cm^3

But let's get back now to our experiments with the air pump. Imagine an air pump which has *no opening* at the bottom, so that there would be no means of escape for the air. If the air in the pump is now compressed, its volume becomes ever .... (smaller/greater) the greater the pressure gets.

→ 68

17 1 m^3 = 1,000 L

Advice on performing this calculation is given at 56 .

18 1 L = 1,000 cm^3

Advice on performing this calculation is given at 46 .

19 You appear to have made a mistake.

In the first case we have: 1 bar and 2 L

Thus, 1 x 2 = 2 .

In the second case we have: 2 bar and 1 L

Thus, 2 x 1 = ...

In the third case we have: 4 bar and 0.5 L

Thus, 4 x 0.5 = ...

Finally, in the fourth case we have: 8 bar and 0.25 L

Thus, 8 x 0.25 = ...

In our example therefore multiplication of the pressure with the corresponding volume *always* yields the product ....

All the products thus have the same value. → 43

---

20 Your answer is wrong. Recalculate your result as follows:

When a force of 18×10^5 N is exerted over a surface of 6 m^2, the pressure on each m^2 will be

$$\frac{18 \times 10^5 \text{ N}}{6 \text{ m}^2} = ?\text{ bar}$$

Go back now to 10 .

---

21 The gas in our example has a volume of 10 L at a pressure of 10 bar, since

10 x 10 = 100.

What would happen now if we were to reduce the pressure?

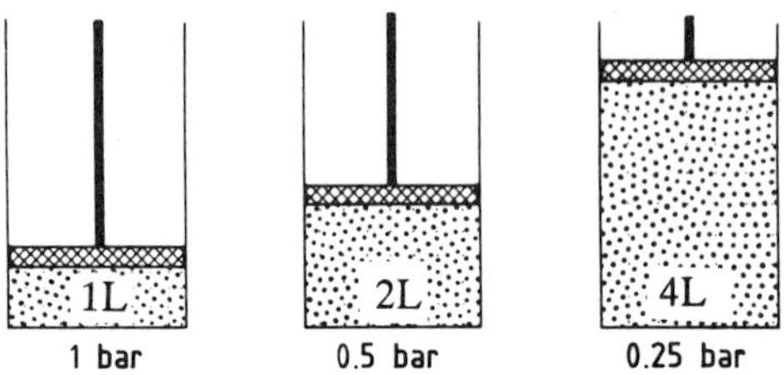

- Nothing would happen → 44
- The volume will be reduced → 8
- The volume will be increased → 35

---

22 Your answer is wrong. Remember here that:

The gas contained in the vessel is at a reduced pressure, i.e. at 0.5 bar. The pressure of the sourrounding air is 1 bar; it is thus greater! If a hole were to be bored in the vessel, air would stream into it from the outside until the pressure difference had been equalized. → 34

---

23 A gas occupying a volume of 80 L at 12.5 bar will be under a pressure of only **0.25 bar** when it occupies a volume of 4 m^3 or 4,000 L.

A distinction is made between

Reduced Pressure
Excess Pressure
Atmospheric Pressure.

The air all around us, i.e. the atmosphere, exerts a pressure on us known as atmospheric pressure. We are not aware of this pressure as our body is so used to it. At great heights the atmospheric pressure is, however, much lower and one then speaks of the air "becoming thinner". For this reason aircraft are fitted out with pressurized cabins in which the pressure needed by our body is maintained at the correct level.

Now, just how great is atmospheric pressure?

At sea level it is around 1 bar. How will the air pressure differ in mountainous regions as compared to that at sea level?

- It will be greater → 36
- It will be smaller → 62
- It will be the same → 11

---

24 The gas has a volume of 200 L at 0.5 bar, for when the pressure is halved the volume will be doubled.

At 1 bar we have: 1 x 100 = 100
At 0.5 bar we have: 0.5 x 200 = 100

We see that in this example too the product of pressure and volume always has the same value.

To generalize, the following relationship between the pressure p and the volume v always holds:

p.v = const.	where const. = a constant number.

The constant number assumes a different value in differing situations (in our particular examples it has had the values 2 and 100). *For a given situation*, however, its *value remains unchanged* no matter what changes of pressure and volume occur.

You should now be able to solve the following problem:

A gas is enclosed within a volume of 150 L at a pressure of 3 bar. What will the volume of this gas be if it is compressed by an adjustable piston to a pressure of 9 bar?

- 25 L → 7
- 50 L → 45
- 100 L → 15
- 250 L → 37
- 450 L → 65
- 500 L → 57
- I am unable to calculate it → 77

---

25 You have retained the earlier material we covered very well indeed.

The gases chlorine, hydrogen, nitrogen and oxygen occur only in the form of molecules (Cl_2, H_2, N_2 and O_2). For these elements the molecular mass is always twice the atomic mass.

When performing chemical experiments with these gases it is often insufficient merely to know their molecular masses. We frequently need to determine accurately the amount of gas utilized in a particular experiment. In the case of solids and liquids this may be conveniently done by *weighing out* the amounts required. Liquid substances can also be *measured out* by *volume,* e.g. by taking (say) one liter.

For a variety of reasons that we shall go into later, the weighing or measuring out of gases is often quite difficult.

Quantities of solid and liquid substances are determined by .... whereas the volume of a liquid is fixed by .... → 32

---

26 You appear to have made a mistake.

In the first case we have: a volume of 2 L at 1 bar

Thus, 1 x 2 = 2

In the second case we have: 4 L at 0.5 bar

Thus, 0.5 x 4 = 2

In the third case we have: 8 L at 0.25 bar

Thus, 0.25 x 8 = ...

In the fourth case we have: 16 L at 0.125 bar

Thus, 0.125 x 16 = ...

In our example therefore multiplication of the pressure with the corresponding volume always yields the product ....

All the products thus have the same value.

Repeat now from [43] .

---

27 Your answer is wrong. Remember here that:

The gas contained in the vessel is at a reduced pressure, i.e. at 0.5 bar. The pressure of the surrounding air is 1 bar; it is thus greater! If a hole were to be bored in the vessel, air would stream into it from the outside until the pressure difference had been equalized. → [34]

---

28 Your answer is wrong. Recalculate your result as follows:

When a force of 18×10^5 N is exerted over a surface of 6 m^2, the pressure on each m^2 will be

$$\frac{18 \times 10^5 \text{ N}}{6 \text{ m}^2} = ?\text{ bar}$$

Go back now to [10] .

10

29 1 m^3 = 1,000 L

Advice on performing this calculation is given at 56.

30 1 L = 1,000 cm^3

Now calculate how many cm^3 there are in one m^3. → 16.

31 You appear to have made a mistake.

10

In the first case we have: 2 L at 1 bar

Thus, 1 x 2 = 2

In the second case we have: 1 L at 2 bar

Thus, 2 x 1 = ...

In the third case we have: 0.5 L at 4 bar

Thus, 4 x 0.5 = ...

In the fourth case we have: 0.25 L at 8 bar

Thus, 8 x 0.25 = ...

In our example, therefore, multiplication of the pressure with the corresponding volume *always* yields the product .... → 43

32 Quantities (masses) of solid and liquid substances are determined by *weighing out* whereas the volume of a liquid is fixed by *measuring out.*

As we shall soon see, fixing the amount of a gas is much more difficult.

Did you ever pump up a bicycle inner tube?

Sure you did! We shall now take a closer look at the process involved. The air in the pump gets compressed until its pressure becomes so great that it can pass through the valve into the inner tube.

The important thing for us here is the fact that the air can be *compressed.* (In the case of solids and liquids this is by no means so easy!)

When air is compressed its volume will .... (increase/decrease), and the pressure of the compressed air will .... (increase/decrease). → 10

---

33 Your answer is wrong. You have not allowed for the fact that the gases chlorine, hydrogen, nitrogen and oxygen occur *only as diatomic molecules,* i.e. as Cl_2, H_2, N_2 and O_2.

In your own interest you are advised to repeat Programs 7 and 9, preferably before starting again at 1.

---

34 Correct. Air will flow into the vessel until the pressure difference has been equalized.

When we have a gas enclosed within a vessel and the pressure p of that gas is *greater than 1 bar* (as in the figure below where p = 3 bar), the gas is said to have an *excess pressure.*

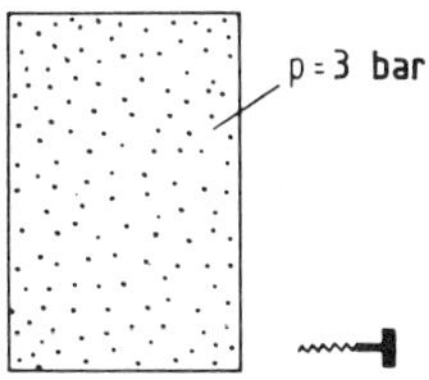

What would happen if a hole were to be drilled in the vessel?

- Air would flow into the vessel → 64
- Some of the enclosed gas would flow out → 82
- Nothing at all would happen → 51

---

35 Your answer is quite right, for when the *pressure* is *reduced* the gas will expand and thus *occupy a larger volume.*

For instance, a gas might have:

A volume of 2 L at a pressure of 1 bar
A volume of 4 L at a pressure of 0.5 bar
A volume of 8 L at a pressure of 0.25 bar
A volume of 16 L at a pressure of 0.125 bar

Multiply together each pressure with its corresponding volume as follows:

Pressure x Volume

1 x 2 =

0.5 x 4 =

etc.

Now compare each of the products you have obtained.

What strikes you about them?

- The products decrease in value → 26
- The products increase in value → 6
- The products all have the same value → 13

---

36 Your answer is wrong. Air or atmospheric pressure decreases with increasing height! In mountainous regions which lie well above sea level the pressure will therefore be .... than that at sea level. → 62

---

37 Your answer is wrong. We shall now solve this problem stepwise, starting at 77 .

---

38 1 L = 1,000 cm^3

Advice on performing this calculation is given at 46 .

---

39 Your answer is quite right, the pressure is indeed 3 bar.

The *volume* of a gas is usually measured in *liters*. For smaller volumes than this one measures the number of cubic centimeters (cm^3) whereas for larger volumes one measures the number of cubic meters (m^3).

Do you know how many liters there are in 1 cubic meter?

- 10 L → 17
- 100 L → 29

- 1000 L → 5
- 10000 L → 50
- I don't know → 56

---

40 Your answer is wrong.

An explanation of this is given at 125 .

---

41 Your answer is correct.

If you hit upon this answer right away, you are progressing well in the Program.

Because the number of moles is the same on both sides of the equation for the following reaction:

H_2	+ Cl_2	→	HCl
22.4 L	22.4 L		2 x 22.4 L

the amount of gas present is not changed in any way. Accordingly, the pressure in the vessel will remain constant and so when is dipped into mercury and the stopcock a is opened nothing will happen.

This should be easy to understand, for we have already said that the gases hydrogen and chlorine consist of molecules.

This observation is, however, *extremely important* as it provides *proof* that the gases hydrogen and chlorine do indeed consist of the molecules H_2 and Cl_2 respectively (and not of the atoms H and Cl).

Let's take a closer look into this.

If hydrogen and chlorine atoms were in fact present, the equation would then assume the form:

H	+ Cl	→	HCl
1 g	35.5 g		36.5 g

and the corresponding volumes under normal conditions would be:

22.4 L	22.4 L	22.4 L

The total volume of the starting materials would thus amount to 44.8 L whereas that for the end product would be 22.4 L. The volume would therefore be decreased by one half (at constant pressure) during this reaction.

But, this clearly did *not* happen in our experiment!
In fact, what we observed in our experiment can be explained *only* if the hydrogen and chlorine are present in the form of *molecules* (H_2 and Cl_2).

The presence of molecules in gases was first recognized in 1811 by the Italian *Avogadro*. He formulated the following law:

10

> Equal volumes of all gases at constant temperature and pressure contain the same number of molecules.

Now copy down this *Law of Avogadro* and learn it by heart.

We have already come across the statement that:

1 mol of any gas under .... occupies a volume of .... L.

Write out the completed version of this statement, and then check what you have written at 91 .

---

42 Your calculation is wrong.

An explanation of this is given at 117 .

---

43 Yes, in all cases multiplication of the pressure and volume gave the same product, namely 2.

This is an important result.

If we take a fixed amount of gas and subject it to various different pressures, we know now that the *product of its pressure and volume will always have the same value.*

In our particular example it was always equal to 2.

Here is a further example. A fixed amount of gas is subjected to the following sequence of pressures:

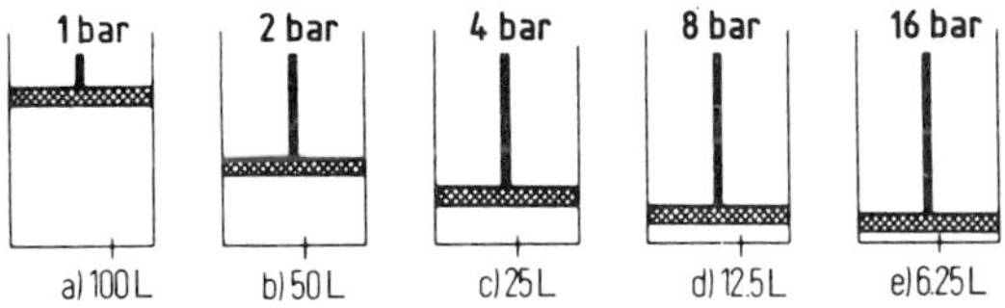

a) At 1 bar the gas has a volume of 100 L

b) At 2 bar the gas has a volume of 50 L

c) At 4 bar the gas has a volume of 25 L

d) At 8 bar the gas has a volume of 12.5 L

e) At 16 bar the gas has a volume of 6.25 L

Here too you should multiply together each pressure with its corresponding volume.

What is the value of each of the products obtained? → 4

---

44 You are obviously not paying attention, for otherwise you would have noticed what changes are occurring in the diagrams.

Read through 21 once again.

---

45 Your answer is quite right.

$p \cdot v = 450$

$9 \cdot 50 = 450$

A gas which occupies a volume of 150 L at 3 bar will have a volume of 50 L at 9 bar.

Here is another problem for you:

A gas occupies a volume of 80 L at 12.5 bar.

What will the pressure be when the same amount of gas occupies a volume of 4 m^3? (This particular problem is not quite straightforward in that you will first have to convert 4 m^3 into liters.)

- 25.0 bar → 80
- 2.5 bar → 12
- 0.25 bar → 23
- 0.025 bar → 9
- I obtained some other pressure, or: I cannot solve this problem → 59

---

46 We asked you:

How many cm^3 are there in one liter?

Remember that one liter is a cube measuring 10 cm x 10 cm x 10 cm.

Perform the calculation now, and then check it at 30 .

---

47 We have already said that 1 bar is the pressure exerted by a force of 10^5 N on a surface of 1 m^2.

$$1 \text{ bar} = \frac{10^5 \text{ N}}{\text{m}^2}$$

Thus, it is important to know what the pressure is on a surface of 1 m^2.

Our problem now is:

What will the pressure in bar be when a force of 18 x 10^5 N is exerted on a surface of 6 m^2?

If the force of 18 x 10^5 N is exerted on a surface of 6 m^2, then on each individual m^2 the pressure will be

$$18 \times 10^5 \text{ N} / 6 \text{ m}^2 \text{ or } \frac{18 \times 10^5 \text{ N}}{6 \text{ m}^2} = ? \text{ bar}$$

Now evaluate this expression. What is your answer? → 39

---

48 How many grams of water will be formed if 44.8 L of hydrogen combine with 22.4 L of oxygen? (Assume normal conditions throughout.)

1. 44.8 L of hydrogen = 2 mol
   22.4 L of oxygen = 1 mol

2. The equation is $2H_2 + O_2 \rightarrow 2H_2O$

   2 mol of H_2 and 1 mol of O_2 yield 2 mol of water

3. 2 mol of water = ... g → [76]

---

[49] Your answer is quite right.

48 g of O_2 = 1.5 mol of O_2 = 33.6 L (1 bar, 0°C)
= 16.8 L (2 bar, 0°C)
73 g of HCl = 2 mol of HCl = 44.8 L (1 bar, 0°C)
= 22.4 L (2 bar, 0°C)

What volume will 56 g of nitrogen occupy at 4 bar and 0 °C?

Check your answer at [66] .

---

[50] $1\ m^3 = 1{,}000\ L$.

Advice on performing this calculation is given at [56] .

---

[51] You are not cooperating too well.

You should now read through the Program again very carefully, starting from [23] .

---

[52] The correct answer is: 22.4 L of hydrogen and 11.2 L of oxygen.

You should now be in a position to answer the following question:

How many grams of water will be formed when 44.8 L of hydrogen combine with 22.4 L of oxygen? (Assume normal conditions throughout.)

- 18g → [95]
- 36 g → [76]
- 54 g → [114]
- I don't know → [48]

53 You chose the answer "I don't know".

It is just as well that you didn't guess, though we would now like to put the following question to you: Couldn't you have managed to arrive at the right answer by really thinking hard about this problem?

Continue your reading at 117 .

54 In order to decide whether a change in pressure has taken place during the course of the reaction in the vessel, it is necessary to know how many moles of gaseous starting materials were present *before* the reaction and how many gaseous end products there are at the *end* of the reaction.

From the equation for the reaction

Cl_2	+ H_2	→	2 HCl
22.4 L	22.4 L		2 x 22.4 L

we see that from 1 mol of chlorine (22.4 L) and 1 mol of hydrogen (22.4 L) 2 mol of hydrogen chloride (44.8 L) are formed.

The number of moles appearing on each side of this equation is the *same,* namely 2. Thus, in this particular reaction the number remains *constant.* Whereas we started out with chlorine and hydrogen molecules, we finish up at the end of the reaction with the same number of hydrogen chloride molecules.

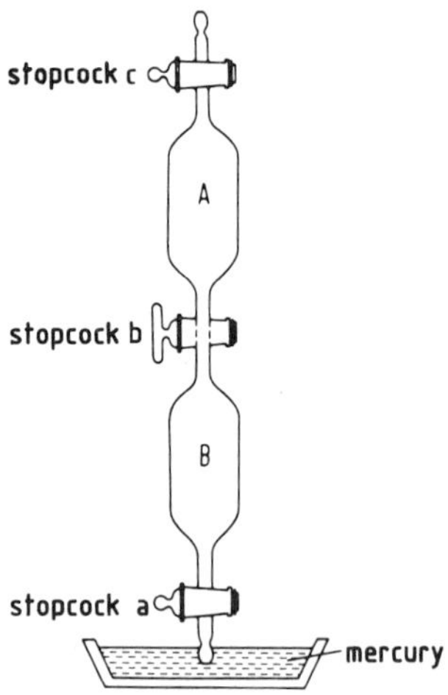

Now what conclusion may be drawn from this observation?

Let's run through the last experiment once again: In A we have chlorine (under normal pressure) and in B we have hydrogen (also under normal pressure). We open stopcock b and irradiate the mixture with UV light. The chlorine and the hydrogen then react together to form hydrogen chloride. Stopcock a is now immersed in a dish of mercury as shown opposite.

What will happen when stopcock a is opened?

- Because of the excess pressure some of the hydrogen chloride will be expelled from the vessel → 118
- As the pressure will not have changed during the reaction, nothing will happen → 41
- Because of the reduced pressure some of the mercury will rise into the lower vessel → 90

---

55 The volume of the starting materials (N_2 + 3 H_2 under normal conditions), V_{start}, is equal to 89.6 L.

The volume of the end product (2 NH_3 under normal conditions), V_{end}, is equal to 44.8 L.

The ratio of the volumes is thus

$$V_{start}/V_{end} = 89.6/44.8$$

This ratio may be simplified by dividing both the numerator and denominator until the smallest possible whole numbers result. Actually, it is much easier to work with small numbers. You had not simplified this ratio sufficiently.

Both 89.6 and 44.8 may be divided by 44.8!

Carry out this division now, and write down the simplified ratio of

$$V_{start}/V_{end} = \ldots\ldots / \ldots\ldots$$

that you obtain. → 94

---

56 We asked you:

How many liters are there in 1 cubic meter?

One liter is a cube having the dimensions:

$$10 \text{ cm} \times 10 \text{ cm} \times 10 \text{ cm} = 1 \text{ L}$$

or: $1 \text{ dm} \times 1 \text{ dm} \times 1 \text{ dm} = 1 \text{ dm}^3 = 1 \text{ L}$
(dm = decimeter, 1 dm = 10 cm, 1 m = 10 dm)

1 cubic meter measures:

$$1 \text{ m} \times 1 \text{ m} \times 1 \text{ m} = 1 \text{ m}^3$$

or: $10 \text{ dm} \times 10 \text{ dm} \times 10 \text{ dm} = 1000 \text{ dm}^3 = 1000 \text{ L}$ → [5]

10

---

[57] Your answer is wrong. We shall solve this problem stepwise, starting at [77] .

---

[58] Normal conditions are:

Pressure = 1 bar (or, more precisely, 1.013 bar)
Temperature = 0 °C

End of Program 10.

---

[59] We shall solve this problem stepwise.

Problem

A gas occupies a volume of 80 L at 12.5 bar. What will its pressure be when the same amount of gas occupies a volume of 4 m^3?

We start by calculating the product $p \cdot v$, that is the constant number. As p = 12.5 bar and v = 80 L, our constant number must equal

$$12.5 \cdot 80 = \mathbf{1000}$$

Before going any further, however, we must first convert 4 m^3 into liters. (Because we obtained our constant number by multiplying together 12.5 *bar* and 80 L, we now have to keep to these units and specify all quantities either in *bar* or *liters*.)

$$1 \text{ m}^3 = 1000 \text{ L}$$

$$4 \text{ m}^3 = 4000 \text{ L}$$

The law states that:

$$p \cdot v = \text{const.}$$

The volume and the constant number are known:

$$v = 4000 \text{ L}$$

$$\text{const} = 1000$$

but the pressure is at present unknown:

$$p = ?$$

Thus we have the equation:

$$p \cdot 4000 = 1000$$

Jot down the number by which 4000 has to be multiplied to produce the result 1000! → 73

---

60 $1 \text{ L} = 1000 \text{ cm}^3$

Advice on performing this calculation is given at 46 .

---

61 Your answer is correct. 5.6 L of hydrogen weigh 0.5 g.

Here is one further practical example based on what you have learnt so far.

How many liters of hydrogen and how many liters of oxygen will be formed under normal conditions if 18 g of water are decomposed by means of direct current in a Hofmann's apparatus?

Some assistance on this is given at 72 .

If, however, you feel sure that you can manage to get the right answer on your own, go instead to 52 .

---

62 Yes, the pressure is *less*.

A gas enclosed within a vessel having a pressure p which is *less* than 1 bar (in the diagram p = 0.5 bar) is said to be under *reduced pressure*.

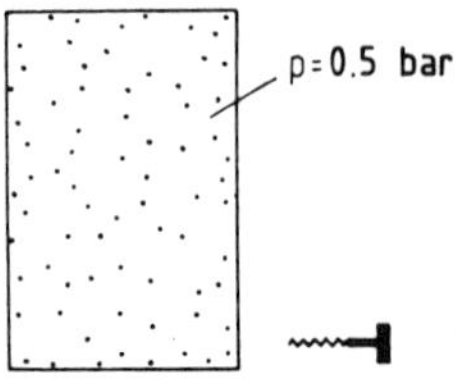

What would happen if a hole were to be drilled in the vessel?

- Air would flow into the vessel → 34
- The gas in the vessel would flow out → 22
- Nothing at all would happen → 27

10

---

63 p · v = *const.*

This law holds, of course, only if we keep the temperature fixed. We have already come across the definition that: 1 mol of any gas under .... occupies a volume of .... L.

Write out the completed version of this definition, and then check your answer at 99 .

---

64 You are not cooperating too well.

You should now go through the Program again very carefully, starting from 23 .

---

65 Your answer is wrong. We shall solve this problem stepwise, starting at 77 .

---

66 The correct answer is: 11.2 L

56 g of N_2 (2 mol) at 1 bar and 0°C = 44.8 L

56 g of N_2 at 4 bar and 0°C = $\frac{44.8\ \text{L}}{4}$ = 11.2 L

Thus, a fourfold increase of pressure reduces the volume to one quarter of its original value. Alternatively, using the formula p · v — const. we obtain the results:

$$1 \cdot 44.8 = 44.8\ (p = 1\ \text{bar},\ v = 44.8\ \text{L})$$
$$4 \cdot\ ? = 44.8\ (p = 4\ \text{bar},\ v = \ ?\ \text{L})$$
$$? = \frac{44.8}{4} = 11.2\ \text{L}$$

What volume will 10 g of hydrogen have at 0 °C and 20 bar?

- 112.0 L → 79
- 56.0 L → 86
- 11.2 L → 107
- 5.6 L → 97
- I obtained a different answer → 115

10

67

Your answer is only partially correct.

An explanation of this is given at 55 .

68

Yes, the greater the pressure the *smaller* the volume.

Let's imagine now a volume of gas equal to 2 L enclosed within a cylinder at a pressure of 1 bar.

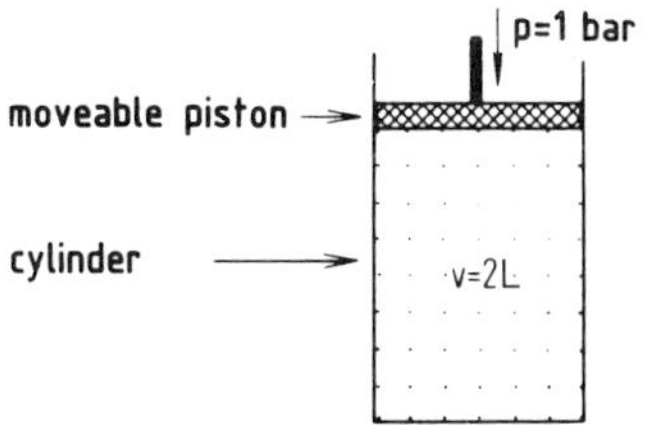

If the pressure is subsequently *doubled* to 2 bar, the piston must move downwards until the volume is *halved*.

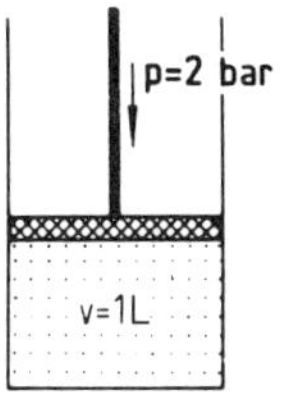

What will the volume be if the pressure is doubled again, i.e. it is increased to 4 bar?

Jot down the new value, and then check your answer at 3 .

---

69 Avogadro's Law states that:

Equal volumes of all gases at the same pressure and temperature contain equal numbers of molecules.

We distinguish between

Reduced pressure
Excess pressure
Atmospheric pressure.

All gases expand when heated and contract again when they are cooled down.

What is meant by normal conditions?

Write out these well-defined conditions, and then check your answer at 58 .

---

70 This question is difficult. You did well not to just guess here. The experimental observation is explained in detail at 54 .

---

71 Both hydrogen and nitrogen occur as diatomic molecules, i.e. as H_2 and N_2.

1 mol of H_2 = 2 g of hydrogen
1 mol of N_2 = 28 g of nitrogen
10 g of H_2 are 5 mol of H_2
56 g of N_2 are 2 mol of N_2

1 mol of any gas under normal conditions occupies a volume of 22.4 L.

Thus, 5 mol of H_2 = ... L
2 mol of N_2 = ... L → 119

---

72 18 g of water are to be decomposed by direct electric current. How many liters of hydrogen and of oxygen will be produced?

We start with the relevant equation. (Note that the equation for the reaction concerned is always the *starting point* for *all calculations* based on a *chemical* process.)

$$2\ H_2O \rightarrow 2\ H_2 + O_2$$

In words this reads: When 2 mol of water are decomposed they yield 2 mol of hydrogen and 1 mol of oxygen.

How many moles of water are there in 18 g of water? If you are able to answer this initial question, solving the problem should then be straightforward.
Check your answer at 52

If you are not sure how to approach this problem, refer to the hints given at 89 .

---

73 The number must be 0.25, since

$$0.25 \times 4000 = 1000$$

The gas is thus under a pressure of 0.25 bar when it occupies a volume of 4000 L = 4 m^3. → 23

---

74 Your answer is wrong. The fact that the color of the chlorine disappeared revealed that the two gases had *reacted* together. If only a mixture of gases were present, the hydrogen would have no more than a diluting effect which would merely reduce the intensity of the green color of the chlorine.

Answer the question at 97 once again.

---

75 In order to make the problem easier to solve, let's suppose that the air is enclosed not within a vessel having a fixed volume, but rather within a cylinder having a moveable piston.

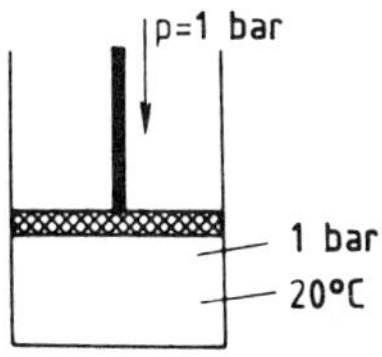

In this first diagram we sketch the cylinder containing air at a pressure of 1 bar.

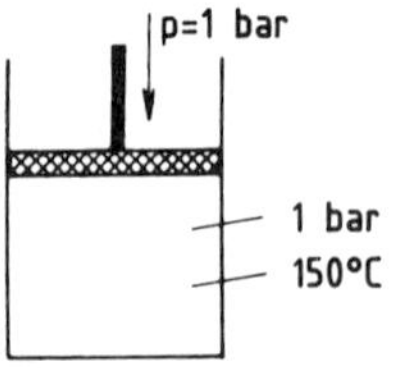

In the second diagram the air has been heated from 20 °C to 150 °C and so it has expanded. The moveable piston has been displaced upwards. Thus, the volume of the enclosed air is now greater.

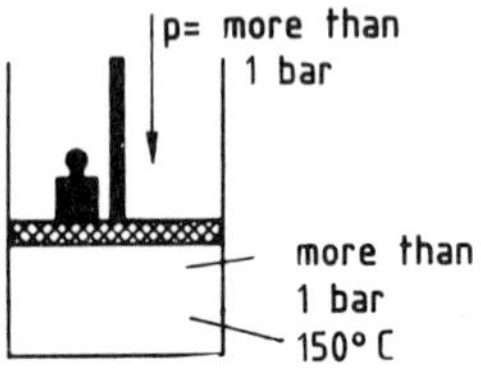

In the third diagram the enclosed air has been brought back to its original volume at 20 °C. This has been done by compressing it, e.g. by adding a weight to the piston. In the process the pressure increases.

10

In view of these explanations, you should now be in a position to answer the question at 96 correctly.

---

76 Your answer is correct. **36 g** of water are formed.

Let's now briefly recapitulate:

The pressure and volume of gases are in a fixed relationship to one another. If the pressure increases the volume will ....

Jot down the missing word, and then go on to 92 .

---

77 We shall solve this problem stepwise.

From our previous examples you should now know that:

$$p \cdot v = \text{const.}$$

Multiplication of the pressure p by the volume v yields a number (called here const.). This number always has the same value for a given quantity of gas enclosed within a vessel, irrespective of the pressure of that gas. Any changes in pressure will, of course, be reflected in changes of volume of the gas.

Now, our problem ran as follows:

A gas is enclosed within a volume of 150 L at a pressure of 3 bar. What will the volume of this gas become if the pressure is increased to 9 bar by the moveable piston?

We start by calculating the number referred to as const. above. We know that $p = 3$ bar and $v = 150$ L.

$$p \cdot v \text{ is thus } 3 \times 150 = 450$$

The number (const.) thus has the value 450.

What will the volume of this gas become when the pressure is 9 bar?

$$p \cdot v = 450$$

$$9 \cdot ? = 450$$

What number when multiplied by 9 will give 450? → [45]

---

[78] *Avogadro*'s Law states that:

Equal volumes of all gases at the same pressure and temperature contain equal numbers of molecules.

In our further discussion we shall be needing some molecular masses. Accordingly, you are asked to jot down the molecular masses of:

Hydrogen
Chlorine
Hydrogen chloride (HCl)
Ammonia (NH_3)

(The atomic masses are H = 1; Cl = 35.5; N = 14.) → [108]

---

[79] Your answer is wrong.

An explanation of this is given at [115] .

---

[80] You appear to have made a mistake.

We shall now solve this problem stepwise, starting at [59] .

---

81 Your answer is only partially correct.

An explanation of this is given at 55 .

82 Quite right.

The gas will flow out until the pressure difference is equalized. (Think here, for instance, of a hole in the tire of a bicycle.)

Now jot down the missing words in the sentences below.

When the pressure is less than 1 bar, we speak of a .... When the pressure is greater than 1 bar, we speak of an .... → 96

83 The calculation may be performed as follows:

71 g of chlorine occupy a volume of 22.4 L
? g of chlorine occupy a volume of 1 L

Thus, $\frac{71}{22.4}$ g of chlorine occupy a volume of 1 L

How much, therefore, will 1 L of chlorine weigh? → 120

84 Your answer is wrong.

An explanation of this is given at 125 .

85 Your answer is quite right.

The pressure inside a *closed* vessel full of air must increase when it is warmed up as the air is unable to expand.

You should note that:

1. All gases expand when they are heated.
2. All gases contract when they are cooled.

Since all gases may be compressed, and because gases expand or contract according to their temperature, the conditions under which gas volumes are measured must always be precisely specified, i.e. the pressure and temperature must be specified. In order that strictly comparable values are obtained, scientists have defined

"*Normal Conditions*" to be:

1.013 bar and 0 °C

Accordingly, whenever a gas volume in measured under *normal conditions* (N.T.P.) the gas will be at a pressure of 1.013 bar and a temperature of 0 °C.

For convenience, the pressure 1.013 bar will in future be rounded off to 1 bar.

Let's now take a closer look at gases and gas reactions. The following statement is of fundamental importance:

> **1 mol of any gas under normal conditions occupies a volume of 22.4 L.**

Make sure you learn this one by heart!

Here are some examples:

1 mol of N_2 (28 g) = 22.4 L of nitrogen at N.T.P.
1 mol of Cl_2 (71 g) = 22.4 L of chlorine at N.T.P.

N.T.P is an abbreviation for the expression "normal temperature and pressure".

With the help of the above statement, see if you can now solve the following problem:

What will the volume of 32 g of oxygen be at N.T.P?

Write out your answer and then check it at 100 .

86 Your answer is wrong.

An explanation of this calculation is given at 115 .

87 The result is

$$\frac{32}{1.43} = 22.4\ \text{L}\ .$$

The law relating to the molar volume of any gas can be applied in many different contexts.

Here is one example:

Suppose we have 5.6 L of hydrogen at N.T.P.. How many grams of hydrogen does this represent?

- 1 g → 110
- 0.5 g → 61
- 0.25 g → 42
- I don't know → 53

---

88 It's a good thing that you didn't guess, just because your figures didn't agree with those given.

A detailed explanation is to be found at 71 .

---

89 18 g of water are to be decomposed. How many liters of H_2 and of O_2 will be formed?

This problem should be approached as follows:

1. The relevant equation is $2\ H_2O \rightarrow 2\ H_2 + O_2$.
   2 mol of water yield 2 mol of hydrogen and 1 mol of oxygen.

2. How many moles of water are there in 18 g?
   The molecular mass of water is 18, i.e. 1 + 1 + 16. Thus, 18 g of water is exactly one mole.

3. 2 mol of water yield 1 mol of O_2.
   Thus, 1 mol of water will yield 0.5 mol of O_2 = 11.2 L.

4. 2 mol of water yield 2 mol of H_2.
   Thus, 1 mol of water will yield 1 mol of H_2 = 22.4 L.

Accordingly, 18 g of water when decomposed will yield .... L of hydrogen and .... L of oxygen (at N.T.P.). → 52

---

90 Your answer is wrong. The gas is *not* at reduced pressure.

An explanation of this is given at 104 .

91 Now check your answer:

1 mol of any gas *under normal conditions* occupies a volume of **22.4** L.

We shall now consider the reaction for the formation of ammonia (NH_3). The relevant equation is:

$$N_2 + 3\ H_2 \rightarrow 2\ NH_3$$

Copy down this equation and write underneath both the starting materials and the end product the volume that each of these gaseous substances will occupy at N.T.P. (Ammonia is a gas at N.T.P.) The actual volumes may be determined from the mole ratios of the substances appearing in the equation. → 122

92 If the pressure increases the volume will *decrease.* Note that the converse also holds, namely:

If the pressure decreases the volume will increase.

This type of relationship may be represented in symbols by the equation:

$$p \cdot v = \dots\dots\dots\dots$$

Write out the full version of this equation, and then check your answer at 63 .

93 Your answer is wrong. The hydrogen chloride is *not* under reduced pressure.

An explanation of this is given at 54 .

94 Your answer is quite right.

In the reaction

$$N_2 + 3\ H_3 \rightarrow 2\ NH_3$$

the ratio of the volume of the starting materials (V_{start}) to that of the end product (V_{end}) is 2 : 1.

The volume of the starting materials (89.6 L) is twice that of the end product (44.8 L).

Our diagram of the cylinder with its moveable piston would now look like this:

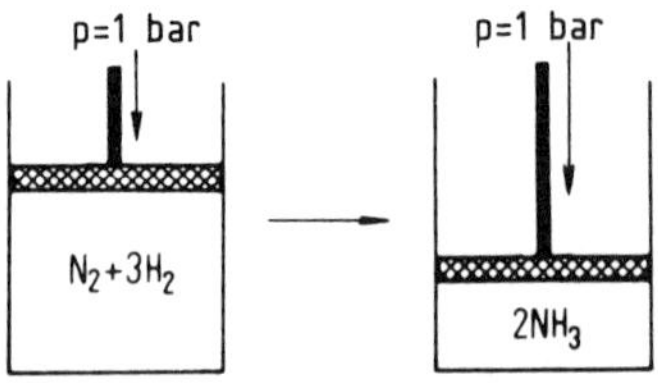

The number of molecules clearly decreases during the reaction; the volume must therefore get smaller. This observation is in accordance with the Law of A..............

Jot down the name, and then check your answer at [78] .

---

[95] Your calculation is wrong.

An explanation of this is given at [48] .

---

[96]

Pressure less than 1 bar:	*Reduced Pressure*
Pressure greater than 1 bar:	*Excess Pressure*

(This particular Program is rather extensive. As you are now roughly half way through it, you might like to take a short pause at this point. If you do, you should note where you got to; as a precaution against losing your place you are advised to insert a bookmark here.)

One further law states that:

All gases expand when they are heated.

What will happen if a gas at 20 °C is heated to 150 °C in a *sealed* vessel? (Note that, because the vessel is sealed, the volume of the gas cannot increase, i.e. *no* expansion is possible.)

- The pressure in the vessel will decrease → 112
- The pressure in the vessel will increase → 85
- The pressure in the vessel will not change → 105
- I don't know → 75

---

97 Your answer is quite right.

10 g of hydrogen at 0 °C and 20 bar will occupy a volume of 5.6 L.

Let's consider now the following experiments:

We start with a glass vessel as shown in the diagram below. This vessel consists of two equal parts A and B, each having a volume of 22.4 L. The two parts are connected together by means of a stopcock and the two ends also have stopcocks. Volume A is filled with chlorine whereas volume B is filled with hydrogen.

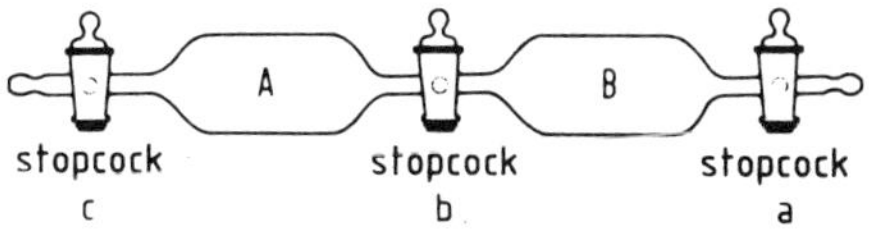

The pressure in both parts of the vessel is equal to atmospheric pressure. All the stopcocks are closed at this stage. The apparatus is now placed under an ultra-violet lamp or exposed to bright sunlight and stopcock b connecting the volumes A and B is opened. Because the greenish-yellow color of the chlorine disappears, we may conclude that the two gases have mixed together after the opening of the middle stopcock and that they have reacted with one another.

What is formed when the chlorine reacts with the hydrogen in this reaction?

- Hydrochloric acid → 121
- Hydrogen chloride → 113
- A mixture of chlorine and hydrogen → 74

98 71 g of chlorine (1 mol) will have a volume of 22.4 L under normal conditions.

1 L of chlorine will thus weigh .... g. → 120

If you find this problem difficult, go instead to 83 .

99 1 mol of any gas under *normal conditions* occupies a volume of 22.4 L.

The volume occupied by one mol of a gas is known as the molar volume. The molar volume of all gases under normal conditions is equal to 22.4 L.

10

This, of course, is in accordance with *Avogadro's Law:* Equal volumes of all gases at .... etc.

Write out now the full version of this law, and then check your answer at 69 .

100 32 g of oxygen are equal to 1 mol of O_2. The Law tells us that 1 mol of oxygen under normal conditions (i.e. at 0 °C and at a pressure of 1 bar) will occupy a volume of **22.4** L.

What will be the volumes occupied by 10 g of hydrogen and by 56 g of oxygen under normal conditions?

10 g of hydrogen	*56 g of oxygen*		
• 22.4 L	22.4 L	→	109
• 112.0 L	44.8 L	→	119
• 224.0 L	89.6 L	→	124
• I obtained other figures		→	88
• I don't know		→	71

101 Your answer is wrong. The hydrogen chloride is *not* under excess pressure.

An explanation of this is given at 54 .

102 The correct answer is: 11.2 L.

1 mol of O_2 at N.T.P. (1 bar and 0 °C)	= 22.4 L
1 mol of O_2 at 2 bar and 0 °C	= 11.2 L
1 mol of O_2 at 4 bar and 0 °C	= 5.6 L

Now try out the following problem:

What will be the volume of 48 g of oxygen and of 73 g of hydrogen chloride at a pressure of 2 bar and a temperature of 0 °C?

48 g of oxygen	*73 g of hydrogen chloride*		
• 22.4 L	22.4 L	→	40
• 11.2 L	11.2 L	→	84
• 33.6 L	44.8 L	→	116
• 16.8 L	22.4 L	→	49
• I obtained different figures		→	125

103 Your answer is only partially correct.

An explanation of this is given at 55 .

104 According to the equation:

$Cl_2 + H_2 \rightarrow 2\ HCl$

1 mol of chlorine (22.4 L) and 1 mol of hydrogen (22.4 L) react together to produce 2 mol of hydrogen chloride (2 x 22.4 = 44.8 L).

The starting materials are:

1 mol of Cl_2	22.4 L
1 mol of H_2	22.4 L
Total	44.8 L

The end product is:

2 mol of HCl 2 x 22.4 L = 44.8 L

The starting materials together thus amount to 2 mol (1 mol of Cl_2 and 1 mol of H_2) and will occupy a volume of 44.8 L (at normal pressure). The end product amounts to 2 mol of HCl which again will occupy a total volume of 44.8 L (under normal conditions).

The number of moles thus does not change during this particular reaction (1 mol of chlorine and 1 mol of hydrogen at the start and 2 mol of hydrogen chloride at the end). This implies that the pressure too will not change. Accordingly, when the bottom part of the apparatus is immersed in mercury and stopcock a is opened, nothing will happen. → 41

10

105 Your ideas here are unfortunately not right.

A detailed explanation of this is given at 75 .

106 The statement tells us that:

The molar volume of any gas under normal conditions is **22.4** L

Make sure you commit this statement to memory!

Use this statement now to solve the following problem:

1.43 g of oxygen at N.T.P. occupy a volume of 1 L.

Calculate the volume occupied by 1 mol = 32 g of oxygen at N.T.P. → 87

107 Your answer is wrong.

An explanation of this calculation is given at 115 .

108 The molecular masses are:

Hydrogen (H_2)	2
Chlorine (Cl_2)	71
Hydrogen chloride (HCl)	36.5
Ammonia (NH_3)	17

Now see if you can answer the following question:

How much will 1 L of chlorine weigh at N.T.P.? We have already said that 71 g of chlorine at N.T.P. occupy a volume of .... L. → 98

---

109 Your answer is wrong. Here one starts by calculating how many moles there are in 10 g of hydrogen and in 56 g of nitrogen.

One mole of any gas under normal conditions occupies a volume of **22.4 L.**

Answer the question at 100 once again.

---

110 Your calculations are wrong.

An explanation is given at 117 .

---

111 Your answer is partially right.

The volume of the starting materials ($N_2 + 3\ H_2$) at N.T.P., V_{start}, is 89.6 L.

The volume of the end product ($2\ NH_3$) at N.T.P., V_{end}, is 44.8 L.

Accordingly, the ratio is

$$V_{start}/V_{end} = 89.6/44.8$$

which is your result.

However, this can be improved upon by reducing the fraction to the smallest possible whole numbers. Only then can one really appreciate the significance of the numbers.

Both 89.6 and 44.8 can be divided by 44.8.

Carry out this division and jot down the new ratio of

$$V_{start}/V_{end} = \ldots\ldots / \ldots\ldots$$

which you obtain. → 94

---

112 Your ideas on this are unfortunately not right.

A detailed explanation of this is given at 75 .

113 Your answer is correct. In both vessels *hydrogen chloride* is formed from the hydrogen and chlorine according to the equation:

$$H_2 + Cl_2 \rightarrow 2\ HCl$$

Hydrogen chloride is a *gaseous* compound at room temperature. The statement below thus applies to hydrogen chloride too:

10

> 1 mol of any gas under normal conditions
> occupies a volume of 22.4 L.

The apparatus we used will contain hydrogen chloride after the reaction has taken place. The end of the vessel at which stopcock a is located is now immersed in a dish of mercury (see diagram). Stopcock b is open whereas stopcocks a and c are closed.

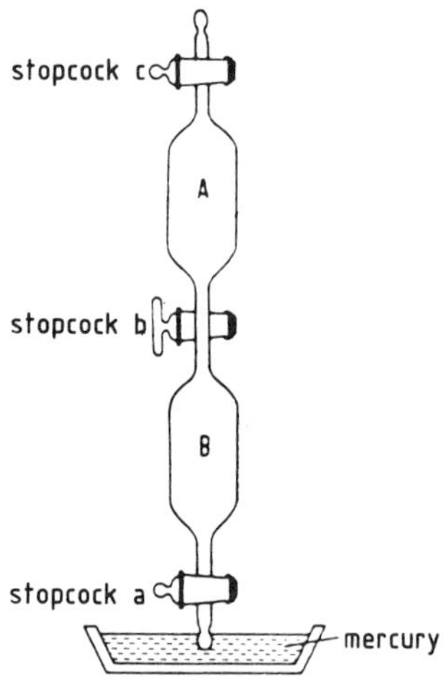

On the basis of the above statement and not forgetting that the relevant equation is

$$H_2 + Cl_2 \rightarrow 2\ HCl$$

it should now be possible for you to answer the question below, although this may not seem very simple.

What will happen when stopcock a is opened?

- Because of the excess pressure some of the hydrogen chloride will be expelled → 101
- Because the pressure has not changed during the reaction nothing will happen → 41
- Because of the reduced pressure some of the mercury will rise up into the lower part of the vessel → 93
- I don't know → 70

---

114 Your calculations are wrong.

An explanation of them is given at 48 .

---

115 What volume will 10 g of hydrogen occupy at 20 bar and 0 °C?

We start by determining the number of mol in 10 g of hydrogen

1 mol of hydrogen = 2 g

10 g of hydrogen are thus $\frac{10}{2} = 5$ mol

> 1 mol of hydrogen under normal conditions occupies a volume of 22.4 L.

5 mol of H_2 under normal conditions (0 °C and 1 bar) will therefore occupy a volume of

$5 \times 22.4 = 112.0$ L.

At 20 bar it becomes: $\frac{112}{20} = \ldots\ldots\ldots\ldots$ L.

Continue your reading at 97 .

---

116 Your answer is wrong.

An explanation of this is given at 125 .

117 The question which you were unable to answer ran as follows:

You are given 5.6 L of hydrogen at N.T.P. How many grams of hydrogen will you have?

At N.T.P. 1 mol (= 2 g) of hydrogen will occupy a volume of 22.4 L. It thus follows that:

22.4 L of hydrogen (H_2) weigh 2 g.

1 L of hydrogen weighs $\frac{2}{22.4}$ g .

5.6 L of hydrogen weigh $\frac{2}{22.4}$ x 5.6 g .

Now work out how much 5.6 L of hydrogen weigh, and then check your answer at 61 .

118 Your answer is wrong. There is *no* excess pressure present.

An explanation of this is given at 104 .

119 Your answer is quite right.

10 g of H_2 = 5 mol of H_2 = 112 L of H_2 (at N.T.P.)
56 g of N_2 = 2 mol of N_2 = 44.8 L of N_2 (at N.T.P.)

If this answer was your first choice you are working well.

Now we said that the volume of a gas is compressed to one half when the pressure on it is doubled.

What volume will 1 mol of O_2 have at 0° C and 2 bar?

Write down your answer, and then check it at 102 .

120

1 L of chlorine weighs $\frac{71}{22.4}$ g = 3.16 g

In this way the mass of one L of any gas under normal conditions can be rapidly determined.

What is the mass of 1 L of hydrogen at N.T.P.? → 123

121

Your answer is not quite right.

Hydrochloric acid is actually the aqueous solution of hydrogen chloride. In the reaction of hydrogen and chlorine, however, ... is formed.

Continue your reading at 113 .

122

Check the following and revise your answer where necessary:

N_2	+ $3\ H_2$	→	$2\ NH_3$
1 x 22.4 L	3 x 22.4 L		2 x 22.4 L
22.4 L	**67.2** L		**44.8** L

Thus, in *this* particular reaction thc volume *decreases* during the course of the reaction.

There are therefore reactions in which the volume remains unchanged (e.g. H_2 + Cl_2 = 2 HCl), reactions in which the volume decreases (e.g. $N_2 + 3\ H_2 = 2\ NH_3$), and those in which the volume increases.

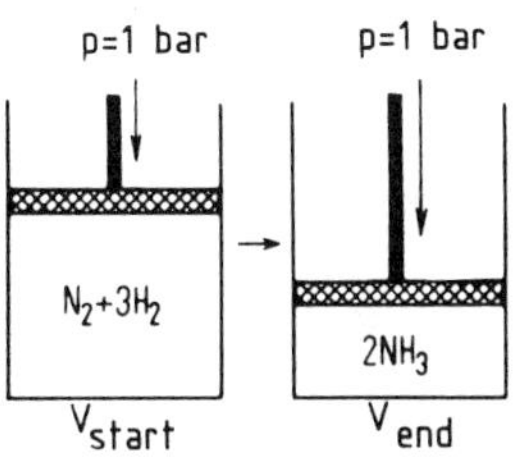

The initial volume of the starting materials at N.T.P. is V_{start}; the final volume of the end products at N.T.P. is V_{end}.

Suppose now that the reaction

$$N_2 + 3\ H_2 \rightarrow 2\ NH_3$$

is carried out in a cylinder with a moveable piston.
(In reality these two gases would react together only at a high temperature and in the presence of a catalyst.)

What will be the ratio V_{start}/V_{end} for this reaction?

V_{start}/V_{end}

- 89.6/44.8 → 111
- 44.8/22.4 → 103
- 11.2/ 5.6 → 67
- 5.6/ 2.8 → 81
- 2 / 1 → 94

123

1 L of hydrogen at N.T.P. weighs $\frac{2}{22.4}$ g = 0.089 g .

Now we know that:

> 1 mol of any gas under normal conditions occupies a volume of 22.4 L.

The volume occupied by *1 mol of a gas* is known as the *molar volume* of that gas.

> The molar volume of any gas under normal conditions is ....

Write out the completed version of this statement, and then check it at 106

124

Your answer is wrong. You have ignored the fact that the gases nitrogen and hydrogen occur as N_2 and H_2 respectively.

Answer the question at 100 once again.

125

We start by calculating the volume occupied by 48 g of oxygen and that occupied by 73 g of hydrogen chloride under normal conditions:

1 mol of O_2 = 32 g; 48 g of O_2 = 1.5 mol of O_2
1 mol of HCl = 36.5 g; 73 g of HCl = 2 mol of HCl

Under normal conditions we have:

48 g of O_2 = 1.5 mol of O_2 = 33.6 L of O_2
73 g of HCl = 2 mol of HCl = 44.8 L of HCl

At 2 bar and 0 °C the volumes will be only half as great:

48 g of O_2 at 2 bar and 0 °C will occupy .... L.
73 g of HCl at 2 bar and 0 °C will occupy .... L.

Continue your reading at 49 .

Program 11

Theoretical Foundations (V)

1

Earlier on we introduced the concepts of:

Acids
Bases
Salts

This Program will be devoted to a discussion of the relationships existing between these three differing types of compound.

We shall start by taking a careful look at *acids.*

11

We know already that included among the important acids are the following:

HCl	= Hydrogen chloride (its aqueous solution is known as hydrochloric acid)
HNO_3	= Nitric acid
H_2SO_4	= Sulfuric acid
H_3PO_4	= Phosphoric acid

Both the names and the formulas of the above acids should be committed to memory.

We also know how one may test for an acid: make use of litmus paper.

Write out the completed version of the sentence below:

Acids have the effect of turning blue litmus .... → 8

2

Nitric acid dissociates in water into *positively* charged hydrogen ions ($H^{\oplus}$) and *negatively* charged nitrate ions ($NO_3^{\ominus}$). The equation for this dissociation takes the form:

$$HNO_3 \xrightarrow{\text{dissociation}} H^{\oplus} + NO_3^{\ominus}$$

Note that the nitrate ion *as a whole* bears the negative charge and not one single part of it.

Dissociation is a very important process. It occurs not only in acids; many other compounds also split up when they are dissolved in water. To be quite sure that you have a good understanding of the significance of the word dissociation, we would now like you to answer the question: What is meant by dissociation?

- The dissolving of a compound in water → 11
- The separation of a mixture by heating it → 21
- The splitting up of a compound into ions when it is dissolved in water → 31

---

3 Nitric acid dissociates according to the equation:

$$HNO_3 \xrightarrow{\text{dissociation}} H^{\oplus} + NO_3^{\ominus}$$

into (positively charged) hydrogen ions and (negatively charged) nitrate ions.

Potassium hydroxide dissociates according to the equation:

$$KOH \xrightarrow{\text{dissociation}} K^{\oplus} + OH^{\ominus}$$

into (positively charged) potassium ions and (negatively charged) hydroxyl ions.

During *neutralization* the (positively charged) *hydrogen ions* of the nitric acid combine with the (negatively charged) *hydroxyl ions* of the potassium hydroxide to form water:

$$H^{\oplus} + OH^{\ominus} \rightarrow H_2O$$

If we now form one new equation from the above three equations by adding together all the right hand sides and all the left hand sides, the resulting equation will assume the form:

$$HNO_3 + KOH + H^{\oplus} + OH^{\ominus} \rightarrow K^{\oplus} + NO_3^{\ominus} + H^{\oplus} + OH^{\ominus} + H_2O$$

The hydrogen and hydroxyl ions appearing on both sides of this equation may now be cancelled out. The equation we are then left with represents the neutralization of nitric acid with potassium hydroxide. This you are asked to write out for yourself.

Check your equation at 16 .

---

4 Your answer is wrong. We stated quite clearly that:

Even in their crystalline form, i.e. in the solid state, salts consist of *ions*!

What type of bonding will therefore be present in salts? → 28

---

5 No. Let's start by writing down the equations for the dissociation of sulfuric acid and sodium hydroxide:

$$H_2SO_4 \rightarrow H^{\oplus} + H^{\oplus} + SO_4^{2\ominus}$$

$$NaOH \rightarrow Na^{\oplus} + OH^{\ominus}$$

During neutralization the hydroxyl ions of the base combine with the hydrogen ions of the acid.

If these two equations are now added together, we obtain the new equation:

$$H_2SO_4 + NaOH \rightarrow H^{\oplus} + H^{\oplus} + SO_4^{2\ominus} + Na^{\oplus} + OH^{\ominus}$$

This new equation, however, has on its right hand side two hydrogen ions but only one hydroxyl ion! There should, of course, be *two* hydroxyl ions present to combine with the *two* hydrogen ions.

Thus, for the neutralization of 1 mol of H_2SO_4 one requires .... mol of NaOH. → 22

---

6 Yes, **1** *mol* of potassium nitrate and **1** *mol* of water are formed.

In figures this means that: 63 g (1 mol) of nitric acid are neutralized by 56 g (1 mol) of potassium hydroxide.

In the process ...... g of potassium nitrate and ...... g of water will be formed.

Calculate the amounts which will be formed, and then check your answers at 14 .

(The relevant atomic masses are: K = 39; N = 14; O = 16)

---

7 Your answer is not quite right. An explanation of this is given at 66 .

---

8 Acids have the effect of turning blue litmus *red.*

This characteristic of acids has previously been used by us to explain what an

11

acid is. Do you remember that the answer to the question: What is an acid? ran as follows:

An acid is a compound which turns blue litmus paper red.

Although this answer is correct as it stands, it in fact does not tell us what an acid really is. All it does is to mention a characteristic of acids without saying anything about their nature.

To understand the essential nature of an acid, it is suggested that you now write out the formulas of hydrogen chloride, sulfuric acid, nitric acid and phosphoric acid.

Then write out your answer to the question:

What element is present in all four of these acids? → 15

---

9 In the previous section we explained in detail what is formed when H_2SO_4 reacts with NaOH, namely water and sodium sulfate.

The relevant equation is:

$$H_2SO_4 + NaOH \rightarrow Na_2SO_4 + H_2O$$

The factors are, however, missing! You should now insert these yourself.
→ 37

---

10 Quite right! Sulfuric acid dissociates in water into two positively charged hydrogen ions ($H^{\oplus}$) and one *doubly* negatively charged sulfate radical ($SO_4^{2\ominus}$).

$$H_2SO_4 \xrightarrow{\text{dissociation}} 2\,H^{\oplus} + SO_4^{2\ominus}$$

Write out now in your own words what is meant by dissociation. → 18

---

11 No. Dissolving the compound in water is certainly a prerequisite for its dissociation, though the word actually refers to something quite different. Remember that not all compounds dissociate when they are dissolved in water!

Go back now to 2 .

12 Here you have simply guessed. Go back to [45] .

13 $NaCl \rightarrow Na^{\oplus} + Cl^{\ominus}$

The negatively charged chlorine ion is called a *chloride* ion. (Hence the name sodium *chloride*, NaCl.)

The other acid radicals have also been given special names, viz.

The acid radical of nitric acid, NO_3, is a nitrate radical.
The acid radical of sulfuric acid, SO_4, is a sulfate radical.
The acid radical of phosphoric acid, PO_4, is a phosphate radical.
The acid radical of carbonic acid, CO_3, is a carbonate radical.

Please try to memorize these names.

Now write out the equation for the dissociation of nitric acid. → [2]

14 **101 g** (1 mol) of potassium nitrate and **18 g** (1 mol) of water are formed.

You should now set up the equation for the neutralization of sulfuric acid with sodium hydroxide:

$$H_2SO_4 + NaOH \rightarrow$$

You will have to think carefully how many moles of NaOH are needed to neutralize 1 mol of H_2SO_4.

Remember that the formula for sulfuric acid contains 2 H atoms which will dissociate into hydrogen ions when the acid is dissolved in water!

How many moles of NaOH will be required to neutralize 1 mol of H_2SO_4?

- 1 mol → [5]
- 2 mol → [22]
- I don't know → [35]

15

Now check your answers:

Hydrogen chloride	HCl (the aqueous solution is called hydrochloric acid)
Nitric acid	HNO_3
Sulfuric acid	H_2SO_4
Phosphoric acid	H_3PO_4

In all these acids the element *hydrogen* is present.
If one were to take away this hydrogen one would be left with the *acid radical.*
Examples of such radicals include:

The acid radical Cl from hydrochloric acid, HCl
The acid radical NO_3 from nitric acid, HNO_3
The acid radical SO_4 from sulfuric acid, H_2SO_4

11

Write down now the acid radicals from carbonic acid, H_2CO_3, and phosphoric acid, H_3PO_4. → 23

16

$$HNO_3 + KOH \rightarrow K^{\oplus} + NO_3^{\ominus} + H_2O$$

If you arrived at this answer without any extra help, your work is really first class.

We have written down the equations for two neutralizations. In both of them there appeared on the right hand side water, a positively charged metal ion and a negatively charged acid radical.

If the experiment corresponding to these equations is now carried out, i.e. an aqueous acid solution is mixed with an aqueous basic solution and the water is subsequently evaporated off, the positively charged metal ions will combine with the negatively charged acid radical to form a *salt.* The latter will remain behind in the vessel as a crystalline deposit.

Although salts consist of *ions,* even when they are in the crystalline state, one frequently omits writing in the relevant charges and sets up neutralization equations as shown in the example below:

$$HNO_3 + KOH \rightarrow KNO_3 + H_2O$$

What type of chemical bonding is present in salts in the solid state?

- Atomic bonding → 4
- Ionic bonding → 28
- I don't know → 36

---

17 You should not forget that:

Sulfuric acid contains two hydrogen atoms which will become positively charged hydrogen ions when the acid dissociates, thus:

$$H_2SO_4 \xrightarrow{\text{dissociation}} 2\,H^{\oplus} + \ldots\ldots\ldots\ldots\ldots$$

The 2 in front of the $H^{\oplus}$ indicates that there are 2 positively charged hydrogen ions, i.e. there will be 2 *positive charges* on the right hand side of the equation. These two *positive* charges must be balanced by a corresponding number of *negative* charges, since the aqueous acid solution as a whole is not charged.

Complete now the equation above for the dissociation of sulfuric acid. How many negative charges are there on the sulfate radical? → 10

---

18 The gist of your answer should be that:

The splitting up of a compound into ions when it is dissolved in water is described as dissociation.

Now set up the equation for the complete dissociation of phosphoric acid (H_3PO_4). → 25

---

19 The full version of the equation is:

HNO_3	+	KOH	$\longrightarrow$	KNO_3	+	H_2O
nitric acid		potassium hydroxide		potassium nitrate		water

This neutralization equation tells us not only which *acid* is neutralizing which *base* but what *salt* is formed in the process. In addition, we also know that an equation gives us information about the actual *quantities* of the substances which are reacting together.

11

When 1 mol of nitric acid is neutralized by 1 mol of potassium hydroxide, the products will be .... of potassium nitrate and .... of water.

What is needed to complete this sentence? → 6

---

20 The correct answers are:

Phenolphthalein
Litmus
Methyl orange

Now copy down the table below and insert the missing colors.

Indicator	With Acids	With Bases
Phenolphthalein	....	....
Methyl orange	....	....
Litmus	....	....

If you are a little uncertain here, go back to 39 . If you can manage to solve this exercise, check your answers at 34 .

---

21 No! You have confused "distillation" with "dissociation".

Read through 2 once again.

---

22 Yes, 2 mol of NaOH.

Sulfuric acid dissociates according to the following equation:

$$H_2SO_4 \rightarrow 2H^{\oplus} + SO_4^{2\ominus}$$

There are therefore 2 hydrogen ions on the right hand side of the equation. Thus, for the neutralization of 1 mol of H_2SO_4 we shall need 2 mol of NaOH.

$$2NaOH \rightarrow 2Na^{\oplus} + 2OH^{\ominus}$$

The hydrogen ions and the hydroxyl ions together form water as follows:

$$2H^{\oplus} + 2OH^{\ominus} \rightarrow 2H_2O$$

When the solution is evaporated, solid sodium sulfate, Na_2SO_4, will remain behind.

$H_2SO_4 + NaOH \rightarrow$

I can manage it on my own → 37

I shall need some assistance → 9

---

23 The acid radical of carbonic acid is CO_3. This may be obtained from the formula for carbonic acid, H_2CO_3, simply by omitting the two hydrogen atoms. The acid radical of phosphoric acid is PO_4. This is similarly obtained by omitting from the formula for phosphoric acid, H_3PO_4, all three of the hydrogen atoms.

Whenever an acid is dissolved in water it splits up into hydrogen and an acid radical. Of course, *both* the hydrogen and the acid radical will be in the form of *ions*. The hydrogen occurs as the *positively* charged hydrogen ion whereas the acid radical occurs as a *negatively* charged ion. This splitting up into ions upon dissolution in water is referred to as *dissociation*.

Accordingly, hydrogen chloride will dissociate in water into positively charged hydrogen ions ($H^{\oplus}$) and negatively charged chlorine ions ($Cl^{\ominus}$) thus:

$$HCl \xrightarrow{\text{dissociation}} H^{\oplus} + Cl^{\ominus}$$

We have already come across the negatively charged chlorine ion ($Cl^{\ominus}$). This is formed when sodium chloride splits up (dissociates) in water.

Write down now the equation for the dissociation of sodium chloride.

$NaCl \rightarrow$ .... + ...

What is the name of the negatively charged $Cl^{\ominus}$ ion again? → 13

---

24 The complete equations are as follows:

For the dissociation of hydrochloric acid:
$HCl \rightarrow H^{\oplus} + Cl^{\ominus}$

For the dissociation of sodium hydroxide:
$NaOH \rightarrow Na^{\oplus} + OH^{\ominus}$

For the combination of hydrogen and hydroxyl ions upon neutralization: $H^{\oplus} + OH^{\ominus} \rightarrow H_2O$

If these three equations are now added together, i.e. all three *left* hand sides and all three *right* hand sides are separately added together, the following new equation is obtained:

$$HCl + NaOH + H^{\oplus} + OH^{\ominus} \rightarrow Na^{\oplus} + Cl^{\ominus} + H^{\oplus} + OH^{\ominus} + H_2O$$

Now, on both sides of this equation we have $H^{\oplus}$ and $OH^{\ominus}$ ions. Accordingly, we may "cancel" these out. We then obtain the equation:

$$HCl + NaOH \rightarrow Na^{\oplus} + Cl^{\ominus} + H_2O$$

11

This is in fact the *equation for the neutralization of an acid by a base.*

Let's consider now a further example of this type of equation.

What would the equation be for the neutralization of nitric acid with potassium hydroxide? Can you manage it on your own or will you need some help?

- I can manage it on my own → 16
- I shall be needing some help → 3

---

25 When phosphoric acid dissociates completely in water, three positively charged hydrogen ions ($H^{\oplus}$) and the triply negative phosphate radical ($PO_4^{3\ominus}$) are formed.

$$H_3PO_4 \xrightarrow{\text{dissociation}} 3\,H^{\oplus} + PO_4^{3\ominus}$$

These specific examples lead us now to a general *definition* for an acid as follows:

> An acid is a compound which forms positively charged hydrogen ions when it dissociates in water.

In fact, it is the positively charged hydrogen ions which are responsible for the acidic properties of acids. Now, we already know that acids may be tested for with blue litmus paper – the paper will turn red. What actually causes this red color in your opinion?

- A reaction of litmus dye with the acid radical → 48

- **A reaction with the positively charged hydrogen ion** → 39
- **I don't know** → 32

---

26 Quite right! The calcium ion bears *two* positive charges. The equation for the dissociation of calcium hydroxide is:

$$Ca(OH)_2 \xrightarrow{\text{dissociation}} Ca^{2\oplus} + 2\ OH^{\ominus}$$

You can now see that both potassium and calcium hydroxides dissociate in aqueous solution to form *hydroxyl ions.*

The same applies, of course, to sodium hydroxide and other metal hydroxides which act as bases. It is thus the *hydroxyl ion* which makes a metal hydroxide a base.

11

Let's go back at this point to our definitions of an acid and a base. Can you still remember both of these definitions? If so, please write them down!

- I can indeed remember them → 41
- I cannot remember them → 50

27

Equation I: $H_2SO_4 + KOH \rightarrow KHSO_4 + H_2O$
Equation II: $KHSO_4 + KOH \rightarrow K_2SO_4 + H_2O$

If equation I is now added to equation II, i.e. the two left hand sides and the two right hand sides are separately added together, the following new equation is obtained:

$$H_2SO_4 + KOH + KHSO_4 + KOH \rightarrow KHSO_4 + H_2O + K_2SO_4 + H_2O$$

The $KHSO_4$ may be cancelled from this equation, and the KOH and the H_2O entries added together. In this way we then arrive at the equation for the complete neutralization of sulfuric acid with sodium hydroxide. The final equation, which you have seen before, is thus:

$$H_2SO_4 + 2\ KOH \rightarrow K_2SO_4 + 2\ H_2O$$

K_2SO_4 ist called potassium sulfate.

$KHSO_4$ is called potassium hydrogen sulfate.

Those salts formed as a result of incomplete neutralization, i.e. those still capable of splitting off hydrogen ions in water, are referred to as

Hydrogen Salts.

Both $KHSO_4$ and $NaHSO_4$ are examples of hydrogen salts.

All *di*basic acids are able to form *two* different types of salts. Because such acids contain two hydrogen atoms, *either one* or *both* of these hydrogen atoms may be replaced by metal ions.

All *mono*basic salts can, of course, form only *one type* of salt. After all, they have only *one hydrogen atom* which may be replaced by a metal ion.

At this point we recall that phosphoric acid is a *tri*basic acid. How many different salts will phosphoric acid be capable of forming when it reacts with sodium hydroxide?

- Two different salts → 7
- Three different salts → 43
- Only one salt → 55
- I don't know → 66

---

28 Quite right, the answer is ionic bonding!

Our equation for neutralization can also be formulated in words as follows:

Acid + Base → Salt + Water

A practical example of this equation is provided by the equation for the neutralization of hydrochloric acid with sodium hydroxide:

HCl	+	NaOH	⟶	NaCl	+	H_2O
hydrochloric acid		sodium hydroxide		sodium chloride (salt)		water

Now complete the following equation:

....	+	KOH	⟶	....	+	....
nitric acid		......		potassium nitrate		....

Check your answers at 19 .

---

29 $2\ KOH + H_2SO_4 \rightarrow$

We first have to decide here what end products will be formed during this neutralization reaction. From the general equation

Acid + Base → Salt + Water

we know that *a salt and water* must appear on the right hand side.

A compound consisting of (negatively charged) acid radicals and (positively charged) metal ions is a salt. In our particular case the acid radical is $SO_4^{2\ominus}$ and the metal ion is $K^{\oplus}$. The acid radical bears two charges and is therefore capable of combining with two potassium ions. The salt thereby formed will have the formula K_2SO_4.

Accordingly, we may now write on the right hand side:

$$\rightarrow \quad K_2SO_4 + H_2O$$

So far then we have the equation:

$$2\ KOH + H_2SO_4 \quad \rightarrow \quad K_2SO_4 + H_2O$$

However, as 2 $H^{\oplus}$ and 2 $OH^{\ominus}$ are combining together here, we shall have to write 2 H_2O on the right hand side.

The final and complete equation is therefore:

$$2\ KOH + H_2SO_4 \quad \rightarrow \quad K_2SO_4 + 2\ H_2O$$

Read [37] once again.

30 Red with acids: *methyl orange* and *litmus*
Red with bases: *phenolphthalein*

State which indicator is: colorless in the presence of acids?
blue in the presence of bases?
orange-yellow in the presence of bases?

Write down the names of the indicators in the correct sequence, and then check your answers at [20] .

31 Quite right! *Dissociation means the splitting up of a compound into ions when it is dissolved in water.*

Now see if you can set up the equation for the dissociation of sulfuric acid, H_2SO_4. How many negative charges will the acid radical bear?

- One charge → 40
- Two charges → 10
- I don't know → 17

---

32 You are obviously not fully conversant with the definition of an acid. It actually runs as follows: An acid is a compound which forms positively charged hydrogen ions when it dissociates in water.
From this definition it is evident that positively charged hydrogen ions are characteristic of all acids. These ions are in fact responsible for the acidic properties of acids, e.g. they cause litmus paper to turn red.

Go back now to 25 .

---

33 The part of a metal hydroxide responsible for it acting as a base, i.e. for it combining with (positively charged) hydrogen ions, is the (negatively charged) *hydroxyl ion.* The two ions in question combine together according to the equation:

$$H^{\oplus} + OH^{\ominus} \rightarrow H\text{–}OH\ (= H_2O)$$

to form *water.*

This equation reveals how it is possible for a base to counteract the effects of an acid or, as we would more commonly say, for it to "neutralize" the acid. The (negatively charged) *hydroxyl ions* of the base simply combine with the (positively charged) *hydrogen ions* of the acid to form (electrically neutral) water molecules.

In the preceding sections we have discussed the following concepts:

The dissociation of acids
The dissociation of bases
The neutralization of acids with bases

These three concepts are so important that we shall go over them again at this point in the Program. By means of suitable equations we will illustrate how these concepts may be applied in practice.

Complete each of the following equations:

The dissociation of hydrochloric acid:
HCl → ...... + ...

The dissociation of sodium hydroxide:
NaOH → ...... + ...

The combination of hydrogen and hydroxyl ions during neutralization:
...... + ... → H_2O

Check your equations at [24] .

34

Indicator	With Acids	With Bases
Phenolphthalein	*colorless*	*red*
Methyl orange	*red*	*orange-yellow*
Litmus	*red*	*blue*

We now know that acids and bases *counteract* each other. Let's consider why this is so.
It is characteristic of all *acids* that positively charged *hydrogen ions* will be present. Now, if bases are able to counteract acids, i.e. have the opposite effect of acids, they have to be capable of rendering the positively charged hydrogen ions of acids "harmless". They must therefore contain some substance which is able to "capture" positively charged hydrogen ions.
Such a substance is termed a *base*. This leads us naturally into the following definition:

> A base is a compound which is capable of combining with positively charged hydrogen ions.

What then will the characteristic feature of a base be?

- It dissolves in water to form an alkali → [67]
- It colors litmus paper blue → [58]
- It can combine with positively charged hydrogen ions → [47]

---

35 No. Let's first write down the equation for the dissociation of sulfuric acid and sodium hydroxide:

$$H_2SO_4 \rightarrow H^{\oplus} + H^{\oplus} + SO_4^{2\ominus}$$

$$NaOH \rightarrow Na^{\oplus} + OH^{\ominus}$$

During neutralization the hydroxyl ions of the base combine with the hydrogen ions of the acid.

If the two equations are now added together, we obtain the new equation:

$$H_2SO_4 + NaOH \rightarrow H^{\oplus} + H^{\oplus} + SO_4^{2\ominus} + Na^{\oplus} + OH^{\ominus}$$

On the right hand side of this equation, however, there are two hydrogen ions but only one hydroxyl ion! There should, of course, be *two* hydroxyl ions present to combine with *two* hydrogen ions.

Thus, for the neutralization of 1 mol of H_2SO_4 one will need .... mol of NaOH. → 22

---

36 We said quite clearly that:

Even in crystalline form, i.e. in the solid state, salts consist of *ions*!

What type of bonding must therefore be present in salts? → 28

---

37 $H_2SO_4 + 2\ NaOH \rightarrow Na_2SO_4 + 2\ H_2O$

The acid radical of sulfuric acid, $SO_4^{2\ominus}$, has two negative charges. When it forms salts, it combines with either *two* metal ions bearing a *single* positive charge (e.g. Na_2SO_4, K_2SO_4) or with *one* metal ion bearing a *doubly* positive charge (e.g. $CaSO_4$).

Now complete the following equations:

$2\ KOH + H_2SO_4 \rightarrow$
$Ca(OH)_2 + H_2SO_4 \rightarrow$

Check your equations at 45 .

---

38 $NaHSO_4 + NaOH \rightarrow Na_2SO_4 + H_2O$

The equation for the preparation of sodium hydrogen sulfate is:

$$NaOH + H_2SO_4 \rightarrow NaHSO_4 + H_2O \qquad \text{(Equation I)}$$

When $NaHSO_4$ reacts with sodium hydroxide the product formed is sodium sulfate:

$$NaOH + NaHSO_4 \rightarrow Na_2SO_4 + H_2O \qquad \text{(Equation II)}$$

In this way we have effectively split the familiar equation

$$H_2SO_4 + 2\ NaOH \rightarrow Na_2SO_4 + 2\ H_2O$$

into two *partial* equations (Equations I and II).

This type of splitting is possible only when the 1 mol of sulfuric acid is not fully neutralized at the outset by adding 2 mol of sodium hydroxide. Instead, the sodium hydroxide has been added stepwise: first 1 mole (Equation I) and subsequently an additional mole (Equation II).

You are now asked to split the following equation:

$$H_2SO_4 + 2\ KOH \rightarrow K_2SO_4 + 2\ H_2O$$

into two partial equations in a way similar to that above.

Do not forget to note down the two equations corresponding to our Equations I and II. Check your answers at 27 .

---

39 Quite right! *It is the positively charged hydrogen ions which cause the litmus paper to turn red.*

This red coloration can in fact be reversed. All that need be done is to allow a few drops of some alkaline solution (e.g. caustic soda) to come into contact with the litmus paper. Its color will then immediately change to blue.
Acids and bases are evidently capable of mutually *counteracting* the effects each produces.

Let's consider now the color changes observed in some other dyes as a consequence of the opposing effects of acids and bases. Dyes which display a marked color change in the presence of acids and bases include:

Phenolphthalein	– colorless with acids
	– red with bases
Methyl orange	– red with acids
	– orange-yellow with bases

Dyes which change their color upon the addition of either acid or base are called *indicators*.

> Dyes which display a marked color change in the presence or acids or bases are called *indicators*.

Which indicators have a red color with acids and which have a red color with bases?

Write out your answer thus:
Red with acids: .... and ....
Red with bases: .... → [30]

---

[40] No. Please remember that:

Sulfuric acid contains two hydrogen atoms which appear as positively charged hydrogen ions when the acid dissociates thus:

$$H_2SO_4 \xrightarrow{\text{dissociation}} 2\ H^{\oplus} + \ldots\ldots$$

The 2 in front of the $H^{\oplus}$ tells us that there are two positively charged hydrogen ions formed, i.e. there will be two *positive charges* on the right hand side of the equation. These two *positive* charges must, of course, be balanced up by a corresponding number of *negative* charges. This is because the aqueous solution of sulfuric acid as a whole bears no electric charge.

Now complete the equation for the dissociation of sulfuric acid above. How many negative charges does the sulfate radical bear? → [10]

---

[41] An acid is a compound which forms positively charged *hydrogen ions* when it dissociates in water.

A base is a compound which is capable of combining with positively charged *hydrogen ions.*

If these two definitions are viewed in conjunction with the fact established earlier that metal hydroxides which act as bases form *hydroxyl ions* when they dissociate in water, the following question should not be too difficult to answer:

What part of a basic metal hydroxide is able to combine with positively charged hydrogen ions, and what is formed in the process?

Write out your answer and then go to [33]

If the question is too difficult for you go to [51] .

---

42 You should not forget that: Calcium hydroxide, $Ca(OH)_2$, contains two OH groups. Each of these groups bears *one* negative charge. In all, there are therefore *two* ngative charges in this molecule.

Now, the number of *positive* charges on the calcium must be equal to the total number of *negative* charges in the molecule, for the molecule as a whole is *not charged electrically*.

How many charges will the calcium ion actually bear? → 26

43 Phosphoric acid forms *three* different salts with sodium hydroxide.

The chemical designations and formulas of these three salts are as follows.

NaH_2PO_4	Sodium **di**hydrogen phosphate	(di = 2)
Na_2HPO_4	**Di**sodium hydrogen phosphate	
Na_3PO_4	**Tri**sodium phosphate	(tri = 3)

Phosphoric acid thus forms *two* different hydrogen salts, namely NaH_2PO_4 and Na_2HPO_4.

In NaH_2PO_4 one of the three hydrogen atoms in phosphoric acid is replaced by an $Na^{\oplus}$.

In Na_2HPO_4 two of the hydrogen atoms in phosphoric acid are replaced by $Na^{\oplus}$ ions.

In Na_3PO_4 three of the hydrogen atoms in phosphoric acid are replaced by $Na^{\oplus}$ ions.

Now complete the right hand side of each of the following equations:

$$NaOH + H_3PO_4 \rightarrow$$
$$2\ NaOH + H_3PO_4 \rightarrow$$
$$3\ NaOH + H_3PO_4 \rightarrow$$

Check your equations at 53 .

44 $NaHSO_4 \rightarrow Na^{\oplus} + H^{\oplus} + SO_4^{2\ominus}$

In water sodium hydrogen sulfate splits up into a sodium ion, a hydrogen ion and a sulfate ion. The hydrogen ion gives the solution its *acid* character; it would for instance turn blue litmus paper red.

If 1 mol of sodium hydrogen sulfate ($NaHSO_4$) reacts with 1 mol of sodium hydroxide, sodium sulfate is obtained thus:

$$NaHSO_4 + NaOH \rightarrow .... + ...$$

Complete this equation, and then go to 38 .

45

$$2\ KOH + H_2SO_4 \rightarrow K_2SO_4 + 2\ H_2O$$

$$Ca(OH)_2 + H_2SO_4 \rightarrow CaSO_4 + 2\ H_2O$$

Did you manage to get both equations right? If so, you did very well.

If you made some mistakes, however, you should in your own interest now go to 29 . There you will find a detailed description of the way in which the first equation should be set up.

Write out the following equation, inserting appropriate factors where necessary:

$$NaOH + H_3PO_4 \rightarrow Na_3PO_4 + H_2O$$

What factors have you inserted?

In front of the NaOH	In front of the H_3PO_4	In front of the Na_3PO_4	In front of the H_2O	
● 2	3	3	2	→ 12
● 3	–	–	–	→ 65
● 3	–	–	3	→ 56
● I find this difficult to do				→ 73

---

46 Let's now take a closer look at each of the five compounds mentioned.

KCl is a *salt*. It is made up of the *metal* ion $K^{\oplus}$ and the *acid radical* $Cl^{\ominus}$ (the chloride ion).

HNO_3 is an *acid*, namely nitric acid. When it dissociates in water it forms hydrogen ions thus: $HNO_3 \rightarrow H^{\oplus} + NO_3^{\ominus}$.

NaCl, sodium chloride, is a salt just like potassium chloride.

NaOH is a *base*. It contains the hydroxide group which in water as the hydroxyl ion is capable of combining with hydrogen ions thus: $OH^{\ominus} + H^{\oplus} \rightarrow H_2O$.

$CaCl_2$, calcium chloride, is – like potassium and sodium chlorides – a salt.

Work your way through the relevant definitions once again, starting from 56 .

---

47 Quite right! *A base is a compound capable of combining with positively charged hydrogen ions.*

Examples of bases include the following:

NaOH	=	Sodium hydroxide
KOH	=	Potassium hydroxide
$Ca(OH)_2$	=	Calcium hydroxide

Now answer the question: What elements are present in all three of these bases? → 59

---

48 Your answer is wrong. Do you remember how we defined an acid? It went as follows:

An acid is a compound which forms positively charged hydrogen ions when it dissociates in water.

The common feature of all acids is thus *not* the acid radical, which in any case differs from acid to acid! The common feature is rather the presence of *positively charged hydrogen ions*. It is the latter, of course, which are responsible for the acidic nature of acids.

If you bear all this in mind, you should now be able to answer the question correctly. → 25

---

49 Potassium hydroxide consists of *positively charged potassium ions* ($K^{\oplus}$) and *negatively charged hydroxyl ions* ($OH^{\ominus}$).

When such metal hydroxides are dissolved in water they dissociate, i.e. they split up into the ions of which they are made. The equation for the dissociation of potassium hydroxide is as follows:

$$KOH \xrightarrow{\text{dissociation}} K^{\oplus} + OH^{\ominus}$$

Now write out the equation for the dissociation of calcium hydroxide, $Ca(OH)_2$, and then answer the question:

How many positive charges does the calcium ion bear?

- One positive charge → 57
- Two positive charges → 26
- I need some help → 42

---

50 You should certainly try to learn the definitions better. Write out the definitions for an acid and a base, inserting the missing words where necessary:

An *acid* is a compound which forms positively charged .... when it dissociates in water.

A *base* is a compound which is capable of combining with positively charged .... → 41 .

---

51 The question was: What part of a basic metal hydroxide is able to combine with positively charged hydrogen ions, and what is formed in the process?

Let's take *sodium hydroxide* as an example of a basic metal hydroxide. What parts or (more precisely) what ions does sodium hydroxide consist of?

Jot down your answer and then go to 60 .

---

52 $H_2SO_4 + NaOH \rightarrow NaHSO_4 + H_2O$

Thus, when 1 mol of sulfuric acid reacts with only 1 mol of sodium hydroxide 1 mol of water and 1 mol of sodium *hydrogen* sulfate are formed.

Now write down the equation for the dissociation of sodium *hydrogen* sulfate.

$NaHSO_4 \rightarrow$ → 44

53

$$NaOH + H_3PO_4 \rightarrow NaH_2PO_4 + H_2O$$

$$2\ NaOH + H_3PO_4 \rightarrow Na_2HPO_4 + 2\ H_2O$$

$$3\ NaOH + H_3PO_4 \rightarrow Na_3PO_4 + 3\ H_2O$$

In each case here a *neutralization* reaction is involved. This is true even when the phosphoric acid is only partially neutralized and hydrogen salts are formed in the process – as in the first two equations.

The three equations indicate how we can prepare three quite different salts by reacting phosphoric acid and sodium hydroxide together in different proportions.

What are these three salts called?

Write out the three names with the appropriate formulas, and then check your answers at 63.

---

54 Your answer is not quite right. An explanation of this is given at 46.

---

55 Your answer is not right. An explanation of this is given at 66.

---

56 The correct version of the equation is:

$$3\ NaOH + H_3PO_4 \rightarrow Na_3PO_4 + 3\ H_2O$$

You should remember that:

A compound consisting of a (negatively charged) acid radical and a (positively charged) metal ion is a salt.

Let's repeat now the definitions of an acid, a base and a salt.

A base is a compound which ....

Write out the completed version of this sentence, and then check your answer at 64.

57 You haven't thought this through carefully enough. Remember that the OH ion always bears *one* negative charge, that calcium hydroxide has *two* hydroxyl groups in it, and that in any molecule the total number of *negative* charges must equal the total number of *positive* charges.

What therefore must the positive charge on the calcium ion be? → 26

58 You are quite right: Aqueous solutions of bases, i.e. alkalis, turn litmus paper blue. This, however, is just *one* specific reaction and *not* the characteristic feature of a base!

Re-read the definition given in 34 on the characteristic feature of all bases. The blue coloration observed with the litmus paper is merely one effect of this fundamental characteristic.

59 In all three of the bases we find:

Oxygen and *Hydrogen*

These two elements occur as the OH *group*. You should already know that the OH group bears a negative charge ($OH^{\ominus}$) and is called the *hydroxyl* ion. In combination with metals it is, however, referred to as a hydroxide, e.g. in the compounds:

Sodium *hydroxide*	NaOH
Potassium *hydroxide*	KOH
Calcium *hydroxide*	$Ca(OH)_2$

Thus, the bases mentioned above consist of a metal atom and one or more hydroxyl groups. Between the metal atom and the hydroxyl group there is an ionic bond, i.e. the metal exists in the form of a positively charged ion whereas the hydroxyl group is a negatively charged ion. Sodium hydroxide, for instance, may be represented as:

$Na^{\oplus}OH^{\ominus}$

(Note that neither the H nor the O of the OH group bears the negative charge but rather the OH group *as a whole*!)

What ions does the base potassium hydroxide consist of? → 49

60 Sodium hydroxide consists of sodium ions, $Na^{\oplus}$, and hydroxyl ions, $OH^{\ominus}$.

Decide now which of these two ions is capable of bonding to positively charged hydrogen ions, and see if you can work out what will be formed.

Jot down your answer and then go to [33] .

If you cannot answer the question go to [68] .

61 Your answer is quite correct.

If we react a polybasic acid with a base, *not necessarily all* of its hydrogen ions become bonded to hydroxyl ions. It is quite feasible for 1 mol of sulfuric acid to react with only 1 mol of sodium hydroxide thus:

$$H_2SO_4 + NaOH \rightarrow NaHSO_4 + \ldots\ldots$$

Complete this equation and then go on to [52] .

62 No. Read [71] carefully again, taking note of the number of hydrogen atoms present in the formulas for sulfuric and phosphoric acids.

Remember that a *mono*basic acid produces *one* hydrogen ion in water!

63 The correct answers are:

Sodium dihydrogen phosphate	NaH_2PO_4
Disodium hydrogen phosphate	Na_2HPO_4
Trisodium phosphate	Na_3PO_4

What type of salts are NaH_2PO_4 and Na_2HPO_4?

Jot down the general name for such salts. → [74]

64 A base is a compound which is capable of combining with positively charged hydrogen ions.

An acid is a compound which ....

11

Write out the completed version of this sentence, and then check your answer at 72 .

---

65 Your answer is not yet complete.

According to you, the equation should read:

$$3\ NaOH + H_3PO_4 \rightarrow Na_3PO_4 + H_2O$$

Now just count up the hydrogen atoms on each side of this equation. You will then soon see that something is wrong. Go back to 45 .

---

66 A *mono*basic acid (HCl, HNO_3) splits off *one* hydrogen ion in water. It is therefore capable of forming only *one* salt with a base such as sodium hydroxide (NaCl, $NaNO_3$).

This is not so in the case of a *di*basic acid (H_2SO_4). An acid of this type splits off *two* hydrogen ions in water.

Two salts will be formed when it reacts with sodium hydroxide:
A hydrogen salt capable of splitting off a further hydrogen ion in water ($NaHSO_4$); and a salt which is not capable of splitting off a hydrogen ion in water (Na_2SO_4).

Phosphoric acid is a *tri*basic acid and is therefore capable of splitting off *three* hydrogen ions in water. Accordingly, when phosphoric acid reacts with sodium hydroxide .... different salts can be formed.

Jot down the missing number, and then continue reading at 43 .

---

67 You are quite right: The aqueous solution of a base is an alkali, though this is not the *characteristic feature* of a base!

Re-read the definition given in 34 on the characteristic feature of all bases.

---

68 You should bear in mind here that particles having the same electric charge will repel each other whereas particles having opposite charges will attract one another.

The question at 60 should now be easy to answer.

---

69 Yes, an *exothermic* reaction is involved here. After initiating the reaction by gentle warming, the mixture of sulfur and iron becomes so hot from the reaction that it begins to glow.

What is the name given to a reaction which gets cooler as it proceeds?

It is an .... reaction. → 76

70 Your answer is wrong. In the formula for sodium sulfate (Na_2SO_4) there are *two* sodium ions. In the formula for sodium phosphate (Na_3PO_4) there are *three* sodium ions.

It follows that these two acid radicals must bear different charges. Read 78 once again.

71 Your answer is correct.

HNO_3 is an acid.
NaOH is a base.
KCl, NaCl and $CaCl_2$ are salts.

Acids which produce *one* hydrogen ion in water are also referred to as *mono*basic acids.

1 mol of a monobasic acid will react with 1 mol of either sodium or potassium hydroxide to form 1 mol of salt and 1 mol of H_2O.
Examples of monobasic acids are HCl and HNO_3.

Now what designations would you apply to:

a) Sulfuric acid (H_2SO_4)? b) Phosphoric acid (H_3PO_4)?

- a) is tribasic and b) is dibasic → 62
- a) is dibasic and b) is tribasic → 87
- a) and b) are both dibasic → 96
- a) and b) are both tribasic → 79

72 An acid is a compound which forms positively charged hydrogen ions when it dissociates in water.

A compound consisting of a .... and an .... is a salt.

Jot down the missing words, and then check your answer at 80 .

---

73 Let's try to find the correct factors by means of a stepwise procedure. The equation without any factors was as follows:

$$NaOH + H_3PO_4 \rightarrow Na_3PO_4 + H_2O$$

Phosphoric acid is to be neutralized. Now H_3PO_4 dissociates in water to form *three* hydrogen ions and the acid radical of phosphoric acid, namely the phosphate ion, which bears *three* negative charges.

11

$$H_3PO_4 \rightarrow 3\ H^{\oplus} + PO_4^{3\ominus}$$

For neutralization, therefore, *three* hydroxyl ions, i.e. 3 NaOH, will be needed.

The equation will thus assume the form:

$$3\ NaOH + H_3PO_4 \rightarrow Na_3PO_4 + H_2O$$

But now we have to look at the number of hydrogen atoms. On the left hand side there are 6 whereas there are only 2 on the right. As the *same* number of hydrogen atoms must appear on each side of the equation, we shall put a 3 in front of the H_2O in order that there will be 6 hydrogen atoms on the right hand side.

$$3\ NaOH + H_3PO_4 \rightarrow Na_3PO_4 + 3\ H_2O$$

You should now count up the number of atoms of each type appearing on the right and left hand sides of this equation. You will then see that the equation is belanced.

Continue your reading at 56 .

---

74 NaH_2PO_4 and Na_2HPO_4 are *hydrogen salts.*

What actually are hydrogen salts?

Write out your answer and then go to 82 .

---

75 The acid radical of sulfuric acid: two negative charges
The acid radical of phosphoric acid: three negative charges

The acid radical of nitric acid: one negative charge
The acid radical of hydrochloric acid: one negative charge

The general equation for neutralization reactions is:

Acid + ... → .... + ...

Write out the complete equation and then go to 90 .

11

76 A reaction which gets cooler as it proceeds is an *endothermic* reaction.

A reaction which gets warmer as it proceeds is an *exothermic* reaction.

Write out these two statements, and give two examples of an exothermic reaction. → 83

77 The reaction of sulfur and iron to form iron sulfide is a strongly *exothermic* reaction. However, the reaction first has to be initiated by heating the mixture. If a test tube containing a mixture of sulfur and iron powder is gently heated, the reaction will begin at the point where the heat is applied. The heat generated from this initial reaction is sufficient to cause the reaction to spread throughout the whole of the mixture. In fact so much heat is liberated that the contents of the test tube become glowing hot.

Thus, the reaction of sulfur with iron to form iron sulfide is an .... (exothermic/endothermic) reaction.

Continue your reading at 69 .

78 $H_2SO_4 + 2\ NaOH \rightarrow Na_2SO_4 + 2\ H_2O$

$H_3PO_4 + 3\ NaOH \rightarrow Na_3PO_4 + 3\ H_2O$

The salts of sulfuric acid are called sulfates; the salts of phosphoric acid phosphates.
Na_2SO_4 is called sodium sulfate.
Na_3PO_4 is called sodium phosphate.

How many negative charges are there on the acid radicals of (a) sulfuric acid and (b) phosphoric acid?

- Both of these radicals have two charges → 94

- Both of these radicals have three charges → [70]
- The sulfuric acid radical has two whereas the phosphoric acid radical has three charges → [61]
- The sulfuric acid radical has three whereas the phosphoric acid radical has two charges → [86]

---

79 Your answer is wrong. Read [71] again carefully and take especial note of the number of hydrogen atoms present in the formulas for sulfuric and phosphoric acids.

Remember that a *mono*basic acid splits off *one* hydrogen ion in water.
→ [71]

11

---

80 A compound consisting of a (negatively charged) acid radical and a (positively charged) metal ion is a salt.

Now take a good look at each of the five compounds below and decide which are acids, which are bases and which are salts!

KCl, HNO_3, $NaCl$, $NaOH$, $CaCl_2$

What types of compound are present?

- One acid, two bases and two salts → [88]
- Two acids, one base and two salts → [95]
- One acid, one base and three salts → [71]
- Two acids, two bases and one salt → [54]

---

81 $NaH_2PO_4 + 2\ NaOH \rightarrow Na_3PO_4 + 2\ H_2O$

From the last two equations we see how the two hydrogen salts of phosphoric acid may be completely neutralized.

Every neutralization reaction is associated with the generation of heat. The amount of heat liberated when dilute acids and alkalis are mixed is, however, scarcely noticeable. Concentrated acids and bases on the other hand should not be mixed together. Such a mixture would become too hot and start to spit out violently.

Reactions which generate heat as they proceed are termed *exothermic* reactions. Conversely, reactions which absorb heat as they proceed are termed *endothermic* reactions.

You no doubt still remember our very first experiment involving the reaction of sulfur with iron to form iron sulfide.

What type of reaction was that?

- An exothermic reaction → 69
- An endothermic reaction → 97
- Neither of these → 89

---

82 The gist of your answer should be that:

Hydrogen salts in water split off hydrogen ions.

$KHSO_4$ is a hydrogen salt which splits off one $H^{\oplus}$ when it is dissolved in water. This hydrogen ion may be neutralized by bases such as KOH.

$$KHSO_4 \rightarrow K^{\oplus} + H^{\oplus} + SO_4^{2\ominus}$$

$$\underbrace{K^{\oplus}+H^{\oplus} + SO_4^{2\ominus}}_{KHSO_4} + \underbrace{K^{\oplus} + OH^{\ominus}}_{KOH} \rightarrow \underbrace{2\,K^{\oplus} + SO_4^{2\ominus}}_{K_2SO_4} + H_2O$$

Now complete the following equation:

$$Na_2HPO_4 + NaOH \rightarrow$$

Check your equation at 91 .

---

83 Examples of *exo*thermal reactions include:

1. Neutralization reactions
2. The formation of iron sulfide from iron and sulfur
3. All combustion reactions

In this Program we have presented you with a whole range of key statements and definitions which your are asked to commit to memory. To help you in this task we shall now repeat them. We start with the following:

Dyes which display a clearly recognizable color change in the presence of either acids or .... are known as ....

Jot down the missing words, and then go to 92.

84

	With Acids	With bases
Litmus	*red*	*blue*
*Phenol*phthalein	*colorless*	red
Methyl*orange*	*red*	orange-*yellow*

Jot down now the number of negative charges on each of the following acid radicals:

The sulfuric acid radical has ....

The phosphoric acid radical has ....

The nitric acid radical has ....

The hydrochloric acid radical has .... → 75

85 Phosphoric acid is a *tribasic* acid.
Hydrochloric acid is a *monobasic* acid.
Sulfuric acid is a *dibasic* acid.
Nitric acid is a *monobasic* acid.

What ions are formed when a hydrogen salt is dissolved in water?

A reaction which gets cooler as it proceeds is an .... reaction.

Conversely: A reaction which gets warmer as it proceeds is an .... reaction.

Write out your answers and then go on to 93.

86 Your answer is wrong. Just check to see how many Na ions there are in the formulas for sodium sulfate and sodium phosphate.

The sodium ion always bears *one* positive charge. It should now be easy for you to answer the question in 78 correctly.

Read through 78 once again.

---

87 Yes, sulfuric acid is a *di*basic acid and phosphoric acid is a *tri*basic acid.

Sulfuric acid splits off *two* hydrogen ions in water. The corresponding number for phosphoric acid is *three* hydrogen ions.

What is the significance of the designations mono*basic,* di*basic* and tri*basic?*

When the various acids are neutralized with a base having *one* hydroxyl group, e.g. NaOH or KOH, the following amounts of base will be required:

1 mol of a **mono***basic* acid (HCl,HNO_3): **1** mol of *base*
1 mol of a **di***basic* acid (H_2SO_4) : **2** mol of *base*
1 mol of a **tri***basic* acid (H_3PO_4) : **3** mol of *base*

11

Now write out once again the equations for the neutralizations of sulfuric and phosphoric acids.

$H_2SO_4 + NaOH \rightarrow Na_2SO_4 +$ ....

$H_3PO_4 + NaOH \rightarrow Na_3PO_4 +$ .... → 78

---

88 Your answer is not quite right.

An explanation of this is given at 46 .

---

89 Your answer is wrong.

An explanation of this is given at 77 .

---

90 Acid + Base → Salt + Water

What is an acid?
What is a base?
What is an alkali?
What type of chemical bond is present in salts?

Write out your answers, and then go to 98 .

---

91 $Na_2HPO_4 + NaOH \rightarrow Na_3PO_4 + H_2O$

Now complete the equation:

$NaH_2PO_4 + 2NaOH \rightarrow$

Check your answer at 81 .

92 Dyes which display a clearly recognizable color change in the presence of either acids or *bases* are known as *indicators*.

Now complete the following table:

	With Acids	With Bases
Litmus	....	....
...... phthalein	....	red
Methyl	....	orange- ......

Check your answers at 84 .

93 Hydrogen salts dissociate when dissolved in water to form metal ions, hydrogen ions and a negatively charged acid radical.

A reaction which gets cooler as it proceeds is an *endothermic* reaction.

Conversely: A reaction which gets warmer as it proceeds is an *exothermic* reaction.

End of Program 11.

94 Your answer is wrong. There are *two* sodium ions in the formula for sodium sulfate (Na_2SO_4).

There are *three* sodium ions in the formula for sodium phosphate (Na_3PO_4).

This means that the two acid radicals must bear different charges.

Read through 78 once again.

95 Your answer is not quite right.

An explanation of this is given at 46 .

96 Your answer is wrong. Read 71 carefully again and take a good look at the number of hydrogen atoms appearing in the formulas for sulfuric and phosphoric acids!

Remember that a *mono*basic acid splits off *one* hydrogen ion in water.

97 Your answer is wrong. An explanation of this is given at 77 .

98 An *acid* is a compound which forms positively charged hydrogen ions when it dissociates in water.

A *base* is a compound which is capable of combining with positively charged hydrogen ions.

An *alkali* is an aqueous solution of a base.

The type of bonding present in salts is *ionic bonding.*

Phosphoric acid is a ...... basic acid.
Hydrochloric acid is a ...... basic acid.
Sulfuric acid is a ...... basic acid.
Nitric acid is a ...... basic acid.

Copy down the above four statements and insert the missing prefixes where indicated. → 85

Program 12

Sulfur and its Compounds (I)

1 Sulfur is an element. It has an atomic mass of 32 and its chemical symbol is S.

Sulfur is a solid and brittle substance. It may be purchased in the form of solid sticks or lumps. In the laboratory, however, it is normally used in the form of a fine powder – as so-called "flowers of sulfur". Sulfur is bright yellow in color, has no luster, conducts heat poorly and electric current not at all.

Is sulfur a metal or a non-metal?

- It is a non-metal → 10
- It is a metal → 20
- I don't know → 26

12

2 You are not working accurately enough. Read through 10 carefully once again.

3 Your answer is wrong. To remove the sulfur from a mixture of sulfur and iron by means of distillation, the mixture would have to be heated up. This, however, would induce a chemical reaction. When heated together, sulfur and iron combine to form iron sulfide and so much heat is generated that the mixture actually gets red hot.

Accordingly, it is necessary to utilize another physical property of sulfur to separate it from the mixture of sulfur and iron.

Details of the physical properties of sulfur are given in 22 .

4 H_2S is reacted with *caustic soda* to prepare sodium sulfide.

See if you can write down this chemical process in the form of an equation!

What type of reaction is involved?

- A neutralization reaction → 50

- A reduction reaction → 33
- An oxidation reaction → 43

---

5 Your answer is wrong. The valence of the sulfur must be *different* in the two oxides of sulfur, i.e. in SO_2 and SO_3. After all, the sulfur is combined with differing amounts of oxygen in each. Oxygen is divalent (cf. O in H_2O or Na_2O!). What is the valence of the sulfur in SO_2 and in SO_3?

Jot down the two valences of the sulfur observed when it combines with oxygen. → 39

---

6 You have obviously not drawn the structural formula of hydrogen sulfide, H_2S, correctly. The structural formula of this compound actually has the form:

12

H–S–H

Structural formulas can be used as an aid to illustrate the valence bonds present in molecules.

If you count up the number of bonds emanating from the sulfur, you will be able to determine the valence of the sulfur in hydrogen sulfide. Now complete the sentence:

The sulfur in hydrogen sulfide has a valence of ....

Continue your reading at 12 .

---

7 The varying forms in which a chemical element can occur are termed its *allotropic modifications* or *allotropes*.

When sulfur is heated up in a test tube it melts. A layer of finely divided, solid sulfur settles out on the cooler parts of the tube. The sulfur has thus been vaporized and then immediately converted back from the vapor state to the solid state. In the process the liquid state is completely bypassed.

The above procedure is used for the industrial purification of sulfur. The resulting purified, finely powdered sulfur is described as *"flowers of sulfur"*.

Write out now the various states of aggregation through which sulfur is passed during its purification:

Solid – .... – .... – ....

Continue your reading at [22] .

---

8 Your answer is not correct.

An explanation of this is given at [40] .

---

9 Your equation is still not right.

Here once again is the incomplete equation:

$$H_2S + O_2 \rightarrow \dot{S} + 2\,H_2O$$

The number of oxygen atoms is the same on both sides, and likewise for the number of sulfur atoms. However, there are 2 x 2 = 4 hydrogen atoms on the right whereas on the left there are only 2. It is thus necessary to take 2 H_2S:

$$2\,H_2S + O_2 \rightarrow S + 2\,H_2O$$

Is the equation correct now?

- Yes → [70]
- No → [61]

---

10 Sulfur is a non-metal.

Sulfur melts when it is heated and forms a yellow liquid which flows very readily. Upon further heating this liquid takes on a darker color and becomes more viscous. Finally, at a temperature of about 400°C, it gets much less viscous again. Thus, in the molten state sulfur exists in differing forms according to the temperature. If the molten sulfur at 400°C is cooled down, the different forms will reappear, though now in reverse sequence. The brown, non-viscous molten sulfur first becomes viscous and then finally turns yellow and gets viscous again. The various forms in which an element can exist are termed its *allotropic modifications* or *allotropes*.

Even in the solid state sulfur can exist in the form of different allotropic modifications. If very hot molten sulfur is slowly poured into water, one obtains a very plastic, easily deformable allotrope of sulfur. If, however, this *plastic form of sulfur* is left for some time at room temperature it begins to lose its plasticity and becomes brittle again. The allotrope which corresponds to room temperature is in fact the brittle one.

Now see if you can answer the following question:

A mixture of brittle and plastic sulfur is kept at room temperature for some time. What will eventually happen?

- The brittle sulfur will change into plastic sulfur → 2
- The plastic sulfur will change into brittle sulfur → 19
- Nothing at all; both forms will exist together → 28

---

11 Your answer is not right. An explanation of this is given at 31 .

---

12 The sulfur atom in hydrogen sulfide is *di*valent:

H–S–H

Hydrogen sulfide is a foul smelling gas which smells like rotten eggs.

Hydrogen sulfide is very poisonous and so the greatest possible care must be taken when working with this gas!

Hydrogen sulfide is an acid. As is the case with all acids, the hydrogen atoms of hydrogen sulfide can dissociate off in water.

See if you can write out the equation for the dissociation of hydrogen sulfide. → 23

---

13 Now check your equation:

$H_2S \rightarrow 2\ H^{\oplus} + S^{2\ominus}$

Clearly, the sulfide ion bears *two* negative charges. See if you can work out how many sodium ions the sulfide ion is capable of combining with when it forms the sodium salt.

Write down the formula for sodium sulfide and then calculate its molecular mass. (Relevant atomic masses: Na = 23, S = 32)

What molecular mass do you obtain?

- 55 → 63

- 78 → 24
- 87 → 38

---

14 Now check your answers:

Sulfur dioxide SO_2

Sulfur trioxide SO_3

It is possible to visualize the valence bonds by drawing the corresponding structural formulas.

For instance, sulfur dioxide has the structural formula O=S=O.

Now write out the structural formula for sulfur trioxide. → 27

---

15 The iron in a mixture of iron and sulfur can be removed with a magnet. In this case, however, the separation of the two substances should be dependent on the physical properties of the *sulfur.* Your answer is thus right only as far as it goes.

Several of the physical properties of sulfur are mentioned in 22 .

---

16 Now check your equation:

FeS	+	2 HCl	→	H_2S	+	$FeCl_2$
iron sulfide		hydro-chloric acid		hydrogen sulfide		iron chloride

Complete the sentence below:

The salts of hydrogen sulfide are termed .... → 49

---

17 Your answer is wrong. The valence of the sulfur must be *different* in the two oxides of sulfur, i.e. in SO_2 and SO_3. After all, the sulfur is combined with differing amounts of oxygen in each. Oxygen is divalent (cf. O in H_2O or Na_2O!).

What then is the valence of the sulfur in SO_2 and in SO_3?

Jot down the two valences of sulfur when it combines with oxygen.

→ [39]

[18] Your answer is not correct. An explanation of this is given at [31]

[19] Your answer is quite right.

The permanent form of sulfur at room temperature is the brittle form. For this reason the plastic form of sulfur gradually changes into the brittle form at room temperature.

Do you remember what term is used to describe the differing forms in which an element can occur?

12

If you do, jot it down and then continue reading at [7] .

If you don't, you should turn now to [10] .

[20] No, metals have *luster* and *conduct* both *heat* and *electric current* well.

Read through [1] once again.

[21] No. Hydrogen sulfide has the formula H_2S. Draw the structural formula corresponding to this chemical formula, and then answer again the question:

What is the valence of sulfur in hydrogen sulfide?

- Two → [12]
- Four → [6]
- Six → [29]

[22] Now check your answer:

Solid – *liquid* – *vapor* – *solid.*

Sulfur is found in the free state in nature.

Sulfur may be extracted by heating sulfur-bearing rock after exclusion of the surrounding air. This procedure enables the sulfur to be collected in molten form. Alternatively, one can force superheated steam into the subterranean sulfur deposits, which is another way of melting the sulfur. The pressure of the steam is sufficient to transport the molten sulfur to the surface.

An excellent solvent for sulfur is carbon disulfide.

Below is a summary of the most important physical properties of sulfur:

Sulfur is a solid and brittle substance. It may be converted into its plastic form (allotrope) temporarily by quenching molten sulfur. Even molten sulfur exists in the form of different allotropic modifications at various temperatures. Sulfur is soluble in carbon disulfide.

You probably recall that in a previous experiment we tried to extract sulfur from a mixture of iron filings and flowers of sulfur in the simplest possible way. How may this be done using the physical properties *of the sulfur?*

(Although this question is not very easy, it should be possible for you to answer after thinking about it carefully).

- Remove the sulfur by means of distillation → 3
- Remove the iron by means of a magnet → 15
- Dissolve the sulfur in carbon disulfide and then remove the iron by filtration → 32
- Melt the sulfur and then remove the iron by filtration → 42

---

23 $H_2S \rightarrow 2H^{\oplus} + S^{2\ominus}$

If the two hydrogen atoms in hydrogen sulfide are replaced by a doubly charged positive iron ion ($Fe^{2\oplus}$), one obtains iron sulfide, FeS, a salt of hydrogen sulfide.

What is the acid radical of hydrogen sulfide called and what charge does it bear?

Jot down your answer and then go on to 34 .

---

24 The molecular mass of sodium sulfide is 78. If you managed to get this right first time you are working well.

Sodium sulfide is used in the chemical industry. As it is not found in nature it has to be produced synthetically.

To prepare it, H_2S is reacted with ....

If you remember here that H_2S is an acid, it shouldn't be too difficult to reach the right answer. → 4

---

25 Your answer is wrong. The valence of the sulfur must be *different* in the two oxides of sulfur, i.e. in SO_2 and SO_3. After all, the sulfur is combined with differing amounts of oxygen in each. Oxygen is divalent (cf. O in H_2O or Na_2O!). What is the valence of the sulfur in SO_2 and in SO_3?

Jot down the two valences of the sulfur observed when it combines with oxygen. → 39

---

26 Make sure that you know the difference between metals and non-metals.

Metals have luster and conduct both heat and electric current well.

Non-metals conduct heat very poorly and electric current not at all.

Re-read 1 .

---

27 Now check your answer:

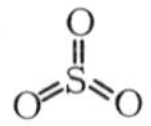

Sulfur trioxide

Sulfur in this molecule has a valence of six.

Sulfur combines with hydrogen to form *hydrogen sulfide.* The molecule of hydrogen sulfide has the formula H_2S.

What is the valence of the sulfur in hydrogen sulfide?

- The valence is two → 12
- The valence is four → 21
- The valence is six → 35

---

28 You are not working accurately enough. Read carefully once again the contents of [10].

---

29 You have obviously not drawn the structural formula of hydrogen sulfide, H_2S, correctly. The structural formula of this compound actually has the form:

H–S–H

Structural formulas can be used as an aid to illustrate the valence bonds present in molecules.

If you count up the number of bonds emanating from the sulfur, you will be able to determine the valence of the sulfur in hydrogen sulfide. Now complete the sentence:

The sulfur in hydrogen sulfide has a valence of ....

Now continue your reading at [12].

---

30 Iron sulfide FeS
Sodium sulfide Na_2S

Large deposits of iron sulfide are found in nature, though the naturally occurring form has *two* sulfur atoms in its formula, i.e. it has the chemical formula FeS_2. This latter compound is known as iron pyrites or just pyrites for short.

In pyrites an ionic bond exists between the (doubly positively charged) iron ions and the (doubly negatively charged) sulfur ions. Actually, the sulfur ions are present as a pair of sulfur atoms and may be represented in the formula for pyrites as follows:

$$Fe^{2\oplus} \quad \left[\begin{array}{c} S \\ | \\ S \end{array}\right]^{2\ominus}$$

Because of their physical appearance, the naturally occurring sulfides are sometimes classified as gravels (e.g. iron pyrites, FeS_2), as blendes (e.g. zinc sulfide or zinc blende, ZnS) or as glances (e.g. lead sulfide or galena, PbS).

For the latter two sulfides write out their normal *chemical* names and formulas.

Now check your answers at [37].

---

31 We shall now work out this equation stepwise.

To start with, here are the substances involved without any factors:

$$H_2S + O_2 \rightarrow SO_2 + H_2O$$

On the right hand side of the equation there are three oxygen atoms whereas on the left hand side there are only two. To equalize the number of oxygen atoms on both sides, we shall write $3O_2$ on the left and $2SO_2$ and $2H_2O$ on the right. Then, at least, the number of oxygen atoms will be the same on each side:

$$H_2S + 3O_2 \rightarrow 2SO_2 + 2H_2O$$

By looking at this equation carefully, it should now be easy to see what factor is missing in front of the H_2S.

12

- This factor is 2 → 57
- This factor is 3 → 45

---

32 Your answer is in fact correct: one dissolves the sulfur in carbon disulfide and filters off the undissolved iron.

When sulfur burns it combines with oxygen:

$$S + O_2 \rightarrow SO_2$$

SO_2 is known as sulfur dioxide. It is called a *di*oxide because there are *two* oxygen atoms in this molecule.

Another oxide of sulfur exists having *three* oxygen atoms in the molecule. Its formula is thus SO_3 and it is known as sulfur *tri*oxide.

During the combustion of sulfur only sulfur dioxide is formed. The further oxidation of sulfur dioxide to sulfur trioxide can be brought about only in the presence of special catalysts.

What valence does the sulfur have in

	SO_2	SO_3		
●	Two	Two	→	5
●	Four	Four	→	17

- Six Six → 25
- Four Six → 39

---

33 No. The reaction of caustic soda with hydrogen sulfide is in fact a *neutralization!*

Now write out the following:

Acid + Base → Salt + Water

H_2S + NaOH →

and complete the second of the two equations. → 50

---

34 The acid radical of hydrogen sulfide is the *sulfide* ion which bears *two* negative charges.

The salts of hydrogen sulfide are known as *sulfides,* e.g. iron sulfide, FeS.

Hydrogen sulfide is a weak and volatile acid.

Less volatile acids, such as hydrochloric and sulfuric acids, are capable of driving hydrogen sulfide out of its salts. Because of its great volatility it is able to escape from the reaction mixture as a gas. This property is in fact utilized for the preparation of hydrogen sulfide in the laboratory.

The incomplete equation for the relevant reaction is:

FeS + 2 HCl → H_2S +

Write out now the complete equation with the name of each substance involved underneath its formula.

Continue your reading at 16 .

---

35 No. Hydrogen sulfide has the formula H_2S. Attempt to draw the corresponding structural formula and then answer the question once again:

What is the valence of the sulfur in hydrogen sulfide?

- Two → 12

- Four → 6
- Six → 29

---

36 Your answer is not correct. An explanation of this is given at 40.

37

ZnS Zinc sulfide
PbS Lead sulfide

Write out completed versions of the three sentences:

FeS_2 is known as .... or ....
ZnS is known as zinc sulfide or ....
PbS is known as lead sulfide or .... → 47

12

---

38 Your answer is wrong.

The chemical formula of sodium sulfide is Na_2S.

The molecular mass of sodium sulfide may be obtained by simply adding up the individual atomic masses thus:

Na = 23

Na = 23

S = 32

Na_2S = ...

What is the molecular mass of sodium sulfide?

Continue your reading at 24.

---

39 When sulfur combines with oxygen it has a valence of either *four* or *six*.

Oxygen has a valence of two.

Jot down now the names and formulas of the two oxides of sulfur we have just described, and then continue reading at 14.

---

40 You have either made a mistake in your calculations or you have taken the wrong formula for sodium sulfide.

To arrive at the correct formula for this salt, we suggest that you first write out the equation for the dissociation of hydrogen sulfide (which has the formula H_2S).

After writing out this equation, check your answer at 13 .

41 We shall now work out this equation stepwise.

To start with, here are the substances involved without any factors:

$$H_2S + O_2 \rightarrow S + H_2O$$

Next we equalize the number of oxygen atoms on the right and left hand sides of the equation by writing in *two* molecules of water on the right side thus:

$$H_2S + O_2 \rightarrow S + 2\ H_2O$$

It should now be easy to see what factors are missing in front of the H_2S and the S. What are these factors?

	In front of H_2S	In front of S	
•	2	3	→ 48
•	2	2	→ 75
•	3	2	→ 9

42 Your answer is wrong. If one wishes to melt the sulfur present in a mixture of sulfur and iron, the mixture must of course be heated up. This, however, would induce a chemical reaction. When heated together, sulfur and iron combine to form iron sulfide and so much heat is generated that the mixture actually gets red hot.

Accordingly, it is necessary to utilize another physical property of sulfur to separate it from the mixture.

Details of the physical properties of sulfur are given in 22 .

43 Your answer is wrong. The reaction of caustic soda with hydrogen sulfide is in fact a *neutralization!*

Now write out the following:

Acid + Base $\rightarrow$ Salt + Water

H_2S + NaOH $\rightarrow$

and complete the second of the two equations. $\rightarrow$ 50

44 Silver is *mono*valent and lead is *di*valent.

Jot down the missing words in the following sentences in the sequence in which the words occur:

The salts of hydrogen sulfide are known as .... In solution these may be tested for by adding solutions of either .... or .... salts. Precipitates which are either .... or .... in color are formed.

Continue your reading at 52 .

45 No. Read 31 again carefully, taking especial note of the number of sulfur atoms on the right and left hand sides.

46 Your answer is wrong. It is not possible for the sulfur to have the same valence in the anhydrides of both sulfurous and sulfuric acids.

In these anhydrides, the sulfur is combined in the one case with *two* oxygen atoms and in the other case with *three!*

Draw out the structural formulas for these two anhydrides, and then check your answers at 77 .

47 FeS_2 is known as *iron pyrites* or *pyrites,* ZnS as zinc sulfide or *zinc blende* and PbS as lead sulfide or *galena.*

Hydrogen sulfide can be ignited. It burns with a blue flame and in the process sulfur dioxide and water are formed. All the factors are missing from the following equation:

$H_2S + O_2 \rightarrow SO_2 + H_2O$

Although at first sight it may not seem too easy, after some thought you should be able to arrive at the factors to be placed in front of all four substances.

	In front of H_2S	In front of O_2	In front of SO_2	In front of H_2O		
●	2	2	2	2	→	11
●	2	2	3	3	→	18
●	2	3	2	2	→	57
●	3	2	2	2	→	67
●	I can't work this out				→	31

48 Your equation is still not right.

Here once again is the incomplete equation:

$H_2S + O_2 \rightarrow S + 2\ H_2O$

The number of oxygen atoms is the same on each side, as is the number of sulfur atoms.

However, we find 2 x 2 = 4 hydrogen atoms on the right as opposed to only 2 on the left.

It is thus necessary to take 2 H_2S:

$2\ H_2S + O_2 \rightarrow S + 2\ H_2O$

Is the equation alright now?

- Yes → 70
- No → 61

49 The salts of hydrogen sulfide are known as *sulfides*.

Write down the chemical formula for sodium sulfide and then calculate its molecular mass. Relevant atomic masses are: Na = 23 and S = 32.

What is your answer?

- 55 → 8
- 78 → 24
- 87 → 36
- I obtained another number → 40

---

50 The preparation of sodium sulfide by reacting hydrogen sulfide with caustic soda involves a neutralization reaction.

Now check your equation against that below:

$$2\ NaOH + H_2S \rightarrow Na_2S + 2\ H_2O$$

Jot down the names and formulas of the two sulfides we have discussed so far, and then continue your reading at 30 .

---

51 No. If the active charcoal is omitted, the oxidation of the hydrogen sulfide to sulfur and water will *not* take place. Although it is possible to get hydrogen sulfide to react with oxygen simply by igniting it, such a reaction results in sulfur dioxide and water being formed.

In fact, only with active charcoal as the *catalyst* is it possible to carry out the oxidation of hydrogen sulfide without any conflagration. The products then formed are sulfur and water thus:

$$2\ H_2S + O_2 \rightarrow 2\ S + 2\ H_2O$$

Will the active charcoal be present at the end of the reaction in the same form as at the beginning?

- Yes → 59
- No → 72
- I don't know → 79

52 Now check your answers:

The salts of hydrogen sulfide are known as *sulfides*. In solution these may be tested for by adding solutions of either *lead* or *silver* salts. Precipitates which are either *brown* or *black* in color are formed.

And now let's take a look at the oxides of sulfur.

Jot down the names and formulas of the two oxides of sulfur which we have already discussed, and then continue reading at 62 .

53 Your answer is not right. An explanation of this is given at 41 .

54 Your answer is wrong. Repeat now from 69 .

55 Check your answers now:

Sodium sulfite	Na_2SO_3
Sodium sulfate	Na_2SO_4
Potassium sulfite	K_2SO_3
Potassium sulfate	K_2SO_4
Potassium sulfide	K_2S

Write out completed versions of the sentences below:

The salts of hydrogen sulfide are called ....
The acid radical is $S^{2\ominus}$.

The salts of sulfurous acid are called ....
The acid radical is ....

The salts of sulfuric acid are called ....
The acid radical is ....

Continue your reading at 74 .

12

56 Your answer is wrong. It is not possible that sulfur has the same valence in the anhydrides of both sulfurous and sulfuric acids.

In these anhydrides, the sulfur is combined in the one case with *two* oxygen atoms and in the other case with *three!*

Draw out the structural formulas of these two anhydrides, and then check your answers at 77 .

57 Quite right! Now see if you can write out again the complete equation for the combustion of hydrogen sulfide to sulfur dioxide and water.

Continue your reading at 66 .

58 Your answer is correct.

Now see if you can set up the two equations for the oxidation of H_2S

a) to SO_2, and

b) to S (using active charcoal as the catalyst).

Continue your reading at 86 .

59 The active charcoal has the role of a *catalyst* in the oxidation of hydrogen sulfide to sulfur. Without the presence of this catalyst, hydrogen sulfide would be attacked by the air only at a much higher temperature. This latter oxidation process, which results in a conflagration, yields sulfur dioxide as the product.

The two relevant equation are once again:

With the use of active charcoal at a low temperature:

$$2\ H_2S + O_2 \rightarrow 2\ S + 2\ H_2O$$

Without the use of active charcoal at a high temperature:

$$2\ H_2S + 3\ O_2 \rightarrow 2\ SO_2 + 2\ H_2O$$

Now we know that hydrogen sulfide smells like rotten eggs. But how can we test for it chemically?

Hydrogen sulfide and its salts may be detected in aqueous solution by the addition of a dissolved lead or silver salt. A precipitate of either lead or silver sulfide is formed ranging in color from brown to black. Even the metals lead and silver themselves turn brownish-black in the presence of hydrogen sulfide. This explains why silver objects become dark brown after standing around for some time: it is caused by the traces of hydrogen sulfide which are always present in the air.

Lead sulfide has the formula PbS and silver sulfide has the formula Ag_2S.

From these formulas you should be able to work out the valences of lead and silver.

- Lead is divalent and silver is monovalent → 44
- Both of these metals are monovalent → 71
- Both of these metals are divalent → 89
- I don't know and I don't want to guess → 78

60

Your answer is wrong. Repeat now from 69 .

61

You are quite right!

The numbers of sulfur atoms on the left and right hand sides of the equation were indeed different.

Here once again is the incomplete equation:

$$2\,H_2S + O_2 \rightarrow 2\,H_2O$$

Write down now the complete equation, and then check your answer at 75 .

62

Sulfur dioxide, SO_2, is a colorless and pungent gas. It acts as a disinfectant. Sulfur trioxide, SO_3, is solid at room temperature and boils at 45°C.

Both sulfur dioxide and sulfur trioxide react with water to form acids. Sulfur dioxide dissolves in water to form sulfurous acid:

$$SO_2 + H_2O \rightarrow H_2SO_3$$

Sulfur trioxide dissolves in water to form sulfuric acid:

$$SO_3 + H_2O \rightarrow H_2SO_4$$

These two acids dissociate in water thus:

$$H_2SO_3 \rightarrow 2H^{\oplus} + SO_3^{2\ominus}$$

$$H_2SO_4 \rightarrow 2H^{\oplus} + SO_4^{2\ominus}$$

Oxides which form acids when they combine with water are termed *acid anhydrides* or simply *anhydrides.*

Write out now completed versions of the statements below:

Sulfur dioxide = the anhydride of ....

Sulfur trioxide = the .... of ....

Continue your reading at 84 .

---

63 Your answer is wrong.

The chemical formula for sodium sulfide is Na_2S.

The molecular mass of sodium sulfide may be obtained by adding together the individual atomic masses thus:

Na	= 23
Na	= 23
S	= 32
Na_2S	= ......

What then is the molecular mass of sodium sulfide? Continue reading at 24

---

64 a) The combustion of H_2S at high temperature:

SO_2 and water are formed.

b) The oxidation of H_2S at low temperature using active charcoal as the catalyst:

S and water are formed.

See if you can write out the two relevant equations, and then continue reading at 86.

---

65 You seem to be a little confused. When testing for hydrogen sulfide or for sulfides in general, i.e. for sulfide ions, with lead or silver salts a brownish-black precipitate of lead or silver sulfide is obtained.

On the other hand sulfur dioxide may be recognized from its smell. What sort of smell does it possess?

- A smell like rotten eggs → 76
- A pungent smell → 90

---

66 Now check your equation:

$$2\ H_2S + 3\ O_2 \rightarrow 2\ SO_2 + 2\ H_2O$$

Thus, after the ignition of hydrogen sulfide, the ensuing combustion (oxidation) process produces sulfur dioxide and water.

However, if a mixture of hydrogen sulfide and air is not ignited but passed over active charcoal instead, the latter will act as a catalyst. Without any flames or conflagration an *incomplete* oxidation of the hydrogen sulfide will take place on the charcoal. Note that no sulfur dioxide will be formed but rather *sulfur* itself!

$$H_2S + O_2 \rightarrow S + H_2O$$

Once again, two factors are missing from this equation.
See if you can work out the appropriate factors.

	In front of H_2S	In front of O_2	In front of S	In front of H_2O	
●	2	1	2	2	→ 75
●	2	2	2	2	→ 92
●	2	3	3	2	→ 53
●	3	2	2	3	→ 82
●	I can't work this out				→ 41

---

67 Your answer is not right. An explanation of this is given in 31.

---

68 No, it is not burnt.

We have already said that the active charcoal acts as a *catalyst* for the process of incomplete combustion of hydrogen sulfide.

A catalyst is a substance which speeds up a chemical reaction but which at the end of the reaction is present in the same form as at the beginning.

Write out this definition and commit it to memory. Continue your reading now at 59

---

69 Your answer is right: the sulfur in the anhydride of sulfurous acid has a valence of *four* whereas in the anhydride of sulfuric acid its valence is *six*.

Compare the relevant structural formulas again:

O=S=O SO_2, the anhydride of sulfurous acid

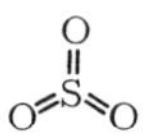
SO_3, the anhydride of sulfuric acid

The salts of sulfurous acid are called *sulfites* whereas the salts of sulfuric acid are called *sulfates*.

What are the acid radicals of sulfurous and sulfuric acids called and what charge does each bear?

Jot down your answers and then go on to 88.

---

70 This equation is still not right!

Take a good look at the number of sulfur atoms on both the right and left hand sides of the equation!

Read through 48 once again.

---

71 Your answer is wrong. An explanation of this is given at 78.

---

72 You are not right here.

We said that active charcoal is a *catalyst* for the process in question.

A catalyst is a substance which speeds up a chemical reaction but which at the end of the reaction is present in the same form as at the beginning.

Write out this definition and commit it to memory. Continue your reading now at 59 .

73 Now check your equations:

$$H_2SO_3 \rightarrow SO_2 + H_2O$$

$$H_2SO_3 \rightarrow 2H^{\oplus} + SO_3^{2\ominus}$$

Most acids have the effect of driving sulfurous acid from its salts, i.e. from the sulfites. As a case in point, we mention the action of hydrochloric acid on sodium sulfite which produces sodium chloride and sulfurous acid. Set up the appropriate equation for this reaction, and then check your answer at 93 .

74 The salts of hydrogen sulfide are called *sulfides*. The acid radical is $S^{2\ominus}$.

The salts of sulfurous acid are called *sulfites*. The acid radical is $SO_3^{2\ominus}$.

The salts of sulfuric acid are called *sulfates*. The acid radical is $SO_4^{2\ominus}$.

What do the formulas Na_2SO_3, K_2SO_4 and Na_2S represent?

- Sodium sulfite, potassium sulfite and sodium sulfide → 60
- Sodium sulfate, potassium sulfate and sodium sulfite → 54
- Sodium sulfite, potassium sulfate and sodium sulfite → 80
- Sodium sulfite, potassium sulfate and sodium sulfide → 85

75 Here is the complete equation:

$$2\ H_2S + 1\ O_2 \rightarrow 2\ S + 2\ H_2O$$

The factor 1 is never written in equations and so the correct version of the equation should read:

$$2\ H_2S + O_2 \rightarrow 2\ S + 2\ H_2O$$

This is the equation for the reaction which occurs when hydrogen sulfide and air are together passed over active charcoal. What part will the active charcoal play in this process?

- The active charcoal will be burnt up → 68
- The active charcoal will act as a catalyst → 59
- The active charcoal could be left out → 51

---

76 You have mixed up the smells of hydrogen sulfide and sulfur dioxide. Write out the following sentence:

Hydrogen sulfide smells like rotten eggs whereas sulfur dioxide has a .... smell.

Continue your reading at 90 .

77

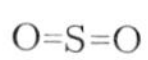

SO_2, the anhydride of sulfurous acid

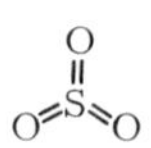

SO_3, the anhydride of sulfuric acid

Compare these two structural formulas, taking especial note of the number of oxygen atoms in each.

By simply counting up the bonds emanating from the sulfur atom in each case, you should now be able to determine the valence of the sulfur in sulfurous acid anhydride and in sulfuric acid anhydride.

Write out the completed version of the sentence below:

In sulfurous acid anhydride sulfur has a valence of .... whereas in sulfuric acid anhydride sulfur has a valence of .... → 69

---

78 Lead sulfide and silver sulfide are salts of hydrogen sulfide. Hydrogen sulfide has the formula H_2S. Hydrogen sulfide dissociates in water according to the equation:

$$H_2S \rightarrow 2\ H^{\oplus} + S^{2\ominus}$$

In lead sulfide, PbS, only one metal ion is bound to the sulfide ion. In silver sulfide, Ag_2S, two metal ions are bound to a single sulfide ion. Accordingly, the two metals must have different valences. In fact, the lead must be ...... valent and the silver ...... valent.

Jot down the valences of the lead and the silver. → 44

---

79 We have already said that the active charcoal acts as a *catalyst* for the process in question.

A catalyst is a substance which speeds up a chemical reaction but which at the end of the reaction is present in the same form as at the beginning.

Write out this definition and commit it to memory.

Continue your reading at 59 .

---

80 Your answer is wrong. Repeat now from 69 .

---

81 Yes, hydrogen sulfide smells like rotten eggs. **Hydrogen sulfide is a very poisonous gas.**

Now let's briefly recapitulate here.

The most important compounds of sulfur are:

Sulfur dioxide
Sulfur trioxide
Sulfurous acid
Sulfuric acid
Sodium sulfide
Sodium sulfite
Sodium sulfate

Copy down these names and write out the chemical formula of each.

Now check your answers at 95 .

---

82 Your answer is not correct. An explanation of this is given in 41 .

83 Check your equations now:

$$Na_2SO_3 + H_2SO_4 \rightarrow Na_2SO_4 + H_2SO_3$$

$$H_2SO_3 \rightarrow H_2O + SO_2$$

$$Na_2SO_3 + H_2SO_4 \rightarrow Na_2SO_4 + H_2O + SO_2$$

This reaction of sulfites may be utilized for their detection. When a sulfite is acidified with HCl or H_2SO_4, gaseous sulfur dioxide is liberated.

How many sulfur dioxide be identified?

- By the black precipitate it forms with lead salts → 65
- By its strong smell of rotten eggs → 76
- By its characteristic pungent smell → 90

84 Now check your answers:

Sulfur dioxide = the anhydride of *sulfurous acid*
Sulfur trioxide = the *anhydride* of *sulfuric acid*

The formula of any acid anhydride can be obtained simply by subtracting water from the formula of the acid in question thus:

H_2SO_3	= sulfurous acid
$- H_2O$	= water
SO_2	= the anhydride of sulfurous acid (sulfur dioxide)

Now carry out the same operation for sulfuric acid, and then check your result at 91 .

85 Your answer is quite right.

Sulfurous acid, H_2SO_3, is a very weak acid. It is unstable and most of it breaks down into its anhydride and water. In aqueous solution sulfurous acid dissociates into ions.

Write down now the equations for the decomposition and for the dissociation of sulfurous acid. Then continue your reading at 73 .

---

86 Now check your equations:

a) $2\ H_2S + 3\ O_2 \rightarrow 2\ SO_2 + 2\ H_2O$

b) $2\ H_2S + O_2 \rightarrow 2\ S + 2\ H_2O$

Sulfur dioxide is also formed during the combustion of sulfur. Further oxidation of the sulfur dioxide to sulfur trioxide becomes feasible only with the aid of special catalysts. The process involved is extremely important in the manufacture of sulfuric acid.

More information on sulfur will be presented in Program 13.

End of Program 12.

---

87 Here are the complete sentences:

Sulfur dioxide and sulfur trioxide are respectively the anhydrides of *sulfurous acid* and *sulfuric acid*. Sulfurous acid is a very *unstable* acid which immediately breaks down into *sulfur dioxide* and water. The sulfur in sulfur dioxide has a valence of *four* whereas in sulfur trioxide its valence is *six*.

There are two possible ways in which hydrogen sulfide can be oxidized, namely:

a) At high temperature without any catalyst; or

b) At low temperature using active charcoal as a catalyst.

What products are formed in reactions a) and b) apart from water?

	In reaction a)	In reaction b)		
•	SO_2	S	→	58
•	S	SO_2	→	94
•	I have simply forgotten		→	64

---

88 The acid radical of sulfurous acid is the *sulfite* ion and this bears two negative charges thus:

$SO_3^{2\ominus}$

The acid radical of sulfuric acid is the *sulfate* ion and this also bears two negative charges:

$SO_4^{2\ominus}$

The salts of sulfurous acid are called sulfites and those of sulfuric acid sulfates.

The salts of hydrogen sulfide are called sulfides.

Particular care should be taken when writing out these salts as it is easy to get them mixed up.

Sulf**ide** Sulf**ite** Sulf**ate**

Copy down the names of the following compounds and give the appropriate chemical formula in each case:

Sodium sulfite, sodium sulfate, potassium sulfite, potassium sulfate, potassium sulfide.

Continue your reading at 55 .

89 Your answer is wrong. An explanation of this is given at 78 .

90 The salts of sulfurous acid, the sulfites, may be detected by acidifying them with HCl or H_2SO_4 and then *smelling* the *characteristically pungent* sulfur dioxide gas which is formed.

The salts of hydrogen sulfide, the sulfides, may be detected in a similar way. If HCl or H_2SO_4 is added to a sulfide, the gas hydrogen sulfide is given off.

What does hydrogen sulfide smell like?

- It is pungent → 76
- Like rotten eggs → 81

91

$$H_2SO_4 \quad = \text{sulfuric acid}$$

$$-\ H_2O \quad = \text{water}$$

$$SO_3 \quad = \text{the anhydride of sulfuric acid (sulfur trioxide)}$$

In the laboratory this process may be forced along simply by heating up concentrated sulfuric acid.

Write down now the structural formulas for the anhydrides of sulfurous and sulfuric acids. You should then be able to determine the valence of the sulfur in the anhydrides of sulfurous and sulfuric acid.

What is the valence of the sulfur in each case?

- In each case it is four → 56
- In each case it is six → 46
- In the anhydride of sulfurous acid it is four whilst in the anhydride of sulfuric acid it is six → 69
- I am unable to write down the relevant structural formulas → 77

92

Your answer is not correct. An explanation of this is given in 41.

93

$$Na_2SO_3 + 2\ HCl \rightarrow 2\ NaCl + H_2SO_3$$

The sulfurous acid formed immediately breaks down into water and sulfur dioxide thus:

$$H_2SO_3 \rightarrow H_2O + SO_2$$

If the two equations are added together we obtain the overall equation:

$$Na_2SO_3 + 2\ HCl \rightarrow 2\ NaCl + H_2O + SO_2$$

We see that, when hydrochloric acid reacts with sodium sulfite, sodium chloride is formed and sulfur dioxide gas is liberated. The latter product may be recognized from its characteristically pungent smell. Other acids are also capable of forcing SO_2 from the salts of sulfurous acid.

Write down now the equation for the reaction of sodium sulfite with sulfuric acid (in terms of two equations which are subsequently added together to yield one overall equation).

Continue your reading at [83] .

---

94 Your answer is wrong.

a) Combustion of H_2S at high temperature:

The products are SO_2 and water.

b) Oxidation of H_2S at low temperature using active charcoal as a catalyst:

12

The products are S and water.

Write out the two relevant equations, and then continue reading at [86] .

---

95 Now check your answers:

Sulfur dioxide	SO_2	Sodium sulfide	Na_2S
Sulfur trioxide	SO_3	Sodium sulfite	Na_2SO_3
Sulfurous acid	H_2SO_3	Sodium sulfate	Na_2SO_4
Sulfuric acid	H_2SO_4		

Write out now completed versions of the sentences below:

Sulfur dioxide and sulfur trioxide are respectively the anhydrides of .... and .... Sulfurous acid is a very .... acid which immediately breaks down into .... and water. The sulfur in sulfur dioxide has a valence of .... whereas in sulfur trioxide its valence is ....

Continue your reading at [87] .

Program 13

Sulfur and its Compounds (II)

---

1 We have already said quite a lot about sulfur and its compounds in the previous Program. It should, therefore, not be too difficult for you to write out the completed version of the text below. See if you are able to insert the correct words.

Sulfur is a ...... -metal. Its color is .... Sulfur is brittle. By pouring molten sulfur into water .... can be produced, although the latter is not stable and slowly changes back into the brittle form of sulfur. Sulfur can be purified by means of ....

The valences of sulfur in its compounds are ......, ... and ...... The most important compound of sulfur with hydrogen is .... having the formula ...... Sulfur forms two oxides, namely .... and .... which have the respective formulas ...... and ......

Continue your reading now at 9 .

---

2 Your answer is wrong. When pyrites is roasted *no* sulfur trioxide is formed but only sulfur dioxide. This was mentioned in the section you have just read. You will need to work a little more carefully!

Read through 48 once again.

---

3 Your answer is wrong. When testing for sulfuric acid or for sulfates with barium chloride one observes the formation of a precipitate!

Read through 59 once again.

---

4 $SO_3 + H_2O \rightarrow H_2SO_4$

This reaction is carried out in *lead* chambers because lead is not attacked by 80% pure sulfuric acid. Iron, for instance, would be totally dissolved away in a short space of time.

The reverse is true for the *Contact Process*. Here sulfur trioxide is introduced into 98% pure sulfuric acid. To ensure that the sulfuric acid remains at 98%, water is carefully added at the same time.

98% pure sulfuric acid does not attack iron. Accordingly, use can be made in this case of iron apparatus, thereby obviating the need to work with expensive lead chambers.

What is the concentration of the sulfuric acid formed in:

a) The Lead Chamber Process ....

b) The Contact Process ....? → 17

---

5 Your answer is not quite right.

Continue your reading now at 29 .

---

6 Now check your equation:

$S + O_2 \rightarrow SO_2$

SO_3 is produced from SO_2 with the aid of catalysts. This will be discussed in greater detail later on.

For the present, we would like to say a little more about the production of SO_2.

Instead of using sulfur itself as the starting material for the manufacture of sulfur dioxide, one can also use sulfur-bearing ores. An ore which is often used for this purpose is pyrites which is plentiful in nature.

What is the actual formula for pyrites?

- FeS → 14
- FeS_2 → 41
- $FeSO_3$ → 24
- $FeSO_4$ → 34
- I don't know → 22

---

7 No. Only 80% sulfuric acid is produced in the Lead Chamber Process.

Repeat now from 13 .

---

8 Your answer is wrong. When pyrites is roasted *no* sulfur trioxide is formed but only sulfur dioxide. This was mentioned in the section you have just read. You will have to work a little more carefully!

Read through 48 once again.

9 Check your answers now:

non-metal
yellow
plastic
distillation
2, 4 and 6
hydrogen sulfide, H_2S
sulfur dioxide, SO_2
sulfur trioxide, SO_3

See if you can also insert the missing words in the sentences below:

SO_2 and SO_3 are respectively the .... of sulfurous and sulfuric acids. The salts of sulfurous acid are called .... To test for these, one adds either sulfuric or hydrochloric acid to their aqueous solution. This results in the formation of ...., which may be recognized by its .... smell. → 18

10 *sulfides*
silver salts
brownish-black

In testing for sulfides, the products formed when the lead or silver salt reacts are respectively:

Lead sulfide	PbS
Silver sulfide	Ag_2S

We shall now continue with our discussion of sulfur and its compounds. The most important sulfur compound is sulfuric acid. This is manufactured in huge quantities in the chemical industry.

What is the chemical formula for the anhydride of sulfuric acid? → 26

11 Your answer is not quite right.

Continue your reading now at 29 .

12 Your answer is wrong.

As the precipitate which forms is both *finely granular and heavy,* it soon settles to the bottom.

Continue your reading now at 43 .

13 Check your equations now:

$$SO_2 + NO_2 \rightarrow SO_3 + NO$$

$$2\ NO + O_2 \rightarrow 2\ NO_2$$

13

The sulfur trioxide obtained in this way then has to be converted into sulfuric acid.

This conversion is brought about in the *Lead Chamber Process* by blowing in steam which produces an approximately 80% solution of sulfuric acid. The oxides of nitrogen escape as gases and may be re-introduced into the lead chambers after being mixed with SO_2 and air.

Write down the equation for the formation of sulfuric acid from sulfur trioxide and water. → 4

14 Your answer is wrong. An explanation of this is given in 22 .

15 Your answer is partially correct.

You should, however, specify in the *formula* the *number* of molecules of water contained in crystalline copper sulfate.

What then should the correct formula be?

- $CuSO_4$ with 5 H_2O → 60
- $CuSO_4 \cdot 5\ H_2O$ → 54

16 Now check your answer:

The solution of sulfur trioxide in sulfuric acid which is produced by the Contact Process is known as *fuming sulfuric acid* or *oleum.*

Concentrated sulfuric acid is a colorless, rather viscous and strongly hygroscopic (water attracting) liquid. **It is very corrosive. Safety goggles must always be worn when working with sulfuric acid.**

The property of the acid to attract and combine with water can be used for the purpose of drying substances. For instance, certain gases may be dried by bubbling them through concentrated sulfuric acid.

Here is one typical example:
The water vapor contained in the air (humidity) can be removed in this way.

When concentrated sulfuric acid and water are mixed together, a great deal of heat is generated. For this reason special precautions have to be taken.

Water must never be added to concentrated sulfuric acid. If this were done, the amount of heat liberated would be so great that some of the water would instantly evaporate causing acid to be sprayed in all directions.

Using key words, make a list of the most important properties of concentrated sulfuric acid. → 23

17

a) In the Lead Chamber Process around 80%.
b) In the Contact Process around 98%.

If, during the Contact Process, SO_3 is introduced into 98% sulfuric acid, *without any water* being *added* at the same time, 100% sulfuric acid will be formed. Further amounts of SO_3 would simply dissolve in the 100% acid. This acid forms a white cloud when exposed to the air and for this reason has been called *fuming sulfuric acid.* In the area of organic chemistry, fuming sulfuric acid is necessary for many of the syntheses carried out.

Which process is used for the production of fuming sulfuric acid?

- The Contact Process → 33
- The Lead Chamber Process → 7

---

18 SO_2 and SO_3 are respectively the *anhydrides* of sulfurous and sulfuric acids. The salts of sulfurous acid are called *sulfites.* To test for a sulfite, one adds either sulfuric or hydrochloric acid to their aqueous solution. This results in the formation of *sulfur dioxide* gas which may be recognized by its *characteristically pungent* smell.

Write down now the equation for the reaction of sodium sulfite with sulfuric acid. → 35

---

19 Your answer is wrong. On the left hand side of the equation

$$4\ FeS_2 + 11\ O_2 \rightarrow 2\ Fe_2O_3 +$$

there are 4 x 2 = 8 sulfur atoms. Remember that in any equation the same number of atoms of each type must appear on both the right and left hand sides.

Re-consider your answer, and then go on to 48 .

13

20 Your answer is quite right.

Do not forget the jingle:

First the water, then the acid
Otherwise things won't be placid!

End of Program 13.

---

21 Your answer is wrong. Sulfides are the salts of hydrogen sulfide!

Note the following carefully:

Hydrogen sulfide, H_2S; the salts are called sulfides.
Sulfurous acid, H_2SO_3; the salts are called sulfites.
Sulfuric acid, H_2SO_4; the salts are called ....

Jot down the missing word, and then continue reading at 40 .

---

22 Pyrites is a sulfide of iron containing two atoms of sulfur thus:

FeS_2

$FeSO_3$ is a salt of sulfurous acid (H_2SO_3).
$FeSO_4$ is a salt of sulfuric acid (H_2SO_4).
FeS is iron sulfide.

What are the correct chemical designations for the four compounds FeS, FeS_2, $FeSO_3$ and $FeSO_4$ in the sequence given?

- Iron sulfide, pyrites, iron sulfate, iron sulfite → 11
- Iron sulfite, pyrites, iron sulfate, iron sulfite → 50
- Iron sulfide, pyrites, iron sulfite, iron sulfate → 41
- Iron sulfite, pyrites, iron sulfate, iron sulfide → 5

23

Your answer should encompass the following points:

Concentrated sulfuric acid is a colorless, viscous and strongly hygroscopic liquid. **It is very corrosive. The great amount of heat generated when the acid is mixed with water means that special precautions must be taken when performing this operation.**

13

To prepare dilute sulfuric acid from concentrated sulfuric acid in the laboratory, one very cautiously pours the required amount of concentrated sulfuric acid into water which should be constantly stirred at the same time.

Never attempt to add water to the acid! And always make sure that protective goggles are worn when performing this operation!

In order to help you remember the correct sequence of events when preparing dilute sulfuric acid, we offer below a rhyming mnemonic:

First the water, then the acid
Otherwise things won't be placid!

(The sequence recommended here arises from the fact that the acid is denser than water and will thus sink to the bottom in water – this ensures that better mixing takes place.)

Write out now completed versions of the following sentences:

Sulfuric acid attracts .... Accordingly, it may be used for the purpose of .... substances. When concentrated sulfuric acid is to be diluted, one starts with the appropriate amount of .... and carefully pours the .... into it.

Continue your reading at 32 .

24 Your answer is wrong.

An explanation of this is given at 22 .

25 Your answer is not correct.

The decomposition of the SO_3 is prevented not by the catalyst but rather by the low temperature of the reaction.

Repeat now from 44 .

26 The anhydride of sulfuric acid has the chemical formula SO_3. It cannot be obtained directly from sulfur by any combustion process. When sulfur burns it is oxidized only to sulfur dioxide.

13

Write out now the equation for the combustion of sulfur to sulfur dioxide.

→ 6

27 Now check your answers:

In the Lead Chamber Process:

The starting material is *sulfur dioxide.*
The oxygen donor is *nitrogen dioxide.*

The end product is *sulfur trioxide.*

Write down the two equations which give a simplified representation of the chemical processes occurring in the lead chambers.

$$SO_2 + NO_2 \rightarrow \ldots\ldots + \ldots\ldots$$

$$NO + \ldots\ldots \rightarrow \ldots\ldots$$

Check your equations at 13 .

28 Check your equation now:

$$CuSO_4 + 5\ H_2O \rightarrow CuSO_4 \cdot 5\ H_2O$$

white blue

One important naturally occurring sulfate is known as *gypsum*. Chemically, gypsum consists of calcium sulfate which contains two moles of water of crystallization per mole of calcium sulfate.

Write down the chemical formula of gypsum, and then continue reading at [38] .

---

[29] You should memorize the following:

Hydrogen sulfide, H_2S;	its salts are called sulfides.
Sulfurous acid, H_2SO_3;	its salts are called sulfites.
Sulfuric acid, H_2SO_4;	its salts are called sulfates.

Read through [22] once again.

---

[30] Yes, the complete equation for the roasting of pyrites is as follows:

$$4\ FeS_2 + 11\ O_2 \rightarrow 2\ Fe_2O_3 + 8\ SO_2$$

In addition to sulfur dioxide an oxide of iron is also formed.

In order to prepare sulfuric acid it is necessary to oxidize SO_2 to SO_3.

There are two industrial processes for doing this, namely:

1. *The Contact Process*

2. *The Lead Chamber Process*

Write out the equation for the oxidation of SO_2 to SO_3 with oxygen. → [44]

---

[31] Your answer is wrong. The formula for copper sulfate containing *no* water of crystallization is $CuSO_4$. But you should have written down the formula for copper sulfate containing 5 moles of water of crystallization per mole of $CuSO_4$.

What is the correct formula for the latter copper sulfate?

- $CuSO_4$ with 5 H_2O → [60]
- $CuSO_4 \cdot 5\ H_2O$ → [54]
- $CuSO_4$ with water of crystallization → [15]

13

32 Now check your answers:

Sulfuric acid attracts *water*. Accordingly, it may be used for the purpose of *drying* substances. When concentrated sulfuric acid is to be diluted, one starts with the appropriate amount of *water* and carefully pours the *acid* into it.

The industrial and economic importance of sulfuric acid can hardly be over-estimated. It is a key product in the manufacture of a wide variety of chemicals.

What are the salts of sulfuric acid called?

- Sulfides → 21
- Sulfites → 56
- Sulfates → 40
- I don't know → 49

33 Yes, fuming sulfuric acid is manufactured by means of the Contact Process. It is also sometimes referred to as *oleum.*

If 100% sulfuric acid has dissolved in it (say) 15% by mass of SO_3, one speaks of 15% oleum.

Now write out the completed version of the sentence below:

The solution of sulfur trioxide in sulfuric acid produced in the Contact Process is known as .... sulfuric acid or ....

Continue your reading at 16 .

34 Your answer is wrong.

An explanation of this is given at 22 .

35 $Na_2SO_3 + H_2SO_4 \rightarrow Na_2SO_4 + H_2O + SO_2$

Write out the completed version of the sentences below:

The salts of hydrogen sulfide are called .... They may be tested for by making use of lead or .... salts. The color of the precipitate they form is ....

Continue your reading at 10 .

36 The missing entries are:

98%
fuming sulfuric acid = oleum

Jot down again now the missing words in the sentence:

Sulfur dioxide may be obtained from the combustion of .... or by the roasting of .... → 59

37 Your answer is wrong. When pyrites is roasted, the product formed is *not* sulfur trioxide but rather sulfur dioxide. This was discussed in the section you have just read. You should read the sections more carefully.

Go back and read through 48 once again.

38 Now check your answer:

Gypsum $CaSO_4 \cdot 2\,H_2O$

When gypsum is heated some of its water of crystallization is driven off. The gypsum is said to burn. If burnt gypsum is mixed with water, it again quickly takes up the water of crystallization which it has lost. In the process it becomes very hard.

Write out now completed versions of the sentences below:

Gypsum has the chemical formula .... When it is heated gypsum loses some of its ....

Continue your reading at 63 .

39 Your answer is not correct.

Oxidation of SO_2 to SO_3 is quite possible without any catalysts, though, such high temperatures have to be used to achieve this, that the SO_3 formed breaks down again to some extent. The oxidation of SO_2 to SO_3 without the aid of catalysts is thus uneconomic.

Repeat now from [44] .

40 The salts of sulfuric acid are called sulfates.

An important example of a sulfate is sodium sulfate, Na_2SO_4. When sodium sulfate crystallizes out from an aqueous solution, one mole of Na_2SO_4 crystallizes together with 10 moles of water. The water contained in crystals of this kind is referred to as *water of crystallization*. Crystalline sodium sulfate is represented by the chemical formula

13

$$Na_2SO_4 \cdot 10\ H_2O.$$

This formula is read as: sodium sulfate *with* ten molecules of water of crystallization.

Another very important sulfate is copper sulfate. This crystallizes out in the form of beautiful blue crystals. The crystals contain water of crystallization in addition to copper sulfate. For every one mole of copper sulfate there are five moles of water of crystallization.

What then will be the chemical formula for crystalline copper sulfate?

- $CuSO_4$ → [31]
- $CuSO_4$ with 5 H_2O → [60]
- $CuSO_4 \cdot 5\ H_2O$ → [54]
- $CuSO_4$ with water of crystallization → [15]

41 Your answer is quite right.

Pyrites is a sulfide of iron containing two atoms of sulfur:

$$FeS_2$$

When heated in the presence of air, pyrites ignites and an exothermic reaction takes place in which it is oxidized. In addition to iron oxide, ... is also formed.

See if you can work out what must be formed. → 48

---

42 $H_2SO_4 + BaCl_2 \rightarrow BaSO_4 + 2\ HCl$

Barium sulfate, $BaSO_4$, settles out as a finely granular white precipitate.

Barium sulfate is also formed by the reaction of barium chloride with potassium sulfate thus:

$$K_2SO_4 + BaCl_2 \rightarrow BaSO_4 + 2\ KCl$$

It is clear that the composition of the precipitate is the same in both reactions. This is because sulfate ions are present in both an aqueous solution of sulfuric acid and in a solution of potassium sulfate. These ions can react with barium ions, $Ba^{2\oplus}$, to form highly insoluble barium sulfate as follows:

$$Ba^{2\oplus} + SO_4^{2\ominus} \rightarrow BaSO_4$$

The appearance of this finely granular, heavy, white precipitate of barium sulfate when barium chloride solution is added, indicates that sulfate ions must be present.

Let's now briefly recapitulate here.

Jot down the missing words in the sentences below:

The oxidation (combustion) of sulfur leads to the production of ....
The process for the manufacture of sulfuric acid is known as the .... → 52

---

43 Your answer is quite right.

When testing for the presence of sulfuric acid and sulfates with barium chloride, one observes the appearance of a finely granular precipitate of barium sulfate. The latter soon settles to the bottom as it is very dense.

$$H_2SO_4 + \ldots\ldots \rightarrow 2\ HCl + BaSO_4$$

$$Na_2SO_4 + BaCl_2 \rightarrow 2\ NaCl + \ldots\ldots$$

Write out the full equation in each case. → 51

44 $2\ SO_2 + O_2 \rightarrow 2\ SO_3$

This reaction requires very high temperatures (over 600 °C) in order for it to proceed at a reasonable speed in industrial processes.

However, at such high temperatures, some of the sulfur trioxide which is formed breaks down again into sulfur dioxide and oxygen.

Write down now the equation for this decomposition of sulfur trioxide.

→ 55

45 The sulfate ion bears two negative charges: $SO_4^{2\ominus}$

Write down the formulas for sodium and calcium sulfates, and then continue your reading at 53 .

13

46 Your answer is quite right.

Now let's move on to the *Lead Chamber Process*. It was the first process in producing sulfuric acid industrially in big amounts. Although the process has lost its importance, it remains interesting because of its chemical reactions.

SO_2, together with air and nitrogen oxides, is pumped into large lead chambers where the reaction takes place. Sulfur trioxide is formed according to the reaction:

$$SO_2 + NO_2 \rightarrow SO_3 + NO$$

Thus, in the process SO_2 is *oxidized* to SO_3 and NO_2 gives up one of its oxygen atoms and is thereby *reduced* to NO.

NO is able to take up oxygen from the air again:

$$2\ NO + O_2 \rightarrow 2\ NO_2$$

In this way, the NO_2 is regenerated and is then available again to oxidize more SO_2. The nitrogen dioxide, NO_2, thus acts as an *oxygen donor.*

Write out the statements below, filling in the gaps where necessary.

The starting material is ....
The oxygen donor is ....

The end product is .... → 27

---

47 *The decomposition of sulfur trioxide into sulfur dioxide and oxygen.*

As sulfur trioxide breaks down to some extent at the high temperatures which are necessary for its manufacture, it is *not* feasible to produce sulfur trioxide economically without the use of catalysts.

With catalysts, however, the SO_3 can be produced at relatively low temperatures (400 – 500 °C) and at a reasonable speed. At such temperatures very little of the SO_3 decomposes.

Let's now take a closer look at the *Contact Process.*

A mixture of sulfur dioxide and air at 400 – 500 °C is passed over *solid* catalysts, known as *contacts*. These bring about the oxidation of the sulfur dioxide to sulfur trioxide. The contacts usually consist of finely-divided platinum or vanadium oxide.

Copy down completed versions of the sentences below:

The Contact Process

The starting material is ....
The catalyst is either .... or ....
Where does the name of the process come from? ....
The end product is .... → 58

---

48 *Sulfur dioxide* and iron oxide are formed. The sulfur dioxide comes off as a gas which can then be used for the synthesis of sulfur trioxide. The iron oxide is left behind.

As this oxidation of pyrites is carried out in large *roasting* ovens, the pyrites is said to be *roasted.*

The incomplete equation for the roasting of pyrites is:

$$4\ FeS_2 + 11\ O_2 \rightarrow 2\ Fe_2O_3 + \ldots\ldots\ldots\ldots$$

What is missing from this equation? Think it out carefully!

- $4\ SO_2$ → 57

- $8\ SO_2$ → 30
- $16\ SO_2$ → 19
- $4\ SO_3$ → 2
- $8\ SO_3$ → 8
- $16\ SO_3$ → 37

---

49 You should memorize the following:

Hydrogen sulfide, H_2S; its salts are called sulfides.
Sulfurous acid, H_2SO_3; its salts are called sulfites.
Sulfuric acid, H_2SO_4; its salts are called ....

Jot down the missing word, and then continue reading at 40 .

---

50 Your answer is not quite right.

Continue your reading at 29 .

---

51 $H_2SO_4 + \mathbf{BaCl_2} \rightarrow 2\ HCl + BaSO_4$

$Na_2SO_4 + BaCl_2 \rightarrow 2\ NaCl + \mathbf{BaSO_4}$

Now write down the three special **dangers** associated with the use of sulfur compounds.

1. **Hydrogen sulfide is very poisonous.**

2. **Sulfuric acid is very corrosive.**

3. **Acid can easily be sprayed about if insufficient care is taken when diluting concentrated sulfuric acid.**

What substance should one start with when diluting sulfuric acid with water?

- The water → 20
- The sulfuric acid → 61

52 The missing words are:

sulfur dioxide

Contact Process

Jot down the missing words in the sentences below:

The Contact Process produces ......% pure sulfuric acid. In the Contact Process a solution of SO_3 in sulfuric acid can also be produced. This solution is known as .... or .... → 36

53 Now check your answers:

Na_2SO_4	$CaSO_4$
sodium sulfate	calcium sulfate

To test for sulfuric acid or for a sulfate, one dissolves the unknown substance in water, acidifies it with hydrochloric acid, and then adds barium chloride solution. If either sulfuric acid or a sulfate is present, an insoluble, finely-grained, white precipitate of barium sulfate will be formed.

Write out now the equation for the reaction of sulfuric acid with barium chloride ($BaCl_2$). Check your answer at 42 .

54 Yes, the formula for crystalline copper sulfate is

$CuSO_4 \cdot 5\ H_2O$

When crystalline copper sulfate is heated, it loses its water of crystallization and changes into anhydrous copper sulfate. The latter material is white in color.

Anhydrous white copper sulfate can be used to test for the presence of water. As soon as it comes into contact with water it turns blue again. Write down the equation for this reaction and indicate the color of each type of copper sulfate involved.

$CuSO_4 + 5\ H_2O \rightarrow$ ....

Continue your reading at 28 .

55

$$2\ SO_3 \rightarrow 2\ SO_2 + O_2$$

This equation is clearly the *reverse* of the equation

$$2\ SO_2 + O_2 \rightarrow 2\ SO_3$$

Reactions like this which can proceed either in a forward or reverse sense are indicated by means of a double arrow ($\rightleftarrows$) thus:

$$2\ SO_2 + O_2 \rightleftarrows 2\ SO_3$$

What is the reaction taking place when we proceed from left to right?
Write out the completed version of the following sentence:

The reaction represents the production of .... from .... and .... → 62

56 Your answer is wrong. Sulfites are the salts of sulfurous acid!

You should memorize the following:

Hydrogen sulfide, H_2S; its salts are called sulfides.
Sulfurous acid, H_2SO_3; its salts are called sulfites.
Sulfuric acid, H_2SO_4; its salts are called ....

Jot down the missing word, and then continue reading at 40 .

57 Your answer is wrong. On the left hand side of the equation

$$4\ FeS_2 + 11\ O_2 \rightarrow 2\ Fe_2O_3 + \ldots.$$

there are in all 4 x 2 = 8 sulfur atoms.

In any equation, the same number of atoms of each type must appear on both the left and right hand sides.

You should now work through this example again. → 48

58 Now check your answers:

In the Contact Process:

The starting material is *sulfur dioxide.*
The catalyst is either *platinum* or *vanadium oxide.*
Where does the name of the process come from? From the solid catalysts used which are known as *contacts.*
The end product is *sulfur trioxide.*

What is the function of the catalysts in the Contact Process?

- They prevent the SO_3 from decomposing → 25
- Without them oxidation of the SO_2 to SO_3 would be impossible → 39
- They make possible the oxidation of the SO_2 to SO_3 at a temperature low enough for the SO_3 to be formed fairly fast but low enough that it is not significantly decomposed again → 46

59 The missing words are:

sulfur, pyrites

To test for sulfuric acid and sulfates, use may be made of barium chloride solution. What does one then observe?

- The appearance of a flaky white precipitate → 12
- The appearance of a fine granular white precipitate which soon settles out → 43
- A change of color → 3

60 Your answer is partially correct.

However, instead of writing the word *"with"* you should have used a point.

Now see if you can write out the correct formula for crystalline copper sulfate. → 54 .

61 Your answer is wrong.

When concentrated sulfuric acid is to be diluted with water, one must start with the water and allow the sulfuric acid to mix into it in a fine stream. Only in this way will the great amount of heat generated be harmlessly dissipated.

Write out now the little couplet we gave you earlier:

First the ...., then the ....
Otherwise things won't be placid!

Continue your reading at 20 .

---

62 The reaction is the production of *sulfur trioxide* from *sulfur dioxide* and *oxygen*.

What will the equation

$$2\ SO_2 + O_2 \rightleftharpoons SO_3$$

represent when it proceeds from right to left?

Write out your answer in full. → 47

---

63 Gypsum has the formula $\mathbf{CaSO_4 \cdot 2\ H_2O}$. When it is burnt the gypsum loses some of its *water of crystallization*.

Do you still remember how many charges there are on the sulfate ion? Write down the formula for the sulfate ion. → 45

Program 14

Theoretical Foundations (VI)

1 In this Program we shall be discussing the subject of titration and a number of related concepts. Titration is an analytical method used to determine accurately the amount of acid or base present in a solution. To understand what follows, it is essential that you are fully conversant with the content of the previous Programs on the theoretical foundations of chemistry. In this regard Program 11 is of especial importance.

Jot down now one example each of a monobasic, a dibasic and a tribasic acid, including both the name and the chemical formula.

→ 26

---

2 Your answer is wrong.

The correct approach to this calculation is outlined in 14.

---

3 Your answer is wrong.

4 g of phosphoric acid and 196 g of water will give 200 g of dilute phosphoric acid when mixed together.

200 g of dilute phosphoric acid contain 4 g of phosphoric acid
100 g of dilute phosphoric acid contain ? g of phosphoric acid

Thus, 100 g of dilute phosphoric acid contain

$$\frac{4 \times 100\ \text{g}}{200} = 2\ \text{g of phosphoric acid}$$

Accordingly, if 100 g of dilute phosphoric acid contain 2 g of phosphoric acid, the solution will be ...... %. → 56

---

4 You have not yet properly understood this method of calculating the equivalent mass. Read through 37 very carefully again.

Do not forget here that phosphoric acid is a *tri*basic acid!

5 *equivalent*

For a sodium hydroxide solution containing 2 gram-equivalents of NaOH per liter, one uses the designation .... → 20

6 You appear to be confused about the basicities of the two acids in question.

Nitric acid, HNO_3, is monobasic and its basicity is **1.**

Carbonic acid, H_2CO_3, is dibasic and its basicity is **2.**

Go back now to 24 .

7 How many grams of NaOH are there in 100 mL of 1 N sodium hydroxide solution?

1 L of 1 N sodium hydroxide solution contains 1 gram-equivalent = 40 g of NaOH.

100 mL represents one tenth of a liter. Thus, 100 mL of 1 N caustic soda contain ......g of NaOH. → 68

8 Your answer is wrong.

Although sulfuric and phosphoric acids happen to have the same molecular mass, they do not have the same equivalent mass. Sulfuric acid is a *di*basic acid whereas phosphoric acid is a *tri*basic acid.

To reach the right answer, you should read through 21 once again.

9 Your answer is wrong.

In the case of nitric acid, the equivalent mass is equal to the molecular mass.

Nitric acid, HNO_3, is a *monobasic acid.* It dissociates in water to yield the acid radical ($NO_3^{\ominus}$) and *one* hydrogen ion ($H^{\oplus}$). Its basicity is thus *one.*

$$\text{The equivalent mass of } HNO_3 = \frac{\text{Molecular mass of } HNO_3}{\text{Basicity}} = \frac{63}{\ldots\ldots} = \ldots\ldots$$

The equivalent mass of nitric acid thus equals......

Continue your reading at [37] .

---

[10] Your answer is wrong. Please note the following fact:

2 gram-equivalents of NaOH in 1 L of solution means that 1 L of this caustic soda solution will contain 2 x 40 g = 80 g of NaOH.

The concentration of the solution in grams per liter is thus known to us.

2 gram-equivalents NaOH/liter = 80 g of NaOH/liter

You should have been able to work this out for yourself.

It would now be advisable for you to start at the beginning of this Program again. → [71]

---

[11] The missing words are:

buret

indicator

End of Program 14.

---

[12] You should have filled out the flow chart as shown below.

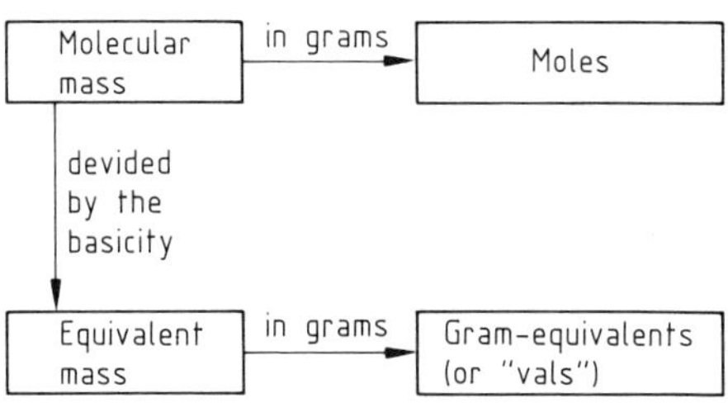

In order to make clear the advantage of using the concept of "equivalent mass", we shall now discuss in some detail the neutralization of hydrochloric and sulfuric acids.

For the neutralization of 1 mol of hydrochloric acid we require 1 mol of sodium hydroxide. This is evident from the equation for this reaction:

$$HCl + NaOH \rightarrow NaCl + H_2O$$

We can say that: 1 mol of HCl is chemically equal to or *equivalent* to 1 mol of NaOH.

The same does not apply, however, for the neutralization of sulfuric acid with caustic soda. In this case we shall need 2 mol of sodium hydroxide for the complete neutralization of 1 mol of sulfuric acid.

$$H_2SO_4 + 2\ NaOH = Na_2SO_4 + 2\ H_2O$$

We see now that 1 mol of sulfuric acid is chemically equal to or *equivalent* to 2 mol of sodium hydroxide. Alternatively, this may be expressed by saying that 1/2 mol of sulfuric acid is *equivalent* to 1 mol of sodium hydroxide.

Starting with 1 mol (= 1 gram-equivalent) of sodium hydroxide, we would thus be able to neutralize:

1 mol of HCl = 1 gram-equivalent of HCl

1/2 mol of H_2SO_4 = 1 gram-equivalent of H_2SO_4

1 *gram-equivalent of an acid is equivalent to* **1** *gram-equivalent of a base.*

Bearing this discussion in mind, it should now be possible for you to solve the problem below without using the equation or calculating the molecular mass.

How many grams of nitric acid (HNO_3) can be neutralized by 2 mol of sodium hydroxide?

The relevant atomic masses are: H = 1; N = 14; O = 16.

- 31.5 g of nitric acid → 19
- 63 g of nitric acid → 28
- 126 g of nitric acid → 36
- I need some help → 46

---

13 Your answer is wrong. To make up a solution having a given concentration – expressed in grams per liter – a fixed amount of substance is weighed out and then dissolved in the appropriate solvent. Care must be taken to ensure that the amount of solvent used is precisely that required to make up one liter of solution. It thus becomes necessary to measure a *volume*. •

Take another look at the concentrations given and then select another answer at 36 .

14 When mixed, 20 g of sulfuric acid and 980 g of water yield 1000 g of sulfuric acid solution.

1000 g of sulfuric acid solution contain 20 g of sulfuric acid
100 g of sulfuric acid solution contain ? g of sulfuric acid.

100 g of sulfuric acid solution thus contain

$$\frac{20 \times 100\text{ g}}{1000} = 2\text{ g of sulfuric acid}$$

When 100 g of sulfuric acid solution contain 2 g of sulfuric acid, the solution is said to be a 2% one.

Now use the same method to calculate:

The concentration (expressed as a percentage) of a dilute solution of phosphoric acid containing 4 g of phosphoric acid mixed with 196 g of water.

- The phosphoric acid concentration is 19.6% → 3
- The phosphoric acid concentration is 4% → 30
- The phosphoric acid concentration is 2% → 56
- I obtained a different concentration → 80

15 Your answer is wrong. To obtain the equivalent mass of phosphoric acid, one has to *divide* the molecular mass of the acid by its basicity. It has a basicity of three.

Read through 37 once again.

16 Now check your answers:

a) 1 L of 1 N H_2SO_4 is equivalent to 2 L of 0.5 N NaOH
b) 0.5 L of 1 N HCl is equivalent to 0.5 L of 1 N NaOH
c) 1 L of 0.1 N H_2SO_4 is equivalent to 0.1 L of 1 N NaOH

Normal solutions are employed in analytical chemistry for the purpose of carrying out titrations. Titration is an analytic process in which the precise amount of an acid, base or other substance present in a solution is determined. Suppose that we wish to know the concentration of an unknown caustic soda solution as an example. We would then proceed as follows:

100 mL of the unknown solution of dilute caustic soda are accurately measured out and some methyl orange indicator is added. A solution of 1 N hydrochloric acid is then run into the flask containing the caustic soda from a buret. The flask is continually shaken throughout the titration, until the indicator changes color from yellow to orange-red. At this point, one has *neutralized* the solution with the 1 N hydrochloric acid. One can then read off from the buret the exact amount of 1 N hydrochloric acid used. (Note that in this context we use mL (milliliters) as the unit of volume and that 1000 mL = 1 L.)

The amount (in mL) of the 1 N hydrochloric acid used in the titration will be equivalent to the unknown amount of sodium hydroxide.

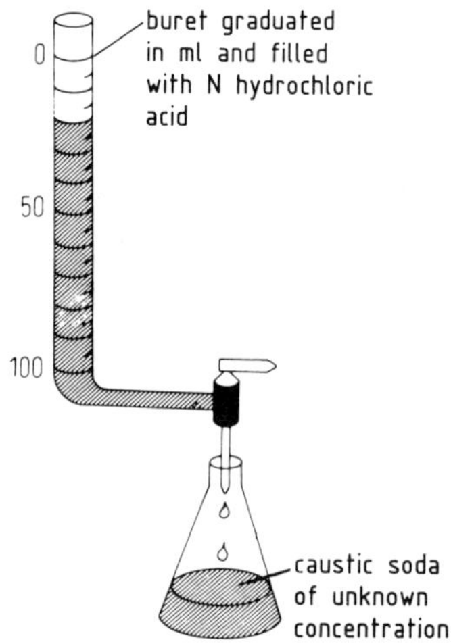

14

If, for example, we require here 100 mL of 1 N hydrochloric acid, we could conclude that the 100 mL of the unknown caustic soda solution would also be 1 N. It would then be easy to calculate the number of grams of NaOH present in the 100 mL of the unknown solution.

See if you can calculate the number of grams of NaOH present in 100 mL of 1 N caustic soda solution. The relevant atomic masses are: Na = 23; O = 16; H = 1.

- It contains 40 g of NaOH → 54
- It contains 4 g of NaOH → 68
- It contains 0.4 g of NaOH → 39
- I cannot calculate it → 7

---

17 Your answer is wrong.

In the case of nitric acid the equivalent mass is equal to the molecular mass.

Nitric acid, HNO_3, is a *monobasic* acid. It dissociates in water to yield the acid radical ($NO_3^{\ominus}$) and *one* hydrogen ion ($H^{\oplus}$). Its basicity is thus *one.*

$$\text{The equivalent mass of } HNO_3 = \frac{\text{Molecular mass of } HNO_3}{\text{Basicity}} = \frac{63}{\ldots\ldots} = \ldots\ldots$$

What is therefore the equivalent mass of nitric acid? → 37

---

18 The correct answer is:

Gram-equivalent

A 1 N standard solution contains 1 gram-equivalent of a substance dissolved in 1 L of solution.

Equal volumes of 1 N solutions are always chemically equal or .... to one another.

Jot down the missing word. → 5

14

---

19 Your answer is wrong. Continue reading now at 46 .

---

20 *2 N caustic soda*

The instrument from which standard solutions flow during a titration is called a ....

Before performing the titration, it is necessary to add an .... to the solution to be titrated so that the end point of the reaction can be determined.

Jot down the two missing words and then check them at 11 .

---

21 Your answer is quite right.

The concentration of the solution is 200 g/liter.

A solution having a standard *normality* is prepared by dissolving a fixed, weighed amount of the substance in question in the solvent so that the final volume of the solution is exactly 1 L. The substance weighed out is specified usually in terms of the number of gram-equivalents it contains.

For instance, if 1 gram-equivalent of a substance is made up to 1 L of solution, we may say that:

The solution is 1 normal

With 2 gram-equivalents/liter: the solution is 2 normal
With 3 gram-equivalents/liter: the solution is 3 normal
With 0.5 gram-equivalents/liter: the solution is 0.5 normal

What amounts of acid would be required in each case to make up 1 L of 1 normal hydrochloric acid, 1 L of 1 normal sulfuric acid and 1 L of 1 normal phosphoric acid?

The relevant molecular masses are: HCl = 36.5; H_2SO_4 = 98; H_3PO_4 = 98.

	HC1	H_2SO_4	H_3PO_4	
•	36.5 g	98 g	98 g	→ 72
•	36.5 g	49 g	32,7 g	→ 53
•	36.5 g	49 g	49 g	→ 8

22 Your answer is wrong.

An explanation of this is given at 59 .

23 You should be more careful in your work. Now read through the explanations given at 73 very carefully.

24 Your answer is quite right.

The equivalent mass of phosphoric acid is obtained by dividing the molecular mass of the acid by 3:

$$\frac{98}{3} = 32.7$$

Now calculate the equivalent masses of nitric acid (HNO_3) and carbonic acid (H_2CO_3).

The relevant atomic masses are: H = 1; N = 14; C = 12; O = 16.

The equivalent masses are:

	HNO_3	H_2CO_3		
●	63	62	→	40
●	31.5	62	→	6
●	63	31	→	52
●	31.5	31	→	64

25 Your answer is quite right.

When a solution having a fixed concentration, expressed as a *mass percentage,* is prepared, it is not necessary to measure out any volumes. Instead, a fixed, weighed amount of the substance in question is dissolved in the solvent. In the simplest case, the amount of solvent taken would be just enough to give a final solution weighing 100 g.

Suppose that 20 g of sulfuric acid are mixed with 80 g of water. The resulting solution would then be 100 g of 20% sulfuric acid.

If instead we mix 20 g of sulfuric acid with 980 g of water, what will then be the mass percentage of the solution obtained?

My calculation indicates that the resulting sulfuric acid solution will be:

- 0.1% → 38
- 0.2% → 47
- 1.0% → 65
- 2.0% → 56
- 10% → 74
- 20% → 83
- 98% → 2
- My answer is different → 14

26 Now check your examples:

Monobasic acids:		Molecular mass
Nitric acid	HNO_3	63
Hydrochloric acid	HCl	36.5
Dibasic acids:		
Sulfuric acid	H_2SO_4	98
Carbonic acid	H_2CO_3	62
Tribasic acid:		
Phosphoric acid	H_3PO_4	98

We have now discussed acids whose radicals bear *differing* negative charges. In addition, we have also covered the concepts of molecular mass and the mole (the molecular mass in grams).

At this point we shall introduce a new concept, namely the *equivalent mass* of a substance.

> *The equivalent mass of an atom or group of atoms is obtained by dividing its atomic or molecular mass by its effective valence.*

To make this concept clearer, we shall now consider the equivalent masses of a number of acids and bases when these react together in neutralization reactions.

HCl is a monobasic acid; it dissociates in water into an acid radical ($Cl^{\ominus}$) and *one* hydrogen ion ($H^{\oplus}$). Its *basicity* or *effective valence* is one.

From the above definition it follows that:

$$\text{The equivalent mass of HCl} = \frac{\text{Molecular mass of HCl}}{\text{Basicity}} = \frac{36.5}{1} = 36.5$$

Note that, in the present context, the effective valence is taken to be equal to the basicity for both acids and bases.

See if you can solve the following problem using an approach similar to that above.

What is the equivalent mass of nitric acid, HNO_3? Its molecular mass is 63.

- The equivalent mass is 31.5 → 9

- The equivalent mass is 126 → 17
- The equivalent mass is 63 → 37

---

27 To neutralize one liter of a 1 N hydrochloric acid solution *or* one liter of a 1 N sulfuric acid solution *or* one liter of a 1 N phosphoric acid solution, we shall need *in each case* one liter of a 1 N solution of sodium hydroxide.

The reason for this will become clearer if we write out the equations for these three neutralizations in a somewhat unusual manner as follows:

I. $HCl + NaOH \rightarrow NaCl + H_2O$

II. $1/2\ H_2SO_4 + NaOH \rightarrow 1/2\ Na_2SO_4 + H_2O$

III. $1/3\ H_3PO_4 + NaOH \rightarrow 1/3\ Na_3PO_4 + H_2O$

In this particular form, the equations reveal exactly what we want to demonstrate. We see now that:

In equation I, one gram-equivalent of hydrochloric acid is neutralized by 1 gram-equivalent of sodium hydroxide.

In equation II, one gram-equivalent of sulfuric acid (1/2 a mole) is neutralized by 1 gram-equivalent of sodium hydroxide.

In equation III, one gram-equivalent of phosphoric acid (1/3 of a mole) is neutralized by 1 gram-equivalent of sodium hydroxide.

Thus, the amounts of acid indicated – 1 mole of hydrochloric acid, 1/2 a mole of sulfuric acid, and 1/3 of a mole of phosphoric acid – are in each case neutralized by **1** mole of sodium hydroxide. This must of course be so, since these amounts of acid correspond exactly in each case to the gram-equivalents.

It should now be clear that equal volumes of normal solutions will always be chemically equal or equivalent to one another.

This is true because the solutions have the same normality, e.g.

1 L of 1 N HCl is equivalent to 1 L of 1 N NaOH.

If the solutions were to have differing normalities, then the equivalent volumes would have to be changed accordingly, e.g.

1 L of 2 N HCl is equivalent to 2 L of 1 N NaOH

1 L of 2 N HCl is equivalent to 4 L of 0.5 N NaOH

1 L of 2 N HCl is equivalent to 0.5 L of 4 N NaOH

See now if you can jot down the missing volumes in each of the statements below:

a) 1 L of 1 N H_2SO_4 is equivalent to ...... L of 0.5 N NaOH

b) 0.5 L of 1 N HCl is equivalent to ...... L of 1 N NaOH

c) 1 L of 0.1 H_2SO_4 is equivalent to ...... L of 1 N NaOH → 16

---

28 Your answer is wrong.

Continue your reading at 46 .

---

29 Your answers are correct if you have written:

The equivalent mass of HCl	=	36.5
1 mol of H_3PO_4	=	98 g
1 gram-equivalent of H_2CO_3	=	31 g
The equivalent mass of HNO_3	=	63

If you made an error in your answers, you should now re-work the Program very thoroughly-starting from 26 – before continuing any further.

If all your answers were correct, copy down now the flow chart below and insert the missing terms. → 12

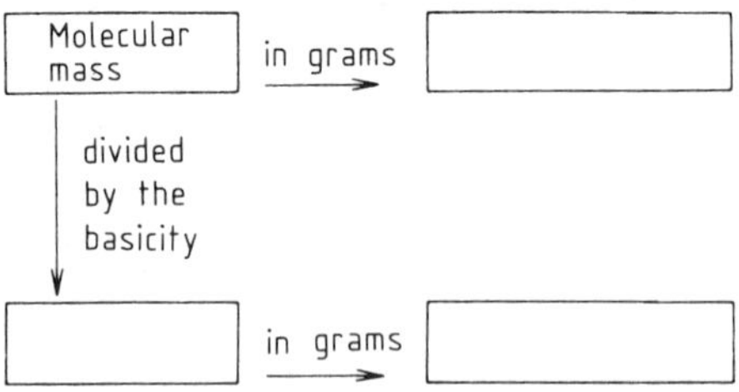

30

Your answer is wrong.

When mixed, 4 g of phosphoric acid and 196 g of water yield 200 g of dilute phosphoric acid.

200 g of dilute phosphoric acid contain 4 g of phosphoric acid

100 g of dilute phosphoric acid contain ? g of phosphoric acid

100 g of dilute phosphoric acid contain $\frac{4 \times 100}{200}$ g

= 2 g of phosphoric acid.

If 100 g of dilute phosphoric acid contain 2 g of phosphoric acid, the solution will be a ......% one. → 56

31

Your answer is wrong. Please note that:

2 gram-equivalents of NaOH in 1 L of solution means

1 L of this caustic soda solution will contain 2 x 40 g = 80 g of NaOH.

This gives us the concentration in grams per liter:

2 gram-equivalents of NaOH/L = 80 g NaOH/L

You should have reached this conclusion yourself! It would be advisable now to work your way through the whole Program again fairly soon. → 71

32

Although you have correctly worked out the molecular mass for each of the three compounds, you appear to have forgotten that $Ca(OH)_2$ has a basicity of *two*.

Make the necessary correction now, and then check the equivalent masses at 41 .

33

You should have been more careful here!

If 3.5 L contain 700 g of sodium sulfate, then 1 L will contain 700 g/3.5 of sodium sulfate.

Now choose the correct answer at 48 .

14

34 Your answer is wrong.

We have already established that equal volumes of normal solutions are chemically equal to one another.

The concept of the normal solution is so important that you would be well advised to work more carefully and accurately here.

You should now read through 27 once again.

35 Your calculation is not correct.

A detailed explanation of this is given at 48 .

36 Your answer is quite right.

2 mol of sodium hydroxide will neutralize 126 g of nitric acid.

Calculations such as those you have just performed are very important in analytical chemistry when determining, for instance, the content or concentration of a solution.

Before discussing how the content of a solution may be determined, however, we must first say something about the concept of "concentration" in the present context.

The *concentration* is a measure of the amount of a substance dissolved in a solution. Thus, we have

A low concentration = a dilute solution
A high concentration = a concentrated solution.

These are, however, rather vague statements about the concentration of a solution. Below we present three *precise* ways of specifying the concentration:

1. The *concentration* can be expressed in *grams per liter (g/L).*

   This is the amount (in grams) of a substance dissolved in one liter of solution.

2. The *concentration* can be expressed as a mass percentage (mass%).

   This is the amount (in grams) of a substance dissolved in 100 g of solution.

3. The *concentration* can be expressed as a normality (N).

> This is the amount (in gram-equivalents) of a substance dissolved in one liter of solution.

Now copy down these three ways of specifying concentration. We shall be discussing each of them in detail later on.

In which of the three ways of specifying concentration is *no* reference made to the volume?

- The concentration expressed in grams per liter → 13
- The concentration expressed as a mass percentage → 25
- The concentration as a normality → 60

---

37 Your answer is quite right.

14

The equivalent mass of nitric acid is 63.

In the case of sulfuric acid, the equivalent mass may be calculated as follows:

H_2SO_4 is a dibasic acid which dissociates in water to form an acid radical ($SO_4^{2\ominus}$) and *two* hydrogen ions ($2H^{\oplus}$). Its basicity is thus equal to *two*.

$$\text{The equivalent mass of } H_2SO_4 = \frac{\text{Molecular mass of } H_2SO_4}{\text{Basicity}} = \frac{98}{2} = 49$$

The molecular mass of phosphoric acid is 98. What will be its equivalent mass?

- Its equivalent mass is 32.7 → 24
- Its equivalent mass is 49 → 4
- Its equivalent mass is 98 → 49
- Its equivalent mass is 294 → 15

---

38 Your answer is wrong.

The correct way to tackle this is described in 14 .

---

39 You appear to have put the decimal point in the wrong place. A detailed description of this calculation is presented in [7].

40 You have not taken the basicity of the carbonic acid into account. Carbonic acid, H_2CO_3, is a dibasic acid, i.e. it has a basicity of two.

Go back now to [24].

41 Your answer is quite right.

The equivalent mass of NaOH $= 40$
KOH $= 56$
$Ca(OH)_2 = 37$

At this point we shall repeat some of the definitions dicussed earlier on.

Concept	Statement	Example	
Atomic mass	*A given dimensionless number*	Sulfur	32
Molecular mass	The sum of the atomic masses; a *given* dimensionless number	H_2SO_4	98
Equivalent mass	The atomic or molecular mass divided by the basicity; a *given* dimensionless number	H_2SO_4	$\frac{98}{2} = 49$
Gram atom	The atomic mass in *grams*	Sulfur	32 g
Gram molecule (Abbreviated: mol)	The molecular mass in *grams*	H_2SO_4	98 g
Gram-equivalent (Abbreviated: g-equiv.)	The equivalent mass in *grams*	H_2SO_4	49 g

The last of these concepts has not yet been introduced here. Make sure that you memorize the very important concept of the gram-equivalent.

Calculate now each of the following quantities:

(The relevant atomic masses are H=1; C=12; Cl=35.5; P=31; N=14; 0=16)

The equivalent mass of HCl =

1 mol of H_3PO_4 =

1 gram-equivalent of H_2CO_3 =

The equivalent mass of HNO_3 = → 29

42

1. The concentration expressed in *grams per liter* (g/L)

2. The concentration expressed as a *mass percentage* (mass%)

3. The concentration expressed as a *normality* (g-equiv/L)

Let's repeat the exercise now using the following example:

A dilute solution of caustic soda contains 2 gram-equivalents of NaOH in 1 L. Calculate what the *concentration* of this solution will be in *grams per liter*. (The equivalent mass of NaOH = 40.)

My answer is:

- 20 g/L → 31
- 40 g/L → 10
- 80 g/L → 71

43

Your answer is wrong.

An explanation of this is given at 59 .

44

Your answer is quite correct.

Obviously, 1 N hydrochloric acid may be replaced by 1 N sulfuric acid or by 1 N phosphoric acid in the titration. Exactly the same amount of acid would be used in each case, as all normal solutions are equivalent to one another.

The method just outlined for determining the concentration of an alkali is known as *alkalimetry*.

Conversely, we may titrate an acid, e.g. hydrochloric acid solution of unknown concentration, with an alkali. In this case the 1 N caustic soda is put into the buret and 100 mL of the unknown hydrochloric acid solution into the flask. The indicator – methyl orange – should not, of course, be forgotten! The caustic soda solution is run out of the buret into the flask until the color of the indicator changes from orange-red to yellow.

Suppose that we use in all 20 ml of 1 N caustic soda solution for the neutralization. The concentration of the hydrochloric acid can then be calculated as follows:

1000 ml of 1 N caustic soda correspond to 1000 ml of 1 N hydrochloric acid containing 36.5 g HCl.

1 ml of 1 N caustic soda corresponds to $\frac{36.5}{1000}$ g of HCl.

20 ml of 1 N caustic soda correspond to $\frac{36.5 \times 20}{1000}$ g of HCl.

When this amount is evaluated we obtain the concentration in g/L. We could, of course, also state the concentration as a *mass percentage*. In the latter case, one would not start out by measuring 100 ml of the unknown acid but rather weigh out 100 g and then titrate that. To keep matters simple, let's suppose that 20 ml of 1 N caustic soda solution were required here. The calculation would then be approached in the same way as above.

See if you are able to complete this calculation. What is the concentration of the hydrochloric acid titrated expressed as a mass percentage?

- 0.73% → 63
- 7.3% → 84
- 0.073% → 76

---

45 Your answer is wrong.

An explanation of this is given at 27 .

---

46 You should now attempt following problem:

How many grams of nitric acid (HNO_3) would be neutralized by 2 mol of sodium hydroxide?

1 gram-equivalent of acid is equivalent to 1 gram-equivalent of base.

2 mol of sodium hydroxide = 2 gram-equivalents of sodium hydroxide.

Thus, 2 gram-equivalents of nitric acid can be neutralized.

Go back now to [12] .

---

47 Your answer is wrong.

The correct way to tackle this calculation is described in [14] .

---

48 After the 1000 g of sodium chloride have been dissolved in water, additional water is added to fill the volumetric flask to the mark. We then have in all 5 L of sodium chloride solution.

5 L of sodium chloride solution contain 1000 g of sodium chloride.
1 L of sodium chloride solution contains ? g of sodium chloride.

1 L of sodium chloride solution contains $\frac{1000 \times 1}{5}$ g = 200 g of sodium chloride.

The concentration of this sodium chloride solution is thus 200 g/L.

14

Now see if you can solve the problem below.

What will be the concentration in g/L of a solution which contains 700 g of sodium sulfate dissolved in 3.5 L?

- The concentration will be 700 g/L → [62]
- The concentration will be 350 g/L → [33]
- The concentration will be 200 g/L → [21]
- The concentration will be 245 g/L → [55]

---

49 Your answer is wrong. 98 is the *molecular mass* of phosphoric acid. To obtain the *equivalent mass*, the molecular mass has to be divided by the basicity of the acid.

You are advised to read through [37] once again and to remember when answering the question there that phosphoric acid is a *tri*basic acid.

---

50 Your answer is wrong. The appearance of a violet coloration when an unknown substance is heated in a flame indicates that potassium is *present* in the substance, i.e. it tells us something about the quality of it. However, it provides us with no information about the *quantity* of the potassium present. We would have to employ special methods to perform a quantitative analysis of the potassium.

Thus, the test for potassium based on a flame colorations is a method of .... analysis. → 70

---

51 Your answer is wrong.

An explanation of this is given at 59 .

---

52 The equivalent mass of HNO_3 is 63.
The equivalent mass of H_2CO_3 is 31.

If you managed to get this answer right first time, you are working well.

Let's now consider how the equivalent mass of a base may be calculated.

14

In water NaOH yields *one* hydroxyl ion ($OH^{\ominus}$) and so has a basicity of **1.**
In water KOH yields *one* hydroxyl ion ($OH^{\ominus}$) and so also has a basicity of **1.**
In water $Ca(OH)_2$ yields *two* hydroxyl ions (2 $OH^{\ominus}$) and so has a basicity of **2.**

Now calculate the equivalent masses of NaOH, KOH and $Ca(OH)_2$.

The relevant atomic masses are: Na = 23 O = 16
K = 39 H = 1
Ca = 40

Make sure you note down these figures!

The equivalent masses are:

	NaOH	KOH	$Ca(OH)_2$		
•	80	112	74	→	23
•	40	56	74	→	32
•	40	56	37	→	41
•	20	28	37	→	66
•	My answers are different			→	73

53 Your answer is quite right.

1 N hydrochloric acid contains in 1 L 36.5/1 = 36.5 g of HCl.
1 N sulfuric acid contains in 1 L 98/2 = 49 g of H_2SO_4.
1 N phosphoric acid contains in 1 L 98/3 = 32.7 g of H_3PO_4.

These three solutions are chemically equal, i.e. equivalent, to one another.

They are equal in the sense that they contain equivalent amounts of acid.

We mean by a normal solution a solution containing 1 gram-equivalent of a substance dissolved in 1 liter of solvent.

Normal solutions are usually designated – in abbreviated notation – as 1 N solutions. Thus, in the case of one normal sulfuric acid we can write 1 N sulfuric acid solution or just 1 N sulfuric acid.

When 73 g (2 mol) of hydrogen chloride are dissolved in water to form 1 L of solution, a two normal or 2 N solution of hydrochloric acid results.

Solutions having various different normalities find numerous applications in research and industry. Solutions which are 5 N, 2 N, 1 N, 0.5 N, 0.1 N and 0.01 N are often used.

The table below provides some examples of the ways in which standard solutions are designated.

Type of Solution	One liter of the Solution contains	Designation of the Solution
3 N solution	3 gram-equivalents	3 normal solution
2 N solution	2 gram-equivalents	2 normal solution
1/2 N solution	1/2 a gram-equivalent	half normal solution
0.5 N solution	0.5 gram-equivalents	0.5 normal solution
1/10 N solution	1/10 of a gram-equiv.	1/10 normal solution
0.1 N solution	0.1 of a gram-equiv.	0.1 normal solution

Note that in this table the 3rd and 4th lines as well as the 5th and 6th lines refer to one and the same solution in each case – only the manner of writing it out is different!

How many moles and how many gram-equivalents of acid are present in one liter of each of the following solutions?

Write out your answers!

- 1 L of 2 N H_2SO_4 contains ...... mol of H_2SO_4 or ...... gram-equivalents of H_2SO_4

- 1 L of 0.5 N HNO_3 contains ...... mol of HNO_3 or ...... gram-equivalents of HNO_3

- 1 L of 5 N HCl contains ...... mol of HCl or ...... gram-equivalents of HCl

Check your answers at [67] .

---

54 You seem to have your decimal point in the wrong place.

A detailed description of this calculation is given at [7] .

---

55 You should take more care over your answers! If 3.5 L contain 700 g of sodium sulfate, then 1 L must contain 700 g/3.5 of sodium sulfate.

Now choose the correct answer at [48] .

---

56 Correct, the solution is a 2% one.

To prepare a solution having a standard concentration, expressed in *grams per liter,* a fixed, weighed amount of substance is dissolved in a solvent (e.g. water) so that the final volume of the solution is exactly 1 liter.

This may be carried out practically as follows:

The substance which is to be dissolved is weighed out and then placed in a 1 liter volumetric flask. The water is then gradually added and the flask continually shaken until the flask is filled precisely to the hair-line with water.

Now see if you can perform the calculation below:

1000 g of sodium chloride are placed in a 5 L volumetric flask and water is then added until the flask is filled to the hair-line mark.

What will be the concentration of this 5 L of sodium chloride solution in grams per liter?

- The concentration will be 100 g/L → [69]

- The concentration will be 200 g/L → [21]

- The concentration will be 500 g/L → 78
- The concentration will be 1000 g/L → 35
- I obtained a different figure → 48

---

57 Your answer is wrong.

We have already established that equal volumes of normal solutions are chemically equal to one another.

The concept of the normal solution is so important that you are advised now to familiarize yourself with it much better.

Read through 27 once again.

---

58 Your answer is wrong.

An explanation of this is given at 27 .

14

---

59 Let's work through the problem here again.

A dissolved substance has a concentration of 200 g/liter.

The density of this solution is 2 g/cm^3.

Now convert this concentration into one expressed as a mass percentage!

A mass percentage means: the quantity (in grams) of a substance dissolved in 100 g of solution.

We approach this problem in the following way:

1 L of solution contains 200 g of dissolved material.
1 L of this solution weighs 2000 g. (The density is 2 g/cm^3.)
2000 g of solution thus contain 200 g of dissolved material.

Hence, 1 g of solution will contain $\frac{200}{2000}$ g of dissolved material,

and 100 g of solution will contain $\frac{200 \times 100}{2000}$ g of dissolved material.

The solution will therefore be a ......% one. → 82

---

60 Your answer is wrong. To prepare a normal solution, 1 gram-equivalent of a substance is dissolved in a solvent (usually water), enough of the solvent being taken that the final volume of the solution will be exactly 1 L. Accordingly, a *volume* has to be measured in this procedure.

Take another look at the concentrations given here and then select a different answer at 36 .

61 The missing words are:

alkalimetry
acidimetry

We have described three different ways of expressing concentration, namely:

1. The concentration expressed in .... (g/L)

2. The concentration expressed as a .... (mass%)

3. The concentration expressed as a .... (gram-equivalents/L)

Jot down the missing words. → 42

62 You should take more care over your answers! If 3.5 L contain 700 g of sodium sulfate, then 1 L must contain 700 g/3.5 of sodium sulfate.

Now choose the correct answer at 48 .

63 Your answer is quite correct.

The procedure for the determination of an acid is known as *acidimetry.*

The procedure for the determination of an alkali is known as ....

Jot down the missing word. → 79

64 Your answer is wrong. In the case of carbonic acid the basicity is 2 whereas nitric acid, HNO_3, has a basicity of 1.

Go back now to 24 .

65 Your answer is wrong.

The correct approach to this problem is outlined in 14 .

66 Your answer is wrong. You have divided all the molecular masses by 2, although it is only $Ca(OH)_2$ which actually has a basicity of 2.

Continue your reading at 73 .

67 Your answers are correct if you wrote:

1 L of 2 N H_2SO_4 contains 1 mol of H_2SO_4 or 2 gram-equivalents of H_2SO_4.

1 L of 0.5 N HNO_3 contains 0.5 mol of HNO_3 or 0.5 gram-equivalents of HNO_3.

1 L of 5 N HCl contains 5 mol of HCl or 5 gram-equivalents of HCl.

A 1 N caustic soda solution contains in one liter 1 gram-equivalent of sodium hydroxide, i.e. 40 g. This solution too is equivalent to other normal solutions. The amount of base contained in one liter of 1 N caustic soda will exactly neutralize the amount of acid contained in one liter of a 1 N acid solution.

How much of a 1 N caustic soda solution would be required to neutralize one liter each of a 1 N hydrochloric acid solution, a 1 N sulfuric acid solution and a 1 N phosphoric acid solution?

- For the 1 N hydrochloric acid 1 L,
  for the 1 N sulfuric acid 2 L,
  for the 1N phosphoric acid 3 L. → 58
- In all three cases 1 L each → 77
- For the 1N hydrochloric acid 1 L,
  for the 1N sulfuric acid 0.5 L,
  for the 1N phosphoric acid 0.33 L. → 45

68 Correct, 100 mL of 1 N caustic soda solution contain 4 g of NaOH.

However, the results of most titrations cannot be so readily calculated. If, for instance, the 100 mL of unknown caustic soda solution had required 60 mL of 1 N hydrochloric acid for neutralization, the calculation would have to be performed as follows:

1000 mL of 1 N hydrochloric acid are equivalent to 1000 mL of 1 N caustic soda solution, which contains 40 g NaOH.

1 mL of 1 N hydrochloric acid thus corresponds to $\frac{40}{1000}$ g NaOH.

60 mL of 1 N hydrochloric acid (our amount) correspond to

$$\frac{40}{1000}\ g \times 60 = \frac{2400}{1000}\ g = 2.4\ g\ NaOH.$$

Thus, the 100 mL sample of the unknown caustic soda solution – requiring 60 mL of a 1 N HCl solution for neutralization – actually contains 2.4 g of NaOH. From this initial result we therefore can conclude that the caustic soda solution contains 24 g NaOH per liter.

Instead of using 1 N HCl for the titration, we could have used 1 N H_2SO_4 solution. How many mL of 1 N sulfuric acid would then have been required for neutralization?

- 60 mL → 44
- 120 mL → 57
- 30 mL → 34

---

69 Your calculation is not correct.

A detailed explanation of it is given at 48 .

---

70 Correct. The test for potassium based on a flame coloration is a method of *qualitative* analysis.

Now write down the answers to the following questions:

1. What is a solution called which contains 1 gram-equivalent in a liter?

2. What do we call all those analytical methods based on measurements using normal solutions?

3. What do we call those analyses in which the precise amounts of either elements in a compound or compounds in a mixture are determined?

Check your answers at 81 .

---

71 Your answer is quite right.

2 gram-equivalents of NaOH in one liter of solution are the same thing as 80 g NaOH/L.

Now we consider one last problem:

A dissolved substance has a concentration of 200 g/L.

The density of its solution is 2 g/cm^3.

Convert this concentration into a mass percentage. What result do you obtain?

- 5% → 43
- 10% → 82
- 20% → 51
- 40% → 22
- I need some help → 59

72 Your answer is wrong. To prepare a normal solution, one gram-equivalent of the substance concerned is dissolved in one liter of solution. You should therefore calculate the equivalents of the three acids.

To reach the right answer you are advised to re-read 21 .

73 We start by calculating the molecular masses:

NaOH:	Atomic mass of Na	=	23
	Atomic mass of O	=	16
	Atomic mass of H	=	1
			40
KOH:	Atomic mass of K	=	39
	Atomic mass of O	=	16
	Atomic mass of H	=	1
			56
$Ca(OH)_2$:	Atomic mass of Ca	=	40
	2 x Atomic mass of O	=	32
	2 x Atomic mass of H	=	2
			74

The equivalent mass of a compound is obtained simply by dividing its molecular mass by its basicity.

	Molecular Mass	Basicity
NaOH	40	1
KOH	56	1
$Ca(OH)_2$	74	2

Now calculate each of the equivalent masses. → 41

---

74 Your answer is wrong.

The correct way to tackle this problem is described in 14 .

---

75 *equivalent mass*

What do we call the equivalent mass in grams?

Jot down your answer and then go to 18 .

---

76 You seem to have the decimal point in the wrong place.

$$\frac{36.5}{1000} \text{g} \times 20 = 0.73 \text{ g of HCl}$$

100 g of the unknown hydrochloric acid contain 0.73 g of HC1.
Thus, the solution is a ......% one. → 63

---

77 Your answer is quite correct.

As this particular topic is so important, we shall be going over it again in different words.

Continue your reading at 27 .

---

78 Your calculation is not correct.

A detailed explanation of this is given at 48 .

79

The determination of an alkali is *alkalimetry*.
The determination of an acid *is acidimetry*.

Both alkalimetry and acidimetry are instances of the general discipline of *volumetric analysis*. This concept is used to describe all those analytical processes which depend upon measurements made with standard solutions.

There are two different types of analytical process:

Qualitative and Quantitative Analysis.

We carry out a *qualitative analysis* to determine which *chemical elements* are present in a given compound or which elements and compounds are present in a mixture.

It is only by means of *quantitative analysis* that we can determine the *relative amounts* of each element in a compound or of each compound in a mixture.

Like all methods of *volumetric analysis* (titrations), both *alkalimetry* and *acidimetry* belong to the domain of *quantitative analysis*.

An illustration of what we have discussed here is provided by the analysis of sea water. The type of investigation which reveals that sea water contains chloride is a *qualitative* method of analysis. But the type of investigation which tells us that sea water contains precisely 3% of sodium chloride is a *quantitative* method of analysis.

Make sure that you know the difference between these two different methods of analysis.

Now answer the question:

When potassium is detected by means of a violet coloration of a flame, what type of analysis is being performed?

- A qualitative type of analysis → 70
- A quantitative type of analysis → 50

80

Your answer is wrong.

When mixed, 4 g of phosphoric acid and 196 g of water yield 200 g of dilute phosphoric acid solution.

200 g of dilute phosphoric acid contain 4 g of phosphoric acid.

100 g of dilute phosphoric acid contain ? g of phosphoric acid.

---

100 g of dilute phosphoric acid contain $\frac{4 \times 100}{200}$g = 2 g

Thus, if 100 g of dilute phosphoric acid solution contain 2 g of phosphoric acid, the solution will be a ......% one. → 56

---

81 The correct answers are:

1. *A normal solution*

2. *Volumetric analysis*

3. *Quantitative analyses*

Jot down the missing words in the sentences below:

So far we have introduced two different methods of volumetric analysis, namely:

The determination of alkalis or ....
The determination of acids or .... → 61

---

82 Correct, the solution is a 10% one.

Let's now briefly recapitulate. The solutions used in volumetric analysis are known as standard solutions.

The molecular mass of a compound divided by its basicity is called the ....

Jot down the missing word. → 75

---

83 Your answer is wrong.

The correct way to tackle this problem is described in 14 .

---

84 You seem to have the decimal point in the wrong place.

$\frac{36.5}{1000}$g x 20 = 0.73 g of HCl

100 g of the unknown hydrochloric acid contain 0.73 g of HCl.

Thus, the solution is a ......% one. → 63

Program 15

Carbon (I)

1 In the next three Programs we shall be discussing one particularly important element: *carbon*. The pure element occurs in three different allotropic modifications, namely as

soot, graphite and diamond.

How would you describe these three different forms of carbon?

- As three different compounds of carbon → 14
- As three different states of aggregation of carbon → 18
- As three different allotropes of carbon → 8

2 Soot, graphite and diamond are the three allotropic modifications or forms of *carbon*. Chemically, they all have the same composition, that is they all consist of pure *carbon*.

15

For this reason they all produce the same combustion product when they burn.

Write out now the completed version of the sentence below:

When the three allotropic forms of carbon are burnt, the same combustion gases are produced.

Check your answer at 31 .

3 No, diamond does *not* conduct electric current.

Read through 36 once again.

4 Your answer is quite correct.

The combustion of carbon involves two stages:

Carbon + Oxygen → Carbon monoxide
Carbon monoxide + Oxygen → Carbon dioxide

Write down the relevant chemical equation for each of these stages, and then check your equations at 27 .

---

5 Your answer is wrong. *Carbon monoxide* is of course *combustible.* In fact, you have just set up the equation for the combustion of carbon monoxide. Carbon monoxide is formed from the *incomplete* combustion of carbon. In the presence of more oxygen it can be burnt to form carbon dioxide.

Write out the two equations once again, and check your answer at 27 .

---

6 No, soot does *not* conduct electric current.

Read through 36 once again.

---

7 No. On an *industrial* scale carbon dioxide is prepared by passing air over red-hot coke. However, this process is totally unsuitable for the production of carbon dioxide in the *laboratory*.

Read through 34 once again.

---

8 Your answer is quite right.

The three different forms of carbon – soot, graphite and diamond – are called more precisely the *allotropes* or *allotropic modifications of carbon.* In all three of these allotropes carbon is present as the pure element. At this point you may recall that sulfur too can assume different forms (allotropic modifications) known as brittle and plastic sulfur.

One of the three forms of carbon is certainly well known to you. We refer to finely divided carbon, such as that formed during the incomplete combustion of carbon compounds. Think here, for instance, of a candle which is not burning properly.

Which allotrope is formed during incomplete combustion?

- Soot → 15
- Graphite → 28
- Diamond → 25

---

9 Your answer is correct. Carbon dioxide is the *anhydride of carbonic acid.*

However, carbon dioxide reacts only to a limited extent with water. The concentration of carbonic acid in water is actually very small. The major part of the carbon dioxide is present in the water as the dissolved gas.

For this reason, the equation for the reaction is written with a double arrow thus:

$$H_2O + CO_2 \rightleftarrows H_2CO_3$$

If this aqueous solution of carbonic acid is heated up, carbon dioxide gas will be driven off.

In this respect carbonic acid thus behaves like sulfurous acid.

In which direction will the reaction proceed when the solution is heated?

- From left to right → 29
- From right to left → 17

10 Your answer is wrong.

Copy down now the completed version of the sentence below:

Soot, graphite and diamond are the three allotropic forms of ....; they are of the same chemical composition as they all consist of pure..............

Check your answers at 2 .

11 $CaCO_3 + H_2SO_4 \rightarrow CaSO_4 + CO_2 + H_2O$

In industry, other processes are used for the production of carbon dioxide, namely:

1. Air is passed over red-hot coke.

$$C + O_2 \rightarrow CO_2$$

2. Chalk is calcined.

Chalk is calcined by roasting it in large ovens:

$$CaCO_3 \xrightarrow{\text{heat}} CaO + CO_2$$

When calcined, calcium carbonate decomposes into .... and ....

Copy down the completed version of this sentence, and then check your answer at 30 .

---

12 Your answer is wrong.

An explanation of this is given at 20 .

---

13 Your answer is quite right.

In the laboratory, carbon dioxide is prepared by the action of hydrochloric acid on calcium carbonate (chalk or marble). In industry, however, carbon dioxide is produced in two quite different ways, namely:

1. By passing .... over red-hot ....
2. By calcining .... → 35

15

---

14 Chemically speaking, soot, graphite and diamond are *pure, elementary forms of carbon.* Your answer is thus wrong.

Read through 1 once again.

---

15 Correct, *soot* is formed during incomplete combustion. It consists of finely divided carbon which, because of its characteristically deep black color, is used as a pigment in paints and elsewhere.

Graphite is found in nature as a mineral, though it can also be synthesized. It is grayish-black in color, rather soft and has an oily luster. Graphite conducts electric current well. Moreover, it is an excellent lubricant and so is used for the purpose of lubricating machinery with moving parts. The so-called lead in pencils is manufactured by firing a mixture of graphite and clay. Because it conducts electric current so well, electrodes can be made from graphite.

Jot down now using key words the most important points concerning soot and graphite, and then check your answers at 36 .

16 Your answer is wrong. Carbon dioxide is certainly the dioxide of carbon. However, this statement tells us nothing about its behavior with water.

Have another good look at the equation in 32 .

17 Yes, when heat is applied the reaction proceeds from right to left. CO_2 is driven out of the water and the H_2CO_3 breaks down again into H_2O and CO_2. In the end the water will contain no more carbonic acid.

Write down now the equation for the formation of carbonic acid from water and carbon dioxide, and then check your equation at 64 .

18 Your answer is wrong. The *state of aggregation* of a substance refers to something else. There are three states of aggregation, namely the *solid, liquid* and *gaseous* states.

For instance, water normally occurs in the liquid state of aggregation. In the form of ice it can occur in the solid state of aggregation and as steam in the gaseous state of aggregation.

Carbon exists as a solid whether in the form of soot, graphite or diamond.

Read through 1 again.

19 The missing word is:

adsorption

From the chemical viewpoint, carbon is not a very reactive element. It forms compounds only with difficulty and combines with other elements only at high temperatures.

An important chemical reaction of carbon is its combustion. For this reaction the three allotropic forms of carbon behave differently.

Soot burns slowly and glows for a long time.

Graphite has to be heated with a blower to get it to burn.

Diamond burns only at *very* high temperatures.

What will be formed when each of the three forms of carbon is burnt?

- In all three cases the same combustion gas is formed → 31

- **The combustion gases are different** → 10

---

20 In the name carbon monoxide we have the prefix *mono* and in the name carbon dioxide we have the prefix *di*. Now the prefix *mono* means *one* and the prefix *di* means *two* here.

The prefix thus tells us that in carbon monoxide there is one oxygen atom whereas in carbon dioxide there are two oxygen atoms. Note that such prefixes stand *before* the word to which they refer.
The same applies to the oxides of sulfur: sulfur dioxide is SO_2 and sulfur trioxide is SO_3. The *di* and the *tri* refer in each case to the oxygen.

Go back now to 31 .

---

21 Yes, when *carbon monoxide burns* it combines with the oxygen in the air to form *carbon dioxide:*

15

$$2\ CO + O_2 \rightarrow 2\ CO_2$$

Carbon dioxide is a colorless gas. It is able to support neither combustion nor respiration. For this reason it is capable of suffocating people. Because carbon dioxide is considerably heavier than air, it always collects on the ground and in low places.

Carbon dioxide is used in several different types of fire extinguisher. It acts as an effective extinguishing agent as it spreads itself over the area of the fire and so prevents oxygen from reaching the fire.

Carbon dioxide may be easily liquified. In liquid form it is commercially available in steel bottles.

When sufficiently cooled (−78 °C) carbon dioxide becomes solid. In this form it is used as a refrigerant. Solid carbon dioxide is referred to as *'dry ice'*. The reason for this name is that, as dry ice warms up, no liquid is formed as is the case with ordinary ice. It is actually converted directly from the solid state to the gaseous state when warmed up, i.e. the carbon dioxide sublimes.

This direct transition from the solid state to the gaseous state is termed *sublimation*. In the sublimation process the liquid state is entirely missed out.

Write out now the properties of CO and CO_2 using key words.

Check your answers at 32 .

22 *Soot, graphite* and *diamond*

Diamond too is a mineral. It is transparent and refracts light very strongly. For these latter two reasons it is used in the manufacture of jewellery. Diamond is extremely hard, in fact one of the hardest substances known. It is thus used for cutting glass, for grinding objects and for the tips of special drills. Diamonds are mined principally in South Africa.

Two of the three allotropes of carbon conduct electric current poorly, and only one of them conducts it well. Which of the three do you think conducts it well?

- Diamond → 3
- Soot → 6
- Graphite → 48

23 No. Carbon monoxide and carbon dioxide are two quite different compounds. They must therefore have different molecular masses!

Read through 31 once again.

24 $Na_2CO_3 + 2\ HCl \rightarrow 2\ NaCl + CO_2 + H_2O$

Did you perhaps write 2 NaCl + H_2CO_3 on the right hand side of this equation? This is not really a mistake, though you should not forget that free carbonic acid is unstable and so breaks down into CO_2 and H_2O.

Another important salt of carbonic acid is calcium carbonate, $CaCO_3$.

Write down now the equation for the reaction of calcium carbonate with hydrochloric acid.

Check your equation at 34 .

25 No, diamond is not formed by incomplete combustion. Diamond is a mineral that we shall be discussing in some detail later on.

Go back now to 8 .

15

15

26 Yes, the property of active charcoal of retaining large quantities of gases, liquids or dissolved substances depends upon a *physical* process. The charcoal itself is *not* changed in this process. The substances retained by the active charcoal can be released again, e.g. by heating it up or by rinsing it out.

Thus, active charcoal can retain foreign molecules. This process is described as .... → [19]

---

27 The combustion of carbon proceeds in two stages:

1. $2\ C + O_2 \rightarrow 2\ CO$

2. $2\ CO + O_2 \rightarrow 2\ CO_2$

The complete combustion of carbon can also be represented by *one* overall equation thus:

Carbon + Oxygen → Carbon dioxide

Write down now the chemical equation corresponding to this equation in words, and check your answer at [39] .

---

28 No, graphite is not formed by incomplete combustion. Graphite is a mineral that we shall be discussing in some detail later on.

Go back now to [8] .

---

29 Your answer is wrong.

Read through [9] carefully again.

---

30 When calcium carbonate is heated, it breaks down into *calcium oxide* and *carbon dioxide.*

How is carbon dioxide prepared in the laboratory?

- By calcining chalk → [37]
- By the action of hydrochloric acid on calcium carbonate → [13]
- By passing air over red-hot charcoal → [7]

---

31 When the three allotropic forms of carbon are burnt, the *same combustion gases* are produced. This provides a proof that all three allotropes really consist of pure carbon.

When carbon is burnt the initial product is carbon monoxide, which in the presence of sufficient oxygen gets further oxidized to carbon dioxide.

Write down now the formulas for carbon monoxide and carbon dioxide.

Carbon has an atomic mass of 12 and oxygen an atomic mass of 16. Choose the correct molecular masses for carbon monoxide and carbon dioxide from the list below.

	Carbon monoxide	Carbon dioxide		
•	28	28	→	23
•	28	44	→	4
•	44	28	→	12
•	28	40	→	41

32 Now check your answer:

CO is a *very poisonous,* colorless, combustible gas.

CO_2 is a suffocating, colorless, heavy gas which can be liquified and used in solid form as dry ice for cooling purposes. CO_2 sublimes.

Carbon dioxide is formed during the *complete* combustion of all compounds containing carbon. In some places carbon dioxide is found in mineral water from springs.

Water saturated with carbon dioxide is sour because some of the carbon dioxide combines with the water to form carbonic acid. Carbonic acid has the formula H_2CO_3 and it is formed according to the equation:

$$H_2O + CO_2 \rightarrow H_2CO_3$$

What is the carbon dioxide reacting as in this equation?

- As an oxidizing agent → 45
- As the dioxide of carbon → 16

- As an acid anhydride → 9

---

33 No, in a chemical process the substances involved are fundamentally changed.

The active charcoal is merely retaining gases, liquids or dissolved substances in the same way as a sponge retains water. The retained substances may be separated from the active charcoal by, for instance, heating it up or rinsing it out. The process known as *adsorption* is a .... (physical/chemical) *process.*

Continue your reading at 26 .

---

34 $CaCO_3 + 2\ HCl \rightarrow CaCl_2 + CO_2 + H_2O$

There are very large natural deposits of calcium carbonate in the form of chalk and marble. In fact, whole mountain ranges are made of chalk.

Carbon dioxide is easily prepared from calcium carbonate. In the laboratory one simply pours acid over lumps of chalk or marble. The mixture effervesces and CO_2 is liberated. The above equation describes this reaction when the acid used is hydrochloric acid. Write down now the equation for the reaction of calcium carbonate with sulfuric acid.

Check your equation at 11 .

---

35 Now check your answer:

1. By passing *air* over red-hot *coke.*

2. By calcining *chalk.*

Write out the equations on which the industrial production of CO_2 is based. Check your answers at 43 .

---

36 Now check your answers:

Soot: incomplete combustion, deep black, paint pigment.

Graphite: mineral, grayish-black, soft, luster, conducts current, ınt, lead lubricant, "lead" pencils, electrodes

What are the three allotropic forms of carbon called?

Jot down your answer and then go on to 22 .

---

37 Your answer is wrong. Carbon dioxide is produced *industrially* by calcining chalk. The relevant equation is:

$$CaCO_3 \rightarrow CaO + CO_2$$

The process is carried out in large ovens, since rather high temperatures are required for this particular reaction. Of course, such a process would be completely unsuitable for the preparation of carbon dioxide in the laboratory.

Read through 34 once again.

38

$$H_2CO_3 \rightleftarrows CO_2 + H_2O$$

End of Program 15.

39

$$C + O_2 \rightarrow CO_2$$

15

CO_2 is carbon dioxide.
CO is carbon monoxide.

Carbon monoxide is an extremely insidious poison gas since it is both colorless and odorless. It is in fact the poisonous component of town gas.

Since it is formed during the incomplete combustion of carbon, ovens which do not draw properly for any reason, i.e. ovens which are fed insufficient oxygen, are potential hazards.

Is carbon monoxide combustible?

- **No** → 5
- **Yes** → 21

---

40 Yes, the concentration of the $H^{\oplus}$ ions is only small.

The salts of carbonic acid are called carbonates. Na_2CO_3 is sodium carbonate.

Jot down now the formula for potassium carbonate. → 50

---

41 Your answer is wrong.

An explanation of this is given at 20 .

42 The applications are as follows:

Soot: black pigment.

Graphite: lubricant, manufacture of electrodes, "lead" pencils

Diamond: jewellery, cutting and polishing of hard substances such as glass.

If wood is heated in the absence of air it forms charcoal. In this process gases are liberated which escape along fine capillaries in the wood being charred. The residue left behind is wood charcoal. The latter has an enormous surface area because of its many fine pores and capillaries. It is known as *active charcoal.*

Active charcoal may be pictured as a sponge, though in this case the fine pores and capillaries are not visible. Active charcoal has the ability of retaining large amounts of gases or liquids or of extracting dissolved substances from a solution. It is said that the active charcoal *adsorbs* the foreign molecules and the process itself is described as *adsorption.*

This property of active charcoal is exploited both in the laboratory and by the chemical industry. It is thus utilized to remove impurities from a variety of chemical substances. Even in gas masks, filters containing active charcoal are used. The carbon layer is able to retain the poisonous components and thereby prevent them getting through.

What type of process is this kind of behavior based on?

- A chemical process → 33
- A physical process → 26

43 Now check your equations:

1. $C + O_2 \rightarrow CO_2$

2. $CaCO_3 \xrightarrow{\text{heat}} CaO + CO_2$

Let's now briefly recapitulate:

How many allotropic forms of carbon are there?

- One allotrope → 68
- Two allotropes → 58
- Three allotropes → 53
- Four allotropes → 61

---

44 Your answer is quite right.

Graphite is the allotrope in question.

The three carbon allotropes are combustible, even if this is sometimes rather difficult to accomplish.

Which of the sequences below is correct for the three allotropes of carbon?

Combustible	Combustible with difficulty	Combustible with great difficulty	
● Soot	Diamond	Graphite	→ 66
● Diamond	Soot	Graphite	→ 54
● Graphite	Soot	Diamond	→ 59
● Soot	Graphite	Diamond	→ 62

---

45 Your answer is wrong.

An oxidation is a chemical process in which a substance *combines with oxygen*. Free oxygen, however, is certainly not formed during the reaction of carbon dioxide with water. You are advised to take another good look at the relevant equation. You will find it at 32 .

---

46 Carbon monoxide, CO
Carbon dioxide, CO_2

Which of these two oxides is very poisonous?

- Carbon monoxide → 56
- Carbon dioxide → 57

---

47 CO_2 is the *anhydride* of carbonic acid.

Carbonic acid, H_2CO_3, is a very weak acid which is stable only when very dilute and kept at a low temperature. When it is heated up carbonic acid simply decomposes.

What is the equation for this decomposition? → 38

---

48 Your answer is quite right.

Of the three allotropes of carbon only *graphite* is capable of conducting electric current well.

Use keywords to write down now the properties of the three allotropes of carbon. Check your answers at 63 .

15

---

49 Your answer is wrong. Soot has quite different properties.

Soot is dark black and is used as a coloring agent. It is formed during incomplete combustion.

Go back now to 60 .

---

50 Now check your answer:

K_2CO_3 = potassium carbonate

And now to the method of testing for carbonic acid in its salts: as a very volatile acid it is readily driven from its salts by less volatile acids.

Complete the following equation:

$$Na_2CO_2 + 2\ HCl \rightarrow$$

Check your equation at 24 .

---

51 $H_2CO_3 \rightleftarrows 2\ H^{\oplus} + CO_3^{2\ominus}$

This reaction too is written with a double arrow since carbonic acid is a weak acid. As a weak acid it dissociates very little. In aqueous solution therefore there are, in addition to $H^{\oplus}$ and $CO_3^{2\ominus}$ ions, also $HCO_3^{\ominus}$ ions and undissociated H_2CO_3 molecules present.

As a dibasic acid it dissociates in two stages:

Stage 1: $H_2CO_3 \rightleftarrows H^{\oplus} + HCO_3^{\ominus}$

Stage 2: $HCO_3^{\ominus} \rightleftarrows H^{\oplus} + CO_3^{2\ominus}$

$H_2CO_3 \rightleftarrows 2\ H^{\oplus} + CO_3^{2\ominus}$

What will be the concentration of $H^{\oplus}$ ions in an aqueous solution of carbonic acid?

- The concentration of the $H^{\oplus}$ ions will be high → 67
- The concentration of the $H^{\oplus}$ ions will be low → 40

52 Your answer is wrong. Diamond has quite different properties.

Diamond is transparent. It is one of the hardest known substances.

Go back now to 60 .

53 Your answer is quite right.

Carbon is found in the form of *three* allotropes. Write down the names of these allotropes. → 60

54 Your answer is wrong.

Go back now to 44 .

55 In the laboratory, the action of acids on carbonates:

e.g. $Na_2CO_3 + 2\ HCl \rightarrow 2\ NaCl + CO_2 + H_2O$

or $CaCO_3 + H_2SO_4 \rightarrow CaSO_4 + CO_2 + H_2O$

In industry, the calcination of chalk:

$$CaCO_3 \xrightarrow{\text{heat}} CaO + CO_2$$

Now complete the sentence below:

CO_2 is the .... of carbonic acid. → 47

---

56 Yes, carbon monoxide is very poisonous.

Write down the equations for the formation of CO and CO_2 from the combustion of carbon. → 65

---

57 No. CO_2 has a suffocating effect whenever present in large amounts although it is not poisonous. If it were, it could not be used in the form of dry ice or for the manufacture of mineral water.

Continue your reading at 56 .

15

---

58 No, carbon occurs in the form of several allotropes.

Think, for instance, of the uses of the various allotropes as a paint pigment, the material for lead pencils and as jewellery.

Now answer the question in 43 once again.

---

59 Your answer is wrong.

Go back now to 44 .

---

60 *Soot, graphite and diamond.*

One of the allotropes of carbon is soft, has an oily luster and is grayish-black in appearance. It conducts electric current well and is used for the manufacture of "lead" pencils.

Which of the allotropes is referred to here?

- Soot → 49

- Graphite → 44
- Diamond → 52

---

61 No, carbon occurs in the form of several allotropes.

Think, for instance, of the uses of the various allotropes as a paint pigment, the material in lead pencils and as jewellery.

Now answer the question in 43 once again.

---

62 Your answer is quite right.

Soot is easier to burn than graphite. Diamond is the most difficult to burn.

Carbon forms two oxides.

Jot down the names and formulas of these oxides, and then check your answers at 46 .

---

63 *Soot* is dark black in color.

Graphite is grayish-black, soft and has an oily luster. It is a good conductor of electric current.

Diamond is one of the hardest known substances, transparent and sparkling in appearance.

Jot down now in the form of key words the uses which have been found for the three allotropes of carbon.

(The properties listed above may provide some useful hints!)

Check your answers at 42 .

---

64 $H_2O + CO_2 \rightleftarrows H_2CO_3$

Hopefully you didn't forget the double arrow! The double arrow indicates that the reaction can proceed in either direction.

Carbonic acid is a weak and unstable acid. Like all other acids it dissociates.

It breaks up into positively charged hydrogen ions and a doubly negatively charged acid radical, the carbonate ion.

Write out the relevant equation for this dissociation. → [51]

---

[65] Now check your equations:

1. $2\,C + O_2 \rightarrow 2\,CO$

2. $C + O_2 \rightarrow CO_2$

The second equation describes the production of CO_2 by the method of passing air over red-hot coke.

What other process do you know of for the production of carbon dioxide?

Jot down the relevant key words and formulas. → [55]

---

[66] Your answer is wrong.

15

Go back now to [44] .

---

[67] Carbonic acid is a weak acid. It dissociates very little into positively charged hydrogen ions and a negatively charged acid radical.

Your answer is thus wrong. The concentration of the $H^{\oplus}$ ions can only be *small.*

Read through the explanations given in [51] once again.

---

[68] No, carbon occurs in the form of several allotropes.

Think, for instance, of the uses of the various allotropes as a paint pigment, the material for "lead" pencils and as jewellery.

Now answer the question in [43] once again.

Program 16

Carbon (II)

1 In this Program we shall discuss the most important salts of carbonic acid and their preparation.

We shall assume here that you are familiar with the properties of carbonic acid; these were described in some detail in the preceding Program 15.

To refresh your memory, we start with some questions on carbonic acid which were covered previously.

Jot down the chemical formula for carbonic acid, and then check your answer at [9] .

2 Na_2CO_3: sodium carbonate, soda.

Sodium carbonate can be manufactured by three different processes. What are the two processes named after their inventors?

These are the .... Process and the .... Process. → [62]

3 Your answer is wrong. $NaHSO_4$ is certainly a hydrogen salt, though this does not justify the conclusion that an aqueous solution of it will react as an acid.

$NaHCO_3$ is also a hydrogen salt, yet its aqueous solution has an alkaline reaction.

The decisive factor which determines the type of reaction taking place when a hydrogen salt reacts with water is the strength of the acid.

Is sulfuric acid a strong acid?

- Yes → [43]
- No → [70]

4 Your answer is quite right.

The full equation for the second stage of the Leblanc Process is as follows:

$Na_2SO_4 + 2\ C \rightarrow Na_2S + 2\ CO_2$

The *third* and last *stage* of the Leblanc Process is represented by the equation below:

$Na_2S + CaCO_3 \rightarrow Na_2CO_3 + CaS$

What are the *chemical* names of each of the compounds which appear in this equation in sequence from left to right?

- Sodium sulfide, calcium carbonate, soda, calcium sulfide → 19
- Sodium sulfite, calcium carbonate, sodium carbonate, calcium sulfite → 12
- Sodium sulfide, calcium carbonate, sodium carbonate, calcium sulfide → 26

---

5 Your answer is almost correct.

You seem to have forgotten, however, that carbonic acid is unstable and thus breaks down.

Go back now to 33 .

---

6 Your answer is wrong.

An explanation of this is given at 10 .

7

$2\ NaCl + H_2SO_4 \rightarrow Na_2SO_4 + 2\ HCl$

$Na_2SO_4 + 2\ C \rightarrow Na_2S + 2\ CO_2$

$Na_2S + CaCO_3 \rightarrow Na_2CO_3 + CaS$

The third process for the manufacture of soda starts out from caustic soda and carbon dioxide. Write down the relevant equation for this process. → 66

---

8 Your answer is wrong. The reaction of sodium hydrogen carbonate and aqueous hydrochloric acid has just been discussed in some detail. We mentioned that a salt and water are formed as products.

Continue your reading at 20 .

9 H_2CO_3

Which of the statements below applies to carbonic acid?

- It is a weak and unstable acid → 30
- It is a strong and stable acid → 23
- It is a weak but stable acid → 16

10 By a neutralization we mean the reaction of a base with an acid or acid anhydride. A salt and water are formed as products.

$$HCl + NaOH \rightarrow NaCl + H_2O$$

$$2\ NaOH + CO_2 \rightarrow Na_2CO_3 + H_2O$$

However, in the reaction

$$NaHCO_3 + HCl \rightarrow NaCl + H_2O + CO_2$$

there is no base at all involved. Rather what has happened is that the hydrochloric acid has driven the weaker and more volatile carbonic acid from its salt. This equation could also be written in two steps as follows:

$$NaHCO_3 + HCl \rightarrow NaCl + H_2CO_3$$

H_2CO_3 is very unstable and immediately breaks down further:

$$H_2CO_3 \rightarrow CO_2 + H_2O$$

When added together, these two equations yield the overall equation:

$$NaHCO_3 + HCl \rightarrow NaCl + H_2O + CO_2$$

We have previously come across other examples of weak, volatile acids being driven from their salts by stronger acids.

For instance, we know that iron sulfide reacts with hydrochloric acid to form hydrogen sulfide:

$$FeS + 2\ HCl \rightarrow FeCl_2 + H_2S$$

and that sodium sulfite reacts with hydrochloric acid to form sulfurous acid, which then readily decomposes further:

$$Na_2SO_3 + 2\ HCl \rightarrow 2\ NaCl + H_2SO_3$$

$$H_2SO_3 \rightarrow SO_2 + H_2O$$

Now complete the following sentence:

When sodium hydrogen carbonate and aqueous hydrochloric acid react together the weaker and more volatile .... is driven from its salt.

Continue your reading at 59 .

---

11 Sodium sulfate should be reduced by carbon to sodium sulfide. In the process there is one by-product formed.

If we write down the relevant equation we obtain:

$$Na_2SO_4 + C \rightarrow Na_2S + ?$$

On the left hand side there are clearly four O atoms and also a carbon which will have to be accounted for on the right hand side of the equation. The carbon is, of course, the substance which reduces the sodium sulfate, i.e. it *removes oxygen* from it, and in the process itself combines with the oxygen. The by-product formed is thus carbon dioxide. If we now insert this by-product into the equation we finish up with:

16

$$Na_2SO_4 + C \rightarrow Na_2S + CO_2$$

This equation is, however, still not in its final form, for on the left hand side we still have two oxygen atoms too many.

We shall therefore have to take twice as much carbon.

Now write out the correct version of this equation. Bear in mind that, for each element appearing in the equation, the *same number* of atoms must appear on both the right and left hand sides of the equation. → 4

---

12 You appear to have confused the salts of sulfurous acid with the salts of hydrogen sulfide.

The salts of sulfurous acid are called sulfites.

The salts of hydrogen sulfide are called sulfides.

You are thus advised to take another good look at 4 .

---

13 Have another good look at the last equation in 33 . All you need do is to write this equation the other way round. You should not forget though that carbonic acid is not stable.

---

14 $NaOH + CO_2 \rightarrow NaHCO_3$

or $NaOH + H_2CO_3 \rightarrow NaHCO_3 + H_2O$

$NaHCO_3$ is called sodium hydrogen carbonate.

When $NaHCO_3$ is heated it decomposes. Write down the relevant equation for this reaction. → 63

---

15 In the first stage we start with sodium chloride and sulfuric acid.

The sodium sulfate produced is then reacted with carbon in the second stage. In this reaction sodium sulfide and one by-product are formed.

In the third stage, the sodium sulfide and calcium carbonate are reacted together. In this reaction sodium carbonate and another by-product are formed.

Now set up the three relevant equations, and then check them at 34 .

---

16 Your answer is only partially correct.

Carbonic acid is a weak acid. It decomposes very readily according to the equation:

$$H_2CO_3 \rightarrow CO_2 + H_2O$$

Now write out the completed version of the sentence below:

Carbonic acid is a weak, ... (stable/unstable) acid. → 30

---

17 Now check your answer: $NaHCO_3$

Sodium hydrogen carbonate is a hydrogen salt. In aqueous solution, it dissociates initially into a positively charged sodium ion and a negatively charged hydrogen carbonate ion.

Write out the equation for this dissociation. → 27

---

18 Your answer is wrong.

An explanation of this is given at 11 .

---

19 Your answer is only partially right.

We actually asked you for the *chemical* names of the substances appearing in this equation. The chemical name for Na_2CO_3 is *sodium carbonate* and not soda. You should always make a distinction between the *chemical name* and the *trivial name* of a substance.

Continue your reading at 26 .

---

20 Your answer is quite right.

When heated, sodium hydrogen carbonate decomposes into sodium carbonate, water and carbon dioxide thus:

$$2\ NaHCO_3 \rightarrow Na_2CO_3 + H_2O + CO_2$$

Sodium hydrogen carbonate is used as a medicament for heartburn.

In the stomach of every healthy person there is a small amount of dilute hydrochloric acid. This "stomach acid" plays an important part in the digestion process. If too much of the acid is present, however, the very unpleasant symptoms associated with heartburn are experienced. The ability of sodium hydrogen carbonate to alleviate heartburn rests on the fact that it can react with the excess hydrochloric acid in the stomach. The relevant equation is:

$$NaHCO_3 + HCl \rightarrow NaCl + H_2O + CO_2$$

What type of reaction is this?

- A neutralization → 6
- The driving of a relatively weak and volatile acid from its salt by a stronger and less volatile acid → 59
- I am not sure, and would find some further clarification useful → 10

21 Your answer is wrong. We expressly said that:

Carbonic acid is a *poly*basic acid.

Its chemical formula is H_2CO_3. See if you can now work out the correct designation.

- It is a dibasic acid → 44
- It is a tribasic acid → 37

22 Your answer is wrong. You should decide which of the four substances has to be manufactured by chemical processes, i.e. which does not occur in nature.

If you now feel you can answer the question in 39, you should turn back to it. If not, you will find a detailed explanation at 67.

23 Your answer is wrong. We have mentioned that:

1. Carbonic acid is driven from its salts by stronger and less volatile acids, e.g.

$$Na_2CO_3 + 2\ HCl \rightarrow 2\ NaCl + CO_2 + H_2O$$

2. Carbonic acid breaks down when it is warmed up:

$$H_2CO_3 \rightarrow CO_2 + H_2O$$

Write out these two equations, and then copy down the completed version of the sentence below:

Carbonic acid is a .... (strong/weak), ... (stable/unstable) acid. → 30

24 $2\ KOH + CO_2 \rightarrow K_2CO_3 + H_2O$

K_2CO_3 is called potassium carbonate or potash.

The calcium salt of carbonic acid is found in large quantities in nature. Three different minerals consist of calcium carbonate.

What are the names of these minerals? → 56

25 Your answer is only partially right.

Carbon dioxide is formed when sodium sulfate is reduced with carbon.

However, in any equation the *same number* of each type of atom present must appear on both sides of the equation. Two atoms of carbon will be required to combine with the four oxygen atoms of the sodium sulfate.

Now write out the correct equation, and then check it at 4 .

26 Your answer is quite right.

In the third stage of the Leblanc Process, sodium sulfide is reacted with calcium carbonate to produce sodium carbonate and calcium sulfide. This three-stage process thus finally leads us to the desired end product, sodium carbonate.

To review what we have covered here, you should now write out the three equations for the Leblanc Process, and then check them at 34 .

If you are no longer able to recall what happens in the first two stages, turn instead to 15 .

16

27 $NaHCO_3 \rightarrow Na^{\oplus} + HCO_3^{\ominus}$

The ions interact with the surrounding water, which itself is also dissociated to some extent. With the hydrogen ions from the water, $HCO_3^{\ominus}$ forms a limited amount of carbonic acid. Whereas water and carbonic acid hardly dissociate at all, sodium hydroxide dissociates fully.

$$H_2O \rightleftarrows H^{\oplus} + OH^{\ominus}$$

$$Na^{\oplus} + HCO_3^{\ominus} + H_2O \rightleftarrows Na^{\oplus} + OH^{\ominus} + H_2CO_3$$

Accordingly, there are more $OH^{\ominus}$ ions than $H^{\oplus}$ ions in aqueous solutions of sodium hydrogen carbonate. Such solutions will therefore have a weak alkaline reaction.

This observation applies to any salt made during the reaction of a strong base with a weak acid. This topic will be discussed in detail in Part II of this book.

Certain other hydrogen salts react as acids when they are dissolved in water, e.g. $NaHSO_4$, sodium hydrogen sulfate.

Why does an aqueous solution of $NaHSO_4$ react as an acid?

- Because $NaHSO_4$ is the hydrogen salt of a strong acid → 43
- Because $NaHSO_4$ is a hydrogen salt → 3
- I need some help → 70

28 Your answer is wrong. Na_2SO_4 and Na_2S are *not by-products.*

In fact they are *intermediates* which get used up later on in the process.

Go back now to 41 .

29 $2\ NaOH + CO_2 \rightarrow Na_2CO_3 + H_2O$

For **2** mol of sodium hydroxide we therefore need **1** mol of carbon dioxide.

If we were to take *more* CO_2 than this, that is we took **2** mol of CO_2 for **2** mol of NaOH or **1** mol of CO_2 for **1** mol of NaOH, we would *not* finish up with sodium carbonate. We would then obtain another salt of carbonic acid known as sodium hydrogen carbonate.

Write down the chemical formula for sodium hydrogen carbonate. → 17

30 Your answer is quite right.

Carbonic acid is a *weak, unstable* acid. It is also a polybasic acid.

Which of the designations below applies to carbonic acid?

- It is a monobasic acid → 21
- It is a dibasic acid → 44
- It is a tribasic acid → 37

31 Your answer is wrong. Decide which of the four substances have to be produced by chemical processes, i.e. do not occur naturally.

You should now be able to attempt the question in 39 and come up with the right answer.

If you feel you can, go back to 39. If not, then a detailed explanation is to be found at 67.

---

32 Your answer is quite correct.

In the first stage of the Leblanc Process, sodium chloride is reacted with sulfuric acid to form *sodium sulfate* and hydrochloric acid. The strong and non-volatile sulfuric acid drives the less strong and volatile hydrochloric acid from its salt.

In the *second stage* of the Leblanc Process, the sodium sulfate which has been produced is reduced with carbon. This reduction leads to the formation of sodium sulfide, i.e. oxygen removal takes place at this stage.

Write down now the relevant equation for this reduction. You will then see that, in addition to sodium sulfide, one other by-product is formed.

What is this other by-product?

- CO_2 → 25
- 2 CO → 18
- 2 CO_2 → 4
- I don't know → 11

16

---

33 Your answer is quite right.

As there is an ample supply of CO_2 available, from the moment it is first bubbled in until the solution becomes saturated with it, the salt formed will be the one requiring the greater amount of CO_2 for its formation, namely $NaHCO_3$.

What would be obtained if we were to pass CO_2 into a solution of sodium carbonate? Now, CO_2 dissolved in aqueous solution means that we are effectively adding carbonic acid, for CO_2 and water react together to produce this acid.

$$CO_2 + H_2O \rightleftarrows H_2CO_3$$

$$Na_2CO_3 + H_2CO_3 \rightarrow 2\ NaHCO_3$$

We shall therefore obtain sodium hydrogen carbonate. From this equation it is quite clear that the production of *sodium hydrogen carbonate* requires *more carbonic acid* than is required for the formation of sodium carbonate. Sodium hydrogen carbonate is formed from sodium carbonate and carbonic acid.

When sodium hydrogen carbonate is heated, the process described by the second of the above equations proceeds from right to left.

Write out now the equation for the reaction which occurs when sodium hydrogen carbonate is heated. What are the reaction products formed?

- Sodium carbonate, water and carbonic acid → 13
- Sodium carbonate, water and carbon dioxide → 20
- Sodium carbonate and carbonic acid → 5

---

34 Now check your answers:

Stage I: $2\ NaCl + H_2SO_4 \rightarrow Na_2SO_4 + 2\ HCl$

Stage II: $Na_2SO_4 + 2\ C \rightarrow Na_2S + 2\ CO_2$

Stage III: $Na_2S + CaCO_3 \rightarrow Na_2CO_3 + CaS$

Jot down the name of this three-stage process for the production of sodium carbonate.

It is the .... Process → 41

---

35 Your answer is wrong. By a neutralization is meant a reaction between an acid and a base with the formation of a salt and water. An example of a neutralization is the reaction of hydrochloric acid with caustic soda:

$$HCl + NaOH \rightarrow NaCl + H_2O$$

Now have another look at the equation in 44 .

---

36 $CaCO_3$

The three different minerals consist of calcium carbonate.

Jot down the names of these minerals, and then check your answers at 61 .

---

37 Your answer is wrong. In the case of phosphoric acid, H_3PO_4, we speak of a *tri*basic acid, for when it dissociates it is capable of splitting off three (positively charged) hydrogen ions.

After having a good look at the formula for carbonic acid, go back to 30 .

---

38 Dolomite

End of Program 16.

---

39 Your answer is quite correct.

The starting materials are:

$NaCl$, H_2SO_4, $CaCO_3$ and C.

Which of these starting materials are to be found in nature?

- All of them → 22
- Three of them → 53
- Two of them → 31
- One of them → 47
- None of them → 54

16

---

40 You have stuck rather too rigidly to the definition of neutralization in this case.

Normally, by neutralization we mean a reaction between an *acid* and a *base*. However, if we make use of an *acid anhydride* instead of an acid in a given reaction, we can still speak of a neutralization.

Write out the equation for the formation of sodium carbonate during the complete neutralization of sodium hydroxide with the anhydride of carbonic acid. Check your equation at 29 .

---

41 The *Leblanc* Process

The Leblanc Process is uneconomical as it produces *by-products* which cannot be used. The by-products formed in a chemical process are those substances produced during the course of the process which cannot be used in the manufacture of the main product.

What are the by-products formed in the Leblanc Process?

Take a look at the equations for the three stages of this process!

- CaS → 48
- HCl, CO_2 and CaS → 50
- Na_2SO_4, Na_2S → 28
- Na_2SO_4, Na_2S, HCl, CO_2 and CaS → 55

42

Your answer is wrong. "Until saturated" means that as much CO_2 as possible should be passed in.

The salt which will then be formed must be the one requiring the greater amount of CO_2.

Which salt in fact requires the greater amount of CO_2 for its preparation?

- Na_2CO_3 → 68
- $NaHCO_3$ → 33

43

Yes, because $NaHSO_4$ is the hydrogen salt of a strong acid, its aqueous solution has an acid reaction.

The $HSO_4^{\ominus}$ ion dissociates completely in water:

$$HSO_4^{\ominus} \rightarrow H^{\oplus} + SO_4^{2\ominus}$$

The $H^{\oplus}$ ions are responsible for the acid reaction.

Please note that:
Not all hydrogen salts display an acid reaction when dissolved in water. In fact it is the strength of the acid which determines in each case whether the hydrogen salt will have an acid, neutral or weakly alkaline reaction.

We now return to our discussion on the preparation of sodium carbonate and sodium hydrogen carbonate. Both of these substances can be prepared by passing carbon dioxide into a sodium hydroxide solution.

Write down the two relevant equations, and then check them at 51 .

---

44 Correct, carbonic acid is a *di*basic acid.

The most important salt of carbonic acid is *sodium carbonate*, Na_2CO_3, also known a *soda*. As soda is virtually non-existent in nature, the enormous quantities consumed by the chemical industry have to be produced synthetically. There are three different processes for the manufacture of soda, namely:

1. The Leblanc Process
2. The Solvay Process
3. A process based on the reaction of caustic soda with carbon dioxide.

The first two of these processes have been named after their discoverers.

We start now by considering the *Leblanc Process*. The process was the first to manufacture soda industrially. It was replaced later by the more economical Solvay process, but its chemical reaction remain interesting. The process involves three separate stages. The *first stage* is the reaction of sodium chloride with sulfuric acid thus:

$$2\ NaCl + H_2SO_4 \rightarrow Na_2SO_4 + 2\ HCl$$

What does this equation actually represent?

16

- A neutralization → 35
- The driving of an acid from its salt by a stronger and less volatile acid → 32

---

45 Your answer is quite right.

Normally, by neutralization we mean a reaction between an *acid* and a *base*. However, if we make use of an *acid anhydride* instead of an acid in the reaction, we can still speak of a neutralization.

Write out the equation for the formation of sodium carbonate during the complete neutralization of sodium hydroxide with the anhydride of carbonic acid. Check your equation at 29 .

---

46 When CO_2 is passed into a solution of sodium hydrogen carbonate, nothing at all happens. You must be confusing this with something we discussed earlier on. → 33

47 Your answer is wrong. Decide which of the four substances have to be produced by chemical processes, i.e. do not occur naturally.

You should now be able to attempt the question in 39 and come up with the right answer.

If you feel you can, go back to 39 . If not, then a detailed explanation is to be found at 67 .

48 Your answer is incomplete.

It is not only during the third stage that a by-product is formed, but also during the first and second stages of the process!

Go back now to 41 .

49 Yes, the reaction with caustic soda leads to the formation of sodium hydrogen carbonate thus:

$$NaHCO_3 + NaOH \rightarrow Na_2CO_3 + H_2O$$

Another important carbonate is *potassium* carbonate, K_2CO_3, also known as *potash*. Potassium carbonate may be prepared by passing carbon dioxide into caustic potash solution.

Write out the relevant equation for this reaction, and then check it at 58 .

50 Your answer is quite right.

HCl, CO_2 and CaS are the by-products of the Leblanc Soda Process.

Na_2SO_4 and Na_2S are the *intermediates* formed during the process which get used up later on in the process.

What starting materials are necessary for the Leblanc Soda Process?

- NaCl and H_2SO_4 → 57
- NaCl, H_2SO_4, Na_2SO_4, Na_2S, $CaCO_3$ and C → 64
- NaCl, H_2SO_4, $CaCO_3$ and C → 39

51

1. $2\ NaOH + CO_2 \rightarrow Na_2CO_3 + H_2O$

2. $NaOH + CO_2 \rightarrow NaHCO_3$

According to which of these equations will more CO_2 per mole of NaOH be required?

- According to equation 1 → 68
- According to equation 2 → 60

52

$MgCO_3$

Magnesium carbonate too is found in large quantities in nature. Together with calcium carbonate it forms the mineral *dolomite.*

To revise what we have covered so far, you are asked to write out the completed version of the sentence below:

Carbonic acid is a .... (strong/weak), ... (stable/unstable) acid. → 65

53

You are right.
The following are found in nature:

NaCl,	sodium chloride, table salt
$CaCO_3$,	calcium carbonate, chalk
C,	carbon, coal.

The sulfuric acid, H_2SO_4, however, has to be produced by a chemical process.

Although three of the starting materials used in the Leblanc Soda Process are found in nature and are therefore cheaply available, the overall process is uneconomic because of by-products formed which cannot be used.

Nowadays, soda is manufactured almost exclusively by the more economical *Solvay Process*. This process will be further discussed in Program 19, frame 7.

A *third* process for the manufacture of sodium carbonate essentially involves pumping carbon dioxide into caustic soda:

$$2\ NaOH + CO_2 \rightarrow Na_2CO_3 + H_2O$$

Chemically speaking, does this equation represent a neutralization?

- Yes → 45
- No → 40
- I would rather not guess and would appreciate some further clarification → 69

---

54 Your answer is wrong. Decide which of the four substances have to be produced by chemical processes, i.e. do not occur naturally.

You should now be able to attempt the question in 39 and come up with the right answer.

If you feel you can, go back to 39 . If not, then a detailed explanation is to be found at 67 .

---

55 Your answer is wrong. Some of the compounds selected by you as by-products are in fact used up in the course of the process! Note that products which get used up in a process are *not* by-products.

Go back now to 41 .

---

56 Limestone, marble and chalk.

Large quantities of magnesium carbonate too are found in nature, usually together with calcium carbonate.

What is the name of this composite mineral? → 38

---

57 Your answer is incomplete.

The starting materials Stage I are not the only starting materials used in the process. Materials also have to be introduced at Stages II and III of the process.

Go back now to 50 .

---

58 $2\ KOH + CO_2 \rightarrow K_2CO_3 + H_2O$

Another very important carbonate is *calcium* carbonate, $CaCO_3$. A particularly common mineral form of calcium carbonate is limestone. In fact, it forms the basis of whole mountain ranges and can be obtained by mining operations. A very pure form of limestone is the crystalline and hard material known as *marble*. Calcium carbonate is also found in the form of *chalk*.

Write down now the chemical formula of marble, and then check your answer at 36 .

59 Your answer is quite right.

When sodium hydrogen carbonate and aqueous hydrochloric acid are mixed, the weaker and more volatile *carbonic acid* is driven from its salt. As carbonic acid is very unstable, it immediately breaks down into water and carbon dioxide.

Sodium carbonate may be prepared by heating sodium hydrogen carbonate.

There is also one further way of making sodium carbonate from sodium hydrogen carbonate.

What is this further way based on?

If you write out equations for the three reactions a), b) and c) below, reaching the right answer should be easier.

a) The action of aqueous hydrochloric acid on sodium hydrogen carbonate → 8

b) The action of caustic soda on sodium hydrogen carbonate → 49

c) The passing of CO_2 into a solution of sodium hydrogen carbonate → 46

60 Your answer is quite correct.

To produce $NaHCO_3$, we need *more* CO_2 per mole of NaOH than to produce Na_2CO_3.

Which salt would be obtained if we were to pass CO_2 into a sodium hydroxide solution until the latter is saturated with the gas?

- Na_2CO_3 → [42]
- $NaHCO_3$ → [33]

61

Calcium carbonate is found in the form of *limestone, marble* and *chalk*.

A further important carbonate is magnesium carbonate.

Write down its formula, remembering that magnesium, Mg, is divalent. → [52]

62

The *Leblanc* and *Solvay* Processes

The Leblanc Process has three stages.

Write down the three equations covering the three stages of the Leblanc Process. → [7]

63

$$2\ NaHCO_3 \xrightarrow{\text{heat}} Na_2CO_3 + CO_2 + H_2O$$

By passing CO_2 into a caustic potash solution a salt of carbonic acid is formed.

Write down the relevant equation for this reaction and give both the chemical and trivial name of the salt formed. → [24]

64

Your answer is wrong. Na_2SO_4 and Na_2S are themselves *produced* during the process and subsequently used up again. They *cannot* therefore be starting materials.

Go back now to [50] .

65

Carbonic acid is a *weak, unstable* acid. The most important salt of carbonic acid is Na_2CO_3.

Jot down both the chemical and trivial name of this salt. → [2]

66

$$2\ NaOH + CO_2 \rightarrow Na_2CO_3 + H_2O$$

When CO_2 is passed into a caustic soda solution until the saturation point is reached, the salt formed is not soda but another salt. Write down the

16

relevant equation for this reaction and name the salt formed. Check your answer at [14] .

---

67 We asked you which of the starting materials for the Leblanc Process ($NaCl$, H_2SO_4, $CaCO_3$ and C) are found in nature.

Three of these substances are found in large quantities in the earth's crust.

Only one of these substances has to be produced by means of a chemical process. Think here of the Lead Chamber Process or of the Contact Process.

Now complete the sentences below:

The starting materials found in nature are:

1. ....
2. ....
3. ....

A chemical process is however necessary for the production of ....

Check your answers at [53] .

---

68 Your answer is wrong. Take a good look at the latter two of the equations you have written down.

To produce Na_2CO_3, we shall require for:

2 mol of $NaOH$ 1 mol of CO_2

1 mol of $NaOH$ 1/2 mol of CO_2

To produce $NaHCO_3$, we shall require for:

1 mol of $NaOH$ 1 mol of CO_2

We thus require *twice* the amount in this latter case!

Now complete the sentence below:

To prepare $NaHCO_3$, we require per mole of $NaOH$ .... (more/less) CO_2 than to prepare Na_2CO_3. → [60]

---

69 Normally, by neutralization we mean a reaction between an acid and a base. However, if we make use of an acid anhydride instead of an acid in the reaction, it is still possible to speak of a neutralization.

Write out the equation for the formation of sodium carbonate during the complete neutralization of sodium hydroxide with the anhydride of carbonic acid. Check your equation at 29 .

---

70 Because the ions of a salt will interact with the ions present in water, the possibility exists for the hydrogen ions of weak acids to form the undissociated acid. In such a solution, the $OH^{\ominus}$ ions will predominate and so its reaction will be an alkaline one.

Sulfuric acid, however, is a strong acid which is always fully dissociated in water. The aqueous solution of $NaHSO_4$ will therefore contain more $H^{\oplus}$ ions.

Answer the question in 27 once again.

Program 17

Carbon (III)

1 In the previous Program we discussed the most important salts of carbonic acid. Among others, we mentioned calcium carbonate, large quantities of which occur naturally in the form of limestone.

Limestone is a very important raw material for the chemical industry.

In this Program we shall be concerned primarily with the use of limestone and its by-products in the construction industry. We shall learn, for instance, why mortar gets hard when it sets.

Limestone (or calcite) is the starting material for the manufacture of *quicklime*. Quicklime is produced commercially by *calcining* (roasting) limestone.

You have already seen the equation for this process of calcination. It was in fact introduced in our discussion on carbon dioxide.

See if you can write down now the equation for the calcination of limestone:

$$CaCO_3 \xrightarrow{\text{heat}}$$

Check your equation at 44 .

17

2 You have either used the wrong formulas or your calculation is wrong.

An explanation of this is given at 49 .

3 1. Quicklime is calcium oxide. It is formed during the calcination of calcium carbonate (limestone) thus:

$$CaCO_3 \rightarrow CaO + CO_2$$

2. Slaked lime is calcium hydroxide, $Ca(OH)_2$.

3. Slaked lime is a strong base which is corrosive.

4. a) $Ca(OH)_2 + 2\ HCl \rightarrow CaCl_2 + 2\ H_2O$

   b) $Ca(OH)_2 + H_2SO_4 \rightarrow CaSO_4 + 2\ H_2O$

5. H_2CO_3 $CaCO_3$

   $Ca(HCO_3)_2$ $NaHCO_3$

6. Initially, a precipitate of calcium carbonate is formed. This dissolves again if further CO_2 is bubbled in; during the process calcium hydrogen carbonate, $Ca(HCO_3)_2$, is produced.

7. When such a solution is heated the soluble calcium hydrogen carbonate gets converted into insoluble calcium carbonate (limescale) thus:

$$Ca(HCO_3)_2 \rightarrow CaCO_3 + CO_2 + H_2O$$

8. Hard water contains dissolved salts, especially calcium hydrogen carbonate. A soap solution forms a lather with hard water only with some difficulty. Soft water (rainwater or distilled water) contains virtually no dissolved salts. For this reason such water tastes insipid.

9. Mortar is a mixture of sand and slaked lime, $Ca(OH)_2$.

10. The CO_2 present in air reacts with the calcium hydroxide to form calcium carbonate and water thus:

$$Ca(OH)_2 + CO_2 \rightarrow CaCO_3 + H_2O$$

11. CO_2 is formed during the combustion of coke.

12. Dilute hydrochloric acid is poured over the unknown sample. A gas is formed which produces an effervescence in the sample. The gas can be identified by passing it into a barium hydroxide solution. If the gas is CO_2, a white precipitate of barium carbonate will be formed.

End of Program 17.

---

4 The reaction which takes place between calcium hydroxide and carbon dioxide is a neutralization.

We have already come across such a reaction in our discussion on the reaction of caustic soda with CO_2. Depending on the amount of CO_2 bubbled into the caustic soda solution, either Na_2CO_3 or $NaHCO_3$ is formed.

Using this as your guideline, you should now be able to answer the question in 37 .

17

5 You have obviously not yet fully appreciated the difference between carbonates and hydrogen carbonates.

An explanation of this is given in 49 .

6 The reaction of carbonates with acids may be used as a test for carbonates. Carbonic acid is a very weak and unstable acid. It can be driven from its salts by less volatile and more stable acids as indicated:

$$Na_2CO_3 + 2\ HCl \rightarrow 2\ NaCl + H_2CO_3$$

The carbonic acid formed immediately breaks down as follows:

$$H_2CO_3 \rightarrow CO_2 + H_2O$$

When carbonates react with acids we thus observe the formation of a gas. The gas formed is ....

Jot down the missing word, and then check your answer at 41 .

7 No, acids are compounds which form positively charged hydrogen ions when they dissociate in water.

$Ca(OH)_2$ is a metal hydroxide. No hydrogen ions are formed when metal hydroxides dissociate.

Think, for instance, of what happens in the case of sodium or potassium hydroxide.

Now make another attempt at the question in 20 .

8 You have answered only part of the question.

Calcium hydrogen carbonate is obtained when carbon dioxide is bubbled into an aqueous suspension of calcium carbonate. The relevant equation for this reaction is:

$$CaCO_3 + H_2O + CO_2 \rightarrow Ca(HCO_3)_2$$

We asked you, however, how calcium hydrogen carbonate may be prepared from calcium carbonate and how the former can then be *converted back again* into *calcium carbonate.*

Read through [31] carefully again.

---

9 Your answer is quite right.

Limescale is calcium carbonate which is formed when calcium hydrogen carbonate is heated thus:

$$Ca(HCO_3)_2 \rightarrow CaCO_3 + H_2O + CO_2$$

Whenever hard water is used for industrial purposes, e.g. in boilers, turbines or steam engines, the formation of scale eventually leads to a breakdown in operations. For this reason such water is normally first converted into soft water (or *softened*).

We shall now turn to a discussion of the use of slaked lime (calcium hydroxide) in the construction industry. When buildings are constructed, the gaps between the bricks are filled with *mortar.* After a period of time this mortar becomes hard. *The mortar is said to set.*

What is mortar and why does it harden after some time? You now have sufficient chemical background for us to tackle this very interesting question.

Mortar is a *mixture of slaked lime* (clacium hydroxide) *and sand.* The sand acts basically as a filler and does not take part in the chemical reactions involved.

Write down now the equations for the chemical processes which take place during:

a) The manufacture of quicklime

b) The manufacture of slaked lime

Your answers may be checked by turning to [36] .

---

10 The incomplete equation has two hydrogen atoms on both its left and right hand sides. This means that there are no hydrogens left over for the formation of barium hydrogen carbonate. Accordingly, your answer was wrong.

Have another look at the equation in [40] .

---

11 You have obviously not yet fully appreciated the difference between carbonates and hydrogen carbonates.

You are therefore advised to read through the explanation in 26 with special care.

12 You have obviously not yet fully appreciated the difference between carbonates and hydrogen carbonates.

An explanation of this is to be found in 49 .

13 Your answer is wrong.

An explanation of this is given in 6 .

14 Your answer is wrong.

We already know that:

Hard water contains dissolved salts as impurities. Examples of hard water include the water from springs and household water. Hard water forms an insoluble precipitate with a soap solution. For this reason hard water is not very effective for washing purposes.

Soft water is virtually free from any dissolved impurities. Examples of soft water include rainwater and, in particular, distilled water. Soap is readily soluble in soft water and so soft water is effective for washing purposes.

Continue your reading at 33 .

15 Your answer is wrong. On the left hand side of the equation there is only one barium atom available. It is thus not possible to make use of two barium atoms on the right hand side.

Have another good look at the equation in 40 .

16 Your answer is wrong. Calcium hydrogen carbonate is obtained when carbon dioxide is bubbled into an aqueous suspension of calcium carbonate. The relevant equation for this reaction is:

$$CaCO_3 + H_2O + CO_2 \rightarrow Ca(HCO_3)_2$$

It would, however, be *wrong* to apply heat here. If this is done, *no* $Ca(HCO_3)_2$ would be formed as it is unstable when heated.

Read through the text carefully again starting from [19] .

---

[17] When calcium carbonate is calcined *calcium oxide* is produced. The latter substance can react with water to form *calcium hydroxide.*

Calcium oxide = CaO
Calcium hydroxide = $Ca(OH)_2$

The calcium ion bears a *doubly* positive charge. Accordingly, it is capable of combining with *two* hydroxyl ions.

Calculate now the equivalent mass of calcium hydroxide. The relevant atomic masses are Ca = 40, O = 16, H = 1.

The equivalent mass is:

- 74 → [29]
- 37 → [46]
- 148 → [35]

---

[18] You still do not fully understand the relationship of calcium carbonate to calcium hydrogen carbonate.

The basic notions have been outlined in [4] .

---

[19] Your answer is quite right.

Initially, when carbon dioxide is bubbled into an aqueous suspension of calcium carbonate, the gas reacts with the water to form carbonic acid. The carbonic acid formed then has the effect of converting the calcium carbonate into calcium hydrogen carbonate.

As $CaCO_3$ is fairly insoluble in water whereas $Ca(HCO_3)_2$ is readily soluble in water, the $CaCO_3$ is gradually taken up into solution as $Ca(HCO_3)_2$.

Write out the relevant equation for this reaction, and then check your answer at [43] .

---

20 $CaO + H_2O \rightarrow Ca(OH)_2$

The product is known as slaked lime or, more precisely, as calcium hydroxide.

What is calcium hydroxide?

- A salt → 28
- An acid → 7
- A base → 30

21 Your answer is wrong. Of course calcium hydroxide reacts with carbon dioxide! Calcium hydroxide is in fact a base and carbon dioxide is an acid anhydride.

The way in which these two substances react together is explained in 4 .

22 You have obviously not yet fully appreciated the difference between carbonates and hydrogen carbonates.

An explanation of this is given in 49 .

23 Your answer is only partially right.

For two of the four compounds you have given the *trivial* names instead of the *chemical* names.

Read through 38 once again.

24 Correct, this process is referred to as *setting*.

The water which is gradually given off during the setting reaction of the mortar causes new buildings to remain damp for a long time after they have been built.

The setting process can be speeded up by putting open coke fires into such buildings.

How will this help to speed up the setting process?

- The slowly burning coke will give off carbon dioxide → 39

- The rooms will be heated by the burning coke and so dry out quicker → 51

---

25 Mortar is a mixture of *sand* and *slaked lime* (calcium hydroxide).

When mortar hardens or sets the calcium hydroxide reacts with carbon dioxide, which is always present in small amounts in the air.

Write out the relevant equation for this reaction.

What products are formed during the reaction?

- Calcium carbonate → 45
- Calcium carbonate and water → 34

---

26 There are two series of carbonates, namely the carbonates themselves and the hydrogen carbonates. The chemical formulas for the two calcium carbonates are as follows:

$CaCO_3$	$Ca(HCO_3)_2$
I	II

17

In formula I both of the hydrogen atoms of carbonic acid are replaced by calcium. This formula represents calcium carbonate.

In formula II only one of the two hydrogen atoms in carbonic acid has been replaced by calcium. This formula represents calcium hydrogen carbonate. Note that in hydrogen salts of carbonic acid only one of the two hydrogen atoms is replaced by a metal atom.

You no doubt know that carbonic acid dissociates in two stages. In the first stage the negatively charged hydrogen carbonate ion is formed thus:

$$H_2CO_3 \rightleftarrows H^{\oplus} + HCO_3^{\ominus}$$

As calcium ions bear two positive charges, in calcium hydrogen carbonate two hydrogen carbonate ions are bonded to one calcium ion:

$$Ca^{2\oplus} \begin{matrix} HCO_3^{\ominus} \\ HCO_3^{\ominus} \end{matrix}$$

→ 38

27 Your answer is quite right.

Calcium hydroxide and carbon dioxide react together to form different products depending on the relative amounts of each present at the outset.

1. For a molar ratio of 1:1 the reaction is:

$$Ca(OH)_2 + CO_2 \rightarrow CaCO_3 + H_2O$$

2. For a molar ratio of 1:2 the reaction is:

$$Ca(OH)_2 + 2\ CO_2 \rightarrow Ca(HCO_3)_2$$

What calcium compounds are formed according to these two equations?

Jot down the names of the two compounds formed. → 38

---

28 No, salts are formed by the combination of positively charged metal ions and negatively charged acid radicals.

Examples of salts include:

Sodium chloride	$NaCl$
Sodium nitrate	$NaNO_3$
Potassium sulfate	K_2SO_4
Calcium carbonate	$CaCO_3$

$Ca(OH)_2$ dissociates in water into $Ca^{2\oplus}$ and 2 $OH^{\ominus}$ ions.

We came across a similar dissociation when discussing the metal hydroxides NaOH and KOH.

$Ca(OH)_2$ is also, of course, a metal hydroxide.

Now answer the question in 20 once again.

---

29 Your answer is wrong.

Please copy down the definition below:

The equivalent mass of a substance is its molecular mass *divided by its basicity.*

Go back now to 17 .

17

30 Calcium hydroxide is a base.

Like all bases calcium hydroxide is both alkaline and corrosive.

Take great care when handling calcium hydroxide!

Calcium hydroxide may be obtained from calcium carbonate by a two-stage process.

Now copy down the completed version of the sentence below:

When calcium carbonate is calcined .... is formed, which can react with water to yield ....

Check your answer at 17 .

31 $Ca(HCO_3)_2 \rightarrow CaCO_3 + H_2O + CO_2$

This answer was easy, for all that was required was to write the equation for the formation of $Ca(HCO_3)_2$ from $CaCO_3$ and CO_2 in reverse.

We have already come across a similar decomposition when we discussed the effect of heat on sodium hydrogen carbonate.

$$2\ NaHCO_3 \rightarrow Na_2CO_3 + H_2O + CO_2$$

Starting from calcium carbonate, how may calcium hydrogen carbonate be prepared and then converted back again into calcium carbonate?

- By bubbling carbon dioxide into a calcium carbonate suspension → 8
- By heating a calcium carbonate suspension and then bubbling in carbon dioxide → 16
- By bubbling carbon dioxide into a calcium carbonate suspension and then heating it up → 42

32 Your answer is wrong.

An explanation of this is given in 6 .

33 Your answer is quite right.

Water which contains salts as impurities is referred to as *hard water.* Pure water, such as rainwater or distilled water, is referred to as *soft water.*

We can now go into the cause of hardness in water:

It is due to traces of calcium hydrogen carbonate which are dissolved in the water.

If such water is heated, the calcium hydrogen carbonate starts to decompose as outlined above. An insoluble precipitate is formed which is referred to in the household as limescale.

What will be the chemical composition of limescale?

- It consists of calcium hydrogen carbonate → 18
- It consists of calcium carbonate → 9

34 Your answer is quite correct.

Calcium hydroxide reacts with carbon dioxide in the air to form calcium carbonate and water.

The mortar hardens in this way and will not then react any more. In addition to calcium carbonate, *water* is also formed when mortar sets:

$$Ca(OH)_2 + CO_2 \rightarrow CaCO_3 + H_2O$$

For quite a long time therefore the walls of new buildings remain ....

→ 47

35 Your answer is wrong.

Copy down the sentence below:

The equivalent mass is the molecular mass *divided by the basicity.*

Go back now to 17 .

36 When limestone (calcium carbonate) is calcined quicklime (calcium oxide) is formed:

a) $CaCO_3 \rightarrow CO_2 + CaO$

The addition of water to quicklime yields slaked lime (calcium hydroxide):

b) $CaO + H_2O \rightarrow Ca(OH)_2$

The slaking of lime is associated with the liberation of a great deal of heat.

Write down the completed version of the sentence below:

Mortar is a mixture of .... and ....

Continue your reading at 25 .

37 $Ca(OH)_2 + 2\ HCl \rightarrow CaCl_2 + 2\ H_2O$

The action of carbon dioxide on calcium hydroxide is of special importance. From what we have said so far on calcium hydroxide and carbon dioxide, you should be able to decide which of the three statements below is correct. It will probably be easier to answer the question if you start by writing down the equation for the reaction of calcium hydroxide with carbon dioxide.

- Calcium hydroxide reacts with carbon dioxide in one way only → 18
- Calcium hydroxide can react with carbon dioxide to form two different end products depending on the relative amounts of reactants involved → 27
- Calcium hydroxide does not react with carbon dioxide → 21
- I don't know → 4

38 *Calcium carbonate, calcium hydrogen carbonate.*

Let's now have a look at some chemical formulas and the names we give them.

H_2CO_3 $CaCO_3$ Na_2CO_3 $Ca(HCO_3)_2$

What substances do these formulas represent in the sequence given?

- Carbonic acid, limestone, soda, calcium hydrogen carbonate → 23
- Carbonic acid, calcium carbonate, sodium carbonate, calcium carbonate → 48
- Carbonic acid, calcium carbonate, sodium hydrogen carbonate, calcium hydrogen carbonate → 11
- Carbonic acid, calcium carbonate, sodium carbonate, calcium hydrogen carbonate → 19

---

39 Your answer is quite correct.

This now concludes our discussion of the chemical processes which take place when mortar is used.

We have already said something about *magnesium carbonate.*

Write down the formulas for magnesium hydrogen carbonate and magnesium carbonate, and calculate the molecular mass of each. The relevant rounded atomic masses are: Mg = 24, O = 16, C = 12, H = 1.

What are the two molecular masses?

	Magnesium hydrogen carbonate	*Magnesium carbonate*	
●	85	84	→ 22
●	146	84	→ 53
●	85	108	→ 5
●	146	108	→ 12
●	My figures differ from all those above		→ 2

---

40 $Zn + 2\ HCl \rightarrow ZnCl_2 + H_2$

Hydrogen gas is formed during this reaction.

It should be clear to you that an effervescence observed when acids are added to a substance is *not* absolute proof that a carbonate is present in the substance. It is necessary to prove that the gas given off is actually carbon dioxide.

To do this, the gas formed is passed into an aqueous solution of barium hydroxide. If the gas is indeed CO_2 the following reaction will take place:

$$Ba(OH)_2 + CO_2 \rightarrow \ldots\ldots + H_2O$$

What substance is missing from the right hand side of this equation?

- $Ba(HCO_3)_2$ → [10]
- $BaCO_3$ → [50]
- $2\ BaCO_3$ → [15]

---

41 Correct, *carbon dioxide* is indeed generated.

All carbonates are decomposed in the presence of acids and give off *carbon dioxide.* This is because carbonic acid is a weak and unstable acid which is easily driven from its salts by less volatile and stronger acids.

Write out now the relevant equations for the action of:

1. Hydrochloric acid on calcium carbonate

17

2. Sulfuric acid on sodium carbonate.

Check these equations at [52] .

---

42 Your answer is quite right.

When carbon dioxide is passed into an aqueous suspension of calcium carbonate, the latter is converted into calcium hydrogen carbonate thus:

$$CaCO_3 + H_2O + CO_2 \rightarrow Ca(HCO_3)_2$$

When heated, the reaction proceeds in the opposite sense:

$$Ca(HCO_3)_2 \xrightarrow{\text{heat}} CaCO_3 + H_2O + CO_2$$

The conversion of calcium hydrogen carbonate into calcium carbonate plays a very important part in the water used for household and industrial purposes. When discussing water, you may recall that we differentiated between soft and hard water.

How did we describe water containing salts, e.g. mineral or household water?

- As hard water → 33
- As soft water → 14

---

43 Now check your equation:

$$CaCO_3 + H_2O + CO_2 \rightarrow Ca(HCO_3)_2$$

If the solution of calcium hydrogen carbonate is heated, carbon dioxide will be given off and insoluble calcium carbonate will be formed again.

Write down the equation for this reaction too, and then check your answer at 31 .

---

44 $$CaCO_3 \xrightarrow{\text{heat}} CaO + CO_2$$

Quicklime (CaO) is thus calcium oxide. Quicklime exists as a white, brittle mass which reacts with water and gets very hot in the process. The product formed is *slaked lime.* Can you write down the relevant equation for this reaction?

$$CaO + H_2O \rightarrow$$

(Think here of the reaction of sodium oxide, Na_2O, and water.)

17

Write out the full equation for the slaking of quicklime, and then check your answer at 20 .

---

45 Your answer is only partially right.

From the chemical standpoint, the reaction of calcium hydroxide with carbon dioxide represents a neutralization. In addition to the calcium carbonate, *water* is also formed.

You should now correct your equation and then go on to 34 .

---

46 The equivalent mass equals 74/2 = 37.

Calcium hydroxide reacts with hydrochloric acid to form calcium chloride and water when the neutralization is complete.

$$Ca(OH)_2 + HCl \rightarrow CaCl_2 + H_2O$$

This equation is not a balanced one. Two factors are missing. Write out now the correct version of this equation and then check it at 37 .

---

47 For quite a long time therefore the walls of new buildings remain *damp* (or *wet*).

The reason for this is *not* the water left over from the original mixing of the mortar, as this dries out of the walls relatively fast and normally before the construction work is finished.

What is the hardening process which takes place in mortar referred to as?
→ 24 .

---

48 You have not yet fully appreciated the difference between carbonates and hydrogen carbonates.

You should therefore read the explanation in 26 with special care.

---

49 There are carbonates and hydrogen carbonates.

17

Carbonic acid is a dibasic acid and so dissociates in two stages as follows:

1. $H_2CO_3 \rightleftharpoons H^{\oplus} + HCO_3^{\ominus}$

2. $HCO_3^{\ominus} \rightleftharpoons H^{\oplus} + CO_3^{2\ominus}$

$$H_2CO_3 \rightleftharpoons 2\,H^{\oplus} + CO_3^{2\ominus}$$

During the first stage, hydrogen carbonate ions bearing a single negative charge are formed. During the second stage, carbonate ions bearing a doubly negative charge are formed.

Like calcium, magnesium also forms doubly positively charged ions. These magnesium ions react with carbonate ions to form magnesium carbonate (formula I) and with hydrogen carbonate ions to form magnesium hydrogen carbonate (formula II).

$MgCO_3$	$Mg(HCO_3)_2$
I	II

Calculate the molecular masses of the compounds represented by formulas I and II once again. The relevant atomic masses are Mg = 24, O = 16, C = 12, H = 1. Check your answers at 39 .

---

50 Your answer is quite right. For revision purposes, see if you can now answer the following questions. You should then check your answers in 3 .

1. What is quicklime and how is it produced (give the relevant equation)?

2. What is slaked lime?

3. Why is special care necessary when handling slaked lime?

4. Complete the two equations below:

   a) $Ca(OH)_2 + HCl \rightarrow$

   b) $Ca(OH)_2 + H_2SO_4 \rightarrow$

5. Write out the chemical formulas for carbonic acid, calcium carbonate, calcium hydrogen carbonate and sodium hydrogen carbonate.

6. What is observed when CO_2 is bubbled into a suspension of chalk in water?

7. Why is scale deposited inside kettles at home (give the relevant equation)?

8. What is meant by hard and soft water?

9. What is mortar?

10. What is the process taking place when mortar sets (give the relevant equation)?

11. Why do the walls of new buildings dry out faster if open coke fires are put in them?

12. How can one test for a carbonate? Give a brief answer based on keywords.

Check your answers at 3 .

---

51 Your answer is wrong. Carbon dioxide is necessary for the setting of mortar as may be seen from the relevant equation:

$$Ca(OH)_2 + CO_2 \rightarrow CaCO_3 + H_2O$$

If there is only a limited supply of carbon dioxide available, the reaction cannot proceed very fast – even if heat is used. The slowly burning coke in an open coke fire gives off carbon dioxide, however, which is able to speed up the setting process.

Continue your reading at 39 .

52

1. $CaCO_3 + 2\ HCl \rightarrow CaCl_2 + H_2O + CO_2$

2. $Na_2CO_3 + H_2SO_4 \rightarrow Na_2SO_4 + H_2O + CO_2$

The fact that a substance effervesces is, however, not sufficient to prove the presence of a carbonate. The gas produced must be identified as carbon dioxide to eliminate any possible error.

We mention this here because there are numerous different substances which effervesce in the presence of acids, although they are *not carbonates.* Effervescence from substances *other than carbonates* is thus not caused by CO_2 but by quite *different gases.*

One example of this is provided by the metal zinc. When hydrochloric acid is poured over zinc (zinc = Zn; divalent) the mixture effervesces. What is the gas being produced? Write down the relevant equation, and then check it at 40 .

17

53

Not only are your formulas right, but your calculations are too!

$MgCO_3$ $Mg(HCO_3)_2$

We still want to say something about the test for carbonates.

The test for a carbonate depends on its reaction with acids. What is observed in this reaction?

- The appearance of a color change → 32
- The formation of carbon dioxide → 41
- The separation of a white precipitate → 13
- I really don't know → 6

Program 18

Nitrogen (I)

1 Nitrogen is one of the elements we have already discussed.

Its chemical symbol is N and its atomic mass is 14. Its two principal valences are three and five.

Write down the name and chemical formula of one compound in which nitrogen has a valence of five and one in which nitrogen has a valence of three. → 54

2 NH_4Cl ammonium chloride

$(NH_4)_2S$ ammonium sulfide

Ammonium chloride is an important salt of ammonia. It is formed not only during the neutralization of ammonium hydroxide with hydrochloric acid, but also in the reaction of gaseous hydrogen chloride with gaseous ammonia. We have already come across this particular reaction. White clouds of ammonium chloride are formed during the reaction.

Write out the equation for this reaction and then check your answer at 15 .

3 Your formula for ammonium sulfate is not right. An explanation of this is given in 56 .

4 *Now check your answers:*

1. Three and five.

2. First the *nitrogen* and then the *oxygen.*

3. 1000 volumes of NH_3 dissolve in one volume of water.

4. | | |
   |---|---|
   | Sodium hydroxide | $NaOH$ |
   | Potassium hydroxide | KOH |
   | Calcium hydroxide | $Ca(OH)_2$ |
   | Ammonia | NH_3 |

5. $2\ NH_4OH + H_2SO_4 \rightarrow (NH_4)_2SO_4 + 2\ H_2O$

6. $(NH_4)_2CO_3 \rightarrow 2\ NH_3 + H_2O + CO_2$

7. $(NH_4)_2CO_3 + 2\ NaOH \rightarrow Na_2CO_3 + 2\ NH_3 + 2\ H_2O$

If you made more than one mistake here, you are advised to work your way through the Program again fairly soon.

End of Program 18.

---

5 You appear not to be concentrating properly. We asked you for a nitrogen *compound.*

Read through [25] once again.

---

6 $(NH_4)_2SO_4 + 2\ NaOH \rightarrow Na_2SO_4 + 2\ NH_3 + 2\ H_2O$

The products formed are sodium sulfate, water and ammonia.

The strong and non-volatile base NaOH drives the .... base NH_3 from its .... → [27]

---

7 Your answer is wrong. Apart from CaO all the compounds listed can be thermally decomposed. Complete the equations below for the thermal dissociation of the following substances:

$CuSO_4 \cdot 5\ H_2O \rightarrow$
$NaHCO_3 \rightarrow$
$CaCO_3 \rightarrow$

Check your equations at [37] .

---

8 Ammonium hydroxide is formed.

The strong and non-volatile caustic soda drives the weak and volatile base from its salt.

We have already mentioned that NH_4OH does not exist as such. It occurs only in low concentrations as $NH_4^{\oplus}$ and $OH^{\ominus}$ ions in the presence of NH_3 and H_2O molecules. When heated, it gives off ammonia gas.

Write down the relevant equation. → [29]

18

9 You have ignored the fact that the ammonium ion occurs twice in the ammonium carbonate molecule. → 20

10 Your answer is only partially correct.

When ammonium salts are heated with caustic soda ammonia gas comes off. This may be detected by its smell. However, it is not enough to rely on the smell. The ammonia should also be tested for with litmus paper.

What coloration of the paper will the ammonia induce?

- Damp red litmus paper will turn blue → 51
- Damp blue litmus paper will turn red → 35

11 Your answer is wrong. Do you recall our experiments on the composition of the air?

If one removes the oxygen in the air by means of some suitable chemical reaction, around 4/5th of the air is left over and this portion of the air is not capable of supporting combustion. The portion is, of course, nitrogen.

Copy down the sentence below and insert the missing number.

Air thus consists of up to 4/5th nitrogen, that is ......% is nitrogen. → 24

12 No, ammonium hydroxide is stable only in solution and even there only in the dissociated state. As soon as an aqueous solution of ammonia is heated ammonia gas comes off again.

Ammonia is given off from its heated aqueous solution as .... Jot down the relevant chemical formula. → 22

13 Ammonium carbonate, $(NH_4)_2CO_3$; molecular mass 96

Ammonium sulfate, $(NH_4)_2SO_4$; molecular mass 132

Write down the formulas for ammonium chloride and ammonium sulfide.
→ 2

14 Your answer is only partially correct.

You should remember that CO_2 is carbon dioxide and not carbonic acid. Thus, the thermal decomposition of ammonium carbonate results in the formation of ...... and ....

Continue your reading at 48 .

15 $NH_3 + HCl \rightarrow NH_4Cl$

All ammonium salts have the property of decomposing fairly readily into their components when heated, e.g.

$$NH_4Cl \xrightarrow{\text{heat}} NH_3 + HCl$$

Ammonium chloride, for instance, breaks down into ammonia and hydrogen chloride.

A decomposition of this type, brought about by heating up a chemical compound, is described as a *thermal dissociation.*

Describe the thermal dissociation of ammonium sulfate by means of an equation. → 38

16 Your answer is wrong. There is *no color change* when ammonium chloride reacts with caustic soda. Such a mixture will in fact give off ammonia when heated.

Now complete the equation below:

$$NH_4Cl + NaOH \rightarrow$$

Check your answer at 8 .

17 Your answer is wrong. The nitrogen compound formed during the dry distillation of coal is NH_3. Give the name of this compound.

Continue your reading at 49 .

18 $2\ NH_4Cl + CuO + 2\ HCl \rightarrow 2\ NH_3 + CuCl_2 + 2\ HCl + H_2O$

Those species which appear on both sides of the equation after the addition can be cancelled out. In our case 2 HCl appears on both sides and so the equation may be simplified as follows:

$$2\ NH_4Cl + CuO \rightarrow 2\ NH_3 + CuCl_2 + H_2O$$

From this equation it is apparent that:

The coating of copper oxide (CuO) caused by heating the soldering iron can be removed by bringing it into contact with ammonium chloride (NH_4Cl). The interaction will lead to the formation of ammonia, water and copper chloride. The ammonia and the water evaporate immediately and the copper chloride also sublimes off rapidly. The coating on the soldering iron is removed in this way; it is thus clean and ready for use again at the end of the reaction.

We come now to the last major topic in this Program: how can we test for ammonia or the ammonium ion?

Ammonia is a volatile and weak base. Just as volatile and weak acids are driven from their salts by strong acids of low volatility, ammonia is driven from its salts by strong bases of low volatility. The action of caustic soda on ammonium chloride, for instance, leads to the formation of sodium chloride among other things:

$$NH_4Cl + NaOH \rightarrow NaCl + \ldots\ldots\ldots\ldots$$

After completing this equation you should be able to answer the question below.

What is formed in this reaction in addition to the NaCl?

Jot down the name of the compound formed. → 8

19 We said that hydrogen chloride reacts with the copper oxide to form copper chloride which sublimes off. Your answer was thus wrong.

Return now to 59 and answer the question there again.

20 $(NH_4)_2CO_3$ ammonium carbonate

$(NH_4)_2SO_4$ ammonium sulfate

Calculate now the molecular mass of each of these compounds. The relevant atomic masses are: H = 1, N = 14, O = 16, C = 12 and S = 32.

What two values do you obtain?

- 78 and 114 → 34
- 96 and 114 → 26
- 96 and 132 → 13
- 78 and 132 → 9
- I need some help → 55

---

21 Your formulas for ammonium carbonate and ammonium sulfate are not correct. An explanation is to be found in 56 .

---

22 Yes, ammonia is given off from its heated aqueous solution as NH_3.

To return to the subject of aqueous solutions of ammonia, we know that they contain mainly NH_3 and H_2O molecules and some $NH_4^{\oplus}$ and $OH^{\ominus}$ ions as well as hydrogen ions. The concentration of $H^{\oplus}$ ions is, of course, very low as most of these ions combine with the base NH_3 to form $NH_4^{\oplus}$.

18

How will an aqueous solution of ammonia react toward litmus paper?

- Red litmus paper will turn blue → 41
- Blue litmus paper will turn red → 61

---

23 Ammonium carbonate is obtained according to the following equation:

$$2\ NH_3 + CO_2 + H_2O \rightarrow (NH_4)_2CO_3$$

When read from left to right, this equation represents the production of ammonium carbonate.

When read from right to left, this equation represents the thermal dissociation of ammonium carbonate.

Ammonium carbonate is used as baking powder. The decomposition products of ammonium carbonate are all gaseous at high temperature. The dough gets "blown up" to some extent by the gases which are formed and becomes softer in the process. The products formed are all quite harmless.

What is formed during the decomposition of ammonium carbonate?

- Ammonia and carbonic acid → 32
- Ammonia, carbonic acid and water → 14
- Ammonia, carbon dioxide and water → 48

---

24 Air contains about **80%** of nitrogen. Nitrogen is not combustible and is not capable of supporting breathing or combustion. Chemically speaking, nitrogen is a very unreactive element. It is only under special conditions, e.g. at high temperatures and pressures, that it will combine with other elements.

Nitrogen can be obtained from the air by warming up previously liquefied air. What will happen when this is done?

- The nitrogen will be distilled off first and can thereby be separated from the oxygen → 36
- The oxygen will be distilled off first and liquid nitrogen will remain behind → 45

---

25 Liquid oxygen boils at −183 °C.

Nitrogen is found in numerous naturally occurring compounds:

1. *In saltpeter*
   Saltpeter is a salt found mainly in Chile where it is mined in large quantities. Chemically, saltpeter is sodium nitrate.

2. *In proteins*
   These are found in both plants and animals.

3. *In products of the dry distillation of coal*
   As coal is formed from plants, it is hardly surprising that the products of dry distillation contain nitrogen compounds.

What nitrogen compound is formed during the dry distillation of coal?

- Nitric acid → 17

- Nitrogen → 5
- Ammonia → 49
- Saltpeter → 40

---

26 You have ignored the fact that the ammonium ion occurs twice in the ammonium carbonate molecule. → 20

---

27 The strong and non-volatile base NaOH drives the *weak* and *volatile* base NH_3 from its *compounds (salts).*

Instead of caustic soda one could also use slaked lime for this purpose.

Now complete the following equation:

$$(NH_4)_2SO_4 + Ca(OH)_2 \rightarrow$$

→ 60

---

28 Your answer is wrong. 1 volume of water dissolves 1000 volumes of ammonia.

Remember that 1 L = 1000 cm^3.

Now refer back to 44 .

---

29 $NH_4^{\oplus} + OH^{\ominus} \rightleftarrows NH_3 + H_2O$

As the solubility of NH_3 in water decreases markedly with increasing temperature, NH_3 will be given off when the solution is heated.

The ammonia which comes off can be recognized by its smell, though a better way is to observe the blue coloration it causes in damp red litmus paper.

We may test for ammonium compounds in an unknown substance by adding caustic soda to it and then heating the mixture. If ammonia is given off, we may conclude that the substance contains ammonium compounds.

How can we recognize that ammonia is given off?

- By a coloration in the reaction mixture → 16

- By a blue coloration in damp red litmus paper held in the escaping vapors → 51
- By the characteristic smell of the ammonia → 10

---

30 Your answer is wrong. Apart from CaO all the compounds listed may be thermally decomposed. Now complete the equations below, all of which represent thermal dissociation reactions.

$CuSO_4 \cdot 5\ H_2O \rightarrow$

$NaHCO_3 \rightarrow$

$CaCO_3 \rightarrow$

Check your equations at 37 .

---

31 1 cm^3 of water dissolves 1 L of ammonia.

An aqueous solution of ammonia is thereby formed.

A trivial name for this aqueous solution is *ammoniacal liquor.*

Even with water ammonia reacts as a base. Some of the ammonia molecules dissolved in the water react with the water to form fully dissociated ammonium hydroxide.

See if you can write down the equation for the formation of ammonium hydroxide from ammonia and water. → 52

---

32 No. If you read the equation in 23 from right to left, the required answer should be obvious.

Read through 23 once again.

---

33 A base is a substance which is capable of combining with hydrogen ions.

Thus, ammonia can combine with hydrogen ions. In the process, the uncharged ammonia molecule becomes the positively charged ammonium ion, $NH_4^{\oplus}$, according to the equation:

$$NH_3 + H^{\oplus} \rightarrow NH_4^{\oplus}$$

If one were to bubble ammonia into hydrochloric acid, the ammonia would combine with the hydrogen ions present in the acid to form ammonium ions. There will, of course, also be chloride ions present in such a solution.

Write down now the equation for this reaction. Take as your starting materials ammonia, hydrogen ions and chloride ions. → 43

---

34 You have ignored the fact that the ammonium ion occurs twice in these molecules. → 20

---

35 No, ammonia turns red litmus paper .... → 51

---

36 When liquid air is warmed up the *nitrogen,* which boils at −196 °C, distills off first. The oxygen remains behind.

At what point will the liquid oxygen start to boil?

Jot down its boiling point. → 25

---

37 Quite correct!

CaO (calcium oxide) is thermally stable.

18

$CuSO_4 \cdot 5\ H_2O$ (copper sulfate) loses its water of crystallization when heated:

$$CuSO_4 \cdot 5\ H_2O \rightarrow CuSO_4 + 5\ H_2O$$

$NaHCO_3$ (sodium hydrogen carbonate) decomposes around 100 °C into sodium carbonate, water and CO_2 (cf. the soda production in the Solvay Process):

$$2\ NaHCO_3 \rightarrow Na_2CO_3 + H_2O + CO_2$$

$CaCO_3$ (calcium carbonate) dissociates thermally into CO_2 and CaO (cf. the production of calcined chalk):

$$CaCO_3 \rightarrow CaO + CO_2$$

Ammonium carbonate too decomposes upon heating. It may be prepared by passing ammonia and carbon dioxide into water.

See if you can correctly complete the relevant equation:

$NH_3 + CO_2 + \dots\dots\dots \rightarrow (NH_4)_2CO_3$ Check your answer at 23

38 $(NH_4)_2SO_4 \xrightarrow{heat} 2\ NH_3 + H_2SO_4$

Ammonium sulfate decomposes into ammonia and sulfuric acid.

We now ask you to name the type of chemical process in which a substance is decomposed by heat.

Jot down the type of process concerned, and then check your answer at 62 .

39

1. The copper tip of a soldering iron gets covered with a layer of *copper oxide* when heated.

2. The hot soldering iron is dipped into some *ammonium chloride.*

3. After a short while the copper oxide disappears. The tip of the soldering iron is clean again, i.e. its surface consists of *copper.*

Let's have a look at the chemistry involved in these processes.

1. The formation of the copper oxide layer:

$$2\ Cu + O_2 \rightarrow 2\ CuO$$

2. Contact of the hot soldering iron with ammonium chloride: This involves thermal dissociation of the ammonium chloride.

Write down the equation for the second reaction, and then check your answer at 64 .

40 Your answer is wrong. The nitrogen compound formed during the dry distillation of coal is NH_3. Give the name of this compound and continue reading at 49

41 Yes, red litmus paper turns blue.

An aqueous solution of ammonia reacts as a base because more $OH^{\ominus}$ ions than $H^{\oplus}$ ions are present in it.

The neutralization of sulfuric acid may be described by the following reaction:

$$2\ NH_4OH + H_2SO_4 \rightarrow (NH_4)_2SO_4 + 2\ H_2O$$

Jot down the name of the salt formed in this reaction. → 65

---

42 We have already said that the hydrogen chloride reacts with the copper oxide to form copper chloride, which then sublimes away. Accordingly, your answer was wrong.

Make another attempt at the question in 59 .

---

43 Now check out your answer and correct any mistakes you made:

$$NH_3 + H^{\oplus} + Cl^{\ominus} \rightarrow NH_4^{\oplus} + Cl^{\ominus}$$

Do you remember the name of the $NH_4^{\oplus}$ ion?

Jot it down and then go on to 53 .

---

44 Now check your answers:

a) Ammonium chloride NH_4Cl
b) Ammonium carbonate $(NH_4)_2CO_3$
c) Ammonium sulfide $(NH_4)_2S$
d) Ammonium nitrate NH_4NO_3

18

Did you remember to put in the brackets for both b) and c) above?

Ammonia is extremely soluble in water. At room temperature 1 volume of water will dissolve around 1000 volumes of ammonia.

Now attempt the following question:

How much ammonia will be dissolved by one cm^3 of water?

- 100 cm^3 → 58
- 1 L → 31
- 10 L → 28

---

45 Your answer is wrong. Oxygen boils at −183 °C. The boiling point of nitrogen is −196 °C. When liquid air at −200 °C is slowly warmed up, it begins to boil at −196 °C. The nitrogen begins to boil off. After all the nitrogen has boild off, the temperature of the liquid air will slowly increase again until the temperature − 183 °C is reached, at which all the oxygen will boil off.

Now complete and copy down the sentence below:

When liquid air is warmed up the.............. distills off first. → 36

46 Your answer is wrong. Apart from CaO all the compounds listed may be thermally decomposed. Now complete the equations below, all of which represent thermal dissociation reactions.

$CuSO_4 \cdot 5\,H_2O \rightarrow$
$NaHCO_3 \rightarrow$
$CaCO_3 \rightarrow$

Check your equations at 37 .

47 Your formula for ammonium carbonate is not correct. An explanation of this is to be found in 56 .

48 The thermal dissociation of ammonium carbonate yields ammonia, carbon dioxide and water.

A very interesting utilization of the thermal dissociation of *ammonium chloride* occurs in the process of cleaning soldering irons.

The tip of a soldering iron consists of copper and this becomes coated with a thin layer of copper oxide when the iron is heated. The copper oxide interferes with the soldering process. In fact, the soldering iron must be completely clean when the solder is applied. To make sure of this, the iron is dipped into some ammonium chloride from time to time. The latter has the effect of removing the copper oxide layer very rapidly.

To summarize what we have said so far, you are now asked to complete the following sentences:

1. The copper tip of a soldering iron gets covered with a layer of .... when heated.

2. The hot soldering iron is dipped into some ....

3. After a short while the copper oxide disappears. The tip of the soldering iron is clean again, i.e. its surface consists of ....

Check your answers at 39 .

---

49 Ammonia is formed during the dry distillation of coal.

Ammonia is a colorless gas having a pungent and choking smell. Ammonia can easily be liquefied; it boils at −33 °C.

Ammonia is a base.

Write out now the definition of a base. → 33

---

50 Yes, the soldering iron becomes lighter.

Let's now sum up the partial processes which are taking place into one overall reaction.

$$NH_4Cl \rightarrow NH_3 + HCl$$

$$CuO + 2\ HCl \rightarrow CuCl_2 + H_2O$$

As two molecules of HCl are required in the second equation whereas the first equation has only one such molecule, it will be necessary to double the first equation.

18

$$2\ NH_4Cl \rightarrow 2\ NH_3 + 2\ HCl$$

$$CuO + 2\ HCl \rightarrow CuCl_2 + H_2O$$

$$2\ NH_4Cl + CuO + 2\ HCl \rightarrow$$

Complete now the addition of the two equations, remembering that the two left hand sides are added and then separately the two right hand sides. Check your final equation at 18 .

---

51 Yes, the ammonia which comes off is recognized by the fact that it turns damp red litmus paper blue.

Now complete the following equation:

$$(NH_4)_2SO_4 + NaOH \rightarrow Na_2SO_4 +$$ → 6

52 $NH_3 + H_2O \rightarrow NH_4^{\oplus} + OH^{\ominus}$

As mentioned, this reaction occurs only to a limited extent. In fact, if it is warmed up it proceeds in the reverse direction. Thus, it is possible to drive all the ammonia from an ammonia solution merely by boiling it.

The fact that a reaction can proceed either forward or in the reverse sense is usually indicated by means of a split arrow thus:

$$NH_3 + H_2O \rightleftarrows NH_4^{\oplus} + OH^{\ominus}$$

Make sure you use a split arrow in this equation, and then attempt the question below.

In what form is the ammonia driven off from its aqueous solution?

- In the form of NH_3 → 22
- In the form of NH_4OH → 12

---

53 $NH_4^{\oplus}$ is the ammonium ion.

We can obtain ammonium chloride as a crystalline salt by evaporating an ammonium chloride solution.

When gaseous hydrogen chloride and ammonia vapor react together, white clouds of ammonium chloride are instantly formed.

Write down the equation for this reaction. → 63

---

54 Nitrogen is tervalent in ammonia: NH_3

Nitrogen is pentavalent in nitric acid: HNO_3

Nitrogen is a colorless and odorless gas which can be liquefied. Liquid nitrogen boils at –196 °C.

We already know that the air contains a large amount of nitrogen. How much of the air is actually nitrogen?

- About 20% → 11
- About 80% → 24

55 You must have miscalculated here.

To calculate the molecular mass of ammonium carbonate, $(NH_4)_2CO_3$, proceed as follows:

NH_4 = 18

NH_4 = 18

CO_3 = <u>60</u>

Add up these numbers and you will have the molecular mass of ammonium carbonate. Then use the same method to calculate the molecular mass of ammonium sulfate, $(NH_4)_2SO_4$. The atomic mass of S = 32.

→ 13

56 The formula for carbonic acid is H_2CO_3 and that for sulfuric acid is H_2SO_4.

Both of these acids form acid radicals bearing a double negative charge. Carbonic acid forms the carbonate ion, $CO_3^{2\ominus}$, and sulfuric acid the sulfate ion, $SO_4^{2\ominus}$. The ammonium ion bears a single negative charge, $NH_4^{\oplus}$.

Accordingly, ammonium sulfate consists of two ammonium ions and one sulfate ion. Ammonium carbonate similarly consists of two ammonium ions and one carbonate ion.

18

Now write down the formulas for ammonium carbonate and ammonium sulfate. Check your answers at 20 .

57 Your answer is wrong. Apart from CaO, all the compounds listed may be thermally decomposed. Now complete the equations below, all of which represent thermal dissociation reactions.

$CuSO_4 \cdot 5\ H_2O \rightarrow$

$NaHCO_3 \rightarrow$

$CaCO_3 \rightarrow$

Check your equations at 37 .

58 Your answer is wrong. 1 volume of water dissolves **1000** volumes of ammonia.

Read through [44] once again.

59 $CuO + 2\ HCl \rightarrow CuCl_2 + H_2O$

Copper chloride and water are formed in this reaction.

The water evaporates and the copper chloride sublimes off. The soldering iron will thus be clean again.

What will happen to the mass of the soldering iron if it is continually treated in this way?

- The mass of the soldering iron will increase → [19]
- The mass of the soldering iron will remain constant → [42]
- The mass of the soldering iron will decrease → [50]

60 $(NH_4)_2SO_4 + Ca(OH)_2 \rightarrow CaSO_4 + 2\ NH_3 + 2\ H_2O$

We shall be discusssing the synthesis of ammonia in the next Program.
Now check what you have learnt this far by writing out answers to the questions below.

1. What valences does nitrogen have in compounds?
2. When we distill liquid air, first the .... and then the .... boils off.
3. How many volumes of ammonia under normal conditions will dissolve in one volume of water?
4. What inorganic bases have you come across so far?
5. Write down the equation for the neutralization of an aqueous solution of ammonia by sulfuric acid.
6. Write down the equation for the thermal dissociation of ammonium carbonate, $(NH_4)_2CO_3$.

7. Write down the equation on which the test for the ammonium ion in $(NH_4)_2CO_3$ is based.

Check your answers at 4 .

---

61 No. Blue litmus paper turns red in the presence of acids. An aqueous solution of ammonia, however, reacts as a base since it contains more $OH^{\ominus}$ ions than $H^{\oplus}$ ions. The same color change will, of course, also be observed in litmus paper in the presence of aqueous solutions of NaOH, KOH or $Ca(OH)_2$.

What color change will be observed in litmus paper in an aqueous solution of ammonia? Jot down the color change. → 41

---

62 By *thermal dissociation* we mean the decomposition of a chemical compound by heating it.

To sharpen this concept of thermal dissociation, you are asked to write out the formulas for the following four compounds:

Calcium oxide
Copper sulfate with 5 waters of crystallization
Sodium hydrogen carbonate
Calcium carbonate

How many of the compounds listed here can be decomposed by means of thermal dissociation?

- All four → 46
- Only three → 37
- Only two → 57
- Only one → 7
- None of them → 30

---

63 Now check your equation:

$$NH_3 + HCl \rightarrow NH_4Cl$$

The polybasic acids also form ammonium salts with the base ammonia. As the ammonium ion, $NH_4^{\oplus}$, bears a single positive charge, it will appear more than once in the formulas for salts of polybasic acids. To ensure clear notation, the NH_4 is always put in brackets thus: (NH_4). As an example, the formula for ammonium sulfate is $(NH_4)_2SO_4$.

Write out now the names and formulas for the salts formed by the reaction of ammonia with each of the following acids:

a) Hydrochloric acid
b) Carbonic acid
c) Hydrogen sulfide
d) Nitric acid → 44

---

64 $NH_4Cl \rightarrow NH_3 + HCl$

The hydrogen chloride formed immediately reacts with the copper oxide to produce copper chloride and water.

Write out the relevant equation for this reaction, and then check your answer at 59 .

---

65 *Ammonium sulfate*

An aqueous solution of ammonia can thus be neutralized by acids. In the process salts are formed consisting of ammonium ions and acid radical ions.

Write down now the formulas for ammonium carbonate and ammonium sulfate. Then go on to calculate the molecular mass of each of these compounds. The relevant atomic masses are: H = 1, N = 14, O = 16, C = 12 and S = 32.

What are the respective molecular masses of ammonium carbonate and ammonium sulfate?

- 78 and 114 → 21
- 96 and 114 → 3
- 96 and 132 → 13
- 78 and 132 → 47
- I obtained other figures → 56

Program 19

Nitrogen (II)

---

1 In this Program we shall be discussing the industrial synthesis of ammonia and the importance of ammonia from the commercial standpoint.

To start with, what is the state of aggregation of NH_3 under normal conditions?

- It is solid → 16
- It is liquid → 24
- It is gaseous → 35

---

2 Your answer is wrong. Ammonia is a compound having the chemical formula NH_3.

The hydrogen contained in mixed gas is essential for the synthesis of ammonia.

Write down now the equation for the synthesis of ammonia formed in the Haber Process.

...... + ... → NH_3

Check your equation at 44 .

---

3 No carbon dioxide whatsoever is formed during the manufacture of producer gas and water gas. It cannot therefore be contained in mixed gas.

Read through 42 once again.

---

4 In any industrial synthesis, including of course the synthesis of ammonia, a suitable supply of *raw materials* is indispensable. These are substances which are readily available in nature. In the case of ammonia synthesis we start out from air, water and coal. These raw materials are transformed via a series of *intermediate stages*. The intermediate stages in ammonia synthesis are the water gas and producer gas stages. Chemical transformation of the *intermediate products* formed must result in the production of nitrogen and hydrogen, the .... materials for the synthesis of ammonia. → 18

---

5 Your answer is not quite right.

Sodium hydrogen carbonate does indeed react with caustic soda to form sodium carbonate, though such a process is not an economic one. Too much expensive sodium hydroxide would be required.

The relevant equation is:

$$NaHCO_3 + NaOH \rightarrow Na_2CO_3 + H_2O$$

There is, however, a much more economic way of converting sodium hydrogen carbonate into sodium carbonate.

Make another attempt at the question in 30 .

6 Your answer is wrong. In the Producer Gas Process carbon, i.e. coke, is reacted with air to yield a mixture of CO and N_2.

Go back now to 50 .

7 Your answer is quite right.

Soda may be manufactured by either the Solvay or Leblanc Processes. The Leblanc Process (Program 16, frame 44) is, although chemically very interesting, now out of date and hardly used anymore.

19

The main process in use today is the *Solvay* Process. This is a process which proceeds via several intermediate stages.

Initially, carbon dioxide gas is pumped into an aqueous solution of ammonia until the latter is saturated. Ammonium carbonate is formed in this way.

Write out the relevant equation for this stage. → 20

8 $(NH_4)_2SO_4$, ammonium sulfate

$(NH_4)_2CO_3$, ammonium carbonate

Nitrogen compounds are needed by plants for their synthesis of proteins.

When discussing carbon we said that proteins are necessary for the nourishment of both humans and animals. Proteins are made up of complex organic compounds which contain nitrogen.

Human beings, animals and most plants are not able to extract from the air the nitrogen they need for the synthesis of proteins.

Humans and animals obtain the necessary nitrogen in sufficient amounts from plant and animal sources. Plants on the other hand are – with some exceptions – able to obtain their nitrogen directly from the soil. The supply of nitrogen compounds in the soil is, however, not unlimited.

For this reason it is necessary to add nitrogen to the soil continually. This was done in earlier times simply by mixing dung and manure into the soil. Both of these are rich in nitrogen. The need for artificial fertilizers was made apparent from the research of the German chemist Justus von Liebig (1803–1873). However, it was only after the invention of a process for converting the nitrogen in the air into ammonia, that it became possible to manufacture nitrogen-containing fertilizers on a large scale.

Without the benefits of chemistry, an even greater proportion of humanity would have to suffer from malnutrition and starvation than is the case today.

Now attempt the following questions, all the answers of which are to be found in this section.

1. What do proteins consist of?
2. Where do human beings and animals get the nitrogen which is essential for their nourishment?
3. Where do plants get the necessary nitrogen for the synthesis of proteins?
4. What is the name of the chemist who first introduced the idea of artificial fertilization? → 57

---

9 Your answer reveals that you are not working in sufficient depth. Nitrogen is tervalent, for example, in NH_3, though it has three other valences in addition to this. Continue your reading at 64 .

---

10 Your answer is quite right.

1000 volumes of ammonia dissolve in one volume of water. A solution of ammonia obtained in this way is used as a starting material in the manufacture of soda. We shall be discussing this process very soon.

Soda can be manufactured by either of two processes. What are these processes called?

- The Solvay and Leblanc Processes → [7]
- The Lead Chamber and Contact Processes → [31]

---

[11] Avogadro's Law states that:

Equal volumes of gases at the same temperature and pressure contain equal numbers of molecules.

The equation

$$3\ H_2 + N_2 \rightarrow 2\ NH_3$$

tells us not only that hydrogen and nitrogen combine together to form ammonia. The equation also says something about the number of gram molecules (moles) involved in this reaction. 3 mol of hydrogen react with 1 mol of nitrogen to form 2 mol of ammonia.

Now we know that the molar volume of all gases is 22.4 L under normal conditions.

We can therefore conclude that 3 x 22.4 L of hydrogen react with 22.4 L of nitrogen to form 2 x 22.4 L of ammonia.

1 mol of H_2 (22.4 L)				
				1 mol of NH_3 (22.4 L)
1 mol of H_2 (22.4 L)	+	1 mol of N_2 (22.4 L)	→	
				1 mol of NH_3 (22.4 L)
1 mol of H_2 (22.4 L)				

The question below should now be easy to answer. Which of the three following statements is correct?

- 6 volumes of hydrogen and 2 volumes of nitrogen yield 6 volumes of ammonia → [47]
- 1 volume of hydrogen and 1 volume of nitrogen yield 1 volume of ammonia → [59]

- 3 volumes of hydrogen and 1 volume of nitrogen yield 2 volumes of ammonia → 56

---

12 The structural formula for NO_2 is

$$O{=}N{=}O$$

The nitrogen in NO_2 has a valence of four.

In the third stage NO_2 is reacted with oxygen and water:

$$4\ NO_2 + 2\ H_2O + O_2 \rightarrow 4\ HNO_3$$

Copy down this equation and write under each of the formulas the name of the substance concerned. → 54

---

13 Your answer is quite right.

The *mixed gas* formed from *producer gas* and *water gas* contains carbon monoxide, hydrogen and nitrogen, i.e.

CO, H_2 and N_2

Which of these three gases is not important as a starting material in the synthesis of ammonia?

- Hydrogen → 2
- Carbon monoxide → 27
- Nitrogen → 39

19

---

14 No. Certain concepts need to be clarified. Please read through 4 carefully.

---

15 Your answer is quite right.

Upon heating, sodium hydrogen carbonate decomposes into sodium carbonate and carbon dioxide (thermal dissociation):

$$2\ NaHCO_3 \rightarrow Na_2CO_3 + H_2O + CO_2$$

The carbon dioxide which is given off can be used again in the first stage of the Solvay Process.

What is actually formed during the first stage of the Solvay Process?

- Ammonium carbonate → 32
- Sodium hydrogen carbonate → 48
- Ammonium hydrogen carbonate → 58

---

16 Your answer is wrong. At temperatures below −33 °C ammonia is a liquid whereas under normal conditions it exists as a pungent ....

→ 35

---

17 Your answer is wrong. The *Contact Process* is used for the manufacture of *sulfuric acid.* In the presence of catalysts (contacts) sulfur dioxide is oxidized to sulfur trioxide with oxygen.

Go back now to 50 .

---

18 The *starting* materials are nitrogen and hydrogen.

What raw materials are used in the synthesis of ammonia?

Jot down the names of these raw materials, and then check your answer at 25 .

---

19 Your answer reveals that you are not working in sufficient depth.

The nitrogen in nitric acid has a valence of five, though it has three other valences in addition to this.

Continue your reading at 64 .

---

20 $NH_3 + H_2O + CO_2 \rightarrow NH_4HCO_3$

Sodium chloride is then added to the solution of ammonium hydrogen carbonate obtained. The solution will thus contain $NH_4^{\oplus}$, $HCO_3^{\ominus}$, $Na^{\oplus}$ and $Cl^{\ominus}$ ions. The $Na^{\oplus}$ and $HCO_3^{\ominus}$ ions now precipitate out as virtually insoluble $NaHCO_3$.

See if you can complete the reaction below:

$NH_4HCO_3 + NaCl \rightarrow$

→ 30

---

21 Your answer is correct if you wrote:

$4\ NH_3 + 5\ O_2 \rightarrow 4\ NO + 6\ H_2O$

Now draw the structural formula for NO.

What is the valence of the nitrogen in NO? → 49

---

22 No. *The mixture of carbon monoxide and hydrogen is called water gas.* Mixed gas is something else.

Read through 42 once again.

---

23 No. Although nitrogen certainly has valences of 3 and 5, two other valences it has are missing. → 49

---

24 No. Under normal conditions NH_3 is not liquid. You were probably thinking of an aqueous solution of ammonia.

At room temperature ammonia is a pungent .... → 35

---

25 Correct, the raw materials used in the synthesis of ammonia are *air, water* and *coal.* Starting from these raw materials, one can obtain the gases hydrogen and nitrogen required for the synthesis of ammonia.

We now present a brief review of the material we have covered so far. Please read through it carefully!

Coal and air yield *producer gas,* $CO + N_2$.

Coal and water yield *water gas,* $CO + H_2$.

These two gases are added together to yield *mixed gas,* $N_2 + CO + H_2$.

To oxidize the CO this mixed gas and steam are together passed over catalysts. The product which forms is known as *contact gas,* $N_2 + CO_2 + H_2$.

The CO_2 is washed out of the contact gas with water. We then obtain the mixture of $N_2 + H_2$ required for the *Haber Process.*

The gaseous ammonia produced by the Haber Process may be liquefied by cooling it to below −33 °C. It can then either be stored in gas cylinders (steel bottles) or dissolved in water to form an aqueous solution.

How many cubic meters of NH_3 gas dissolve in one cubic meter of water?

- $100m^3$ → [38]
- $1000m^3$ → [10]
- $10000m^3$ → [46]

---

[26] Your answer is quite correct.

But can you also say what is manufactured in the other four processes we just mentioned?

Use key words to write out the products produced in each of the following four processes:

The Contact Process
The Water Gas Process
The Producer Gas Process

Check your answers at [62] .

---

[27] Yes, the carbon monoxide in mixed gas is not important in the synthesis of ammonia. It is therefore removed before the synthesis takes place.

The simplest way of doing this would be to wash it out with water. However, although CO_2 is readily soluble in water, CO is not soluble. Accordingly, the CO is first oxidized to CO_2 to render it soluble.

This cannot be done simply with air, for then the hydrogen present in mixed gas would also burn.

Instead the carbon monoxide is reacted with *steam* by means of suitable catalysts operating at a high temperature.

See if you can complete the relevant equation:

$CO + H_2O \rightarrow \quad + H_2$ → 40

---

28 Nitrogen cannot have a valence of six.

Have another shot at the question in 54 .

---

29 Your answer is correct, though incomplete as it stands.

Let's take another look at the relevant equation:

$$CO + H_2O \rightarrow CO_2 + H_2$$

A reduction of the water to hydrogen has certainly taken place. At the same time the carbon monoxide has been oxidized to carbon dioxide. Thus, the carbon monoxide takes up the .... which is removed from the water. → 63

---

30 $(NH_4)HCO_3 + NaCl \rightarrow NH_4Cl + NaHCO_3$

As the $NaHCO_3$ is virtually insoluble in water, it crystallizes out and so can be readily removed.

How may sodium hydrogen carbonate be converted into soda in a commercially viable way?

- By heating it up → 15
- By reacting it with carbon dioxide → 41
- By reacting it with sodium hydroxide → 5

---

31 You appear to have confused the processes for the manufacture of soda and sulfuric acid. To manufacture sulfuric acid one uses either the Lead Chamber Process or the Contact Process.

To manufacture soda one uses the Leblanc Process and the .... Process. → 7

---

32 No, in the first stage of the Solvay Process for the manufacture of sodium carbonate, carbon dioxide is pumped into an aqueous solution of ammonia until the saturation point is reached. The product which forms is ammonium hydrogen carbonate.

Write down the relevant equation for this stage. → 20

---

33 Producer gas consists of carbon monixide and *nitrogen.*

The second process for the manufacture of *water gas* involves passing steam over red-hot coke. Under such conditions the water will react with the carbon as follows:

$$H_2O + C \rightarrow CO + H_2$$

The *producer gas* and *water gas* are then added together. The new gas formed is referred to as *mixed gas.*

What will be the components of mixed gas?

- Carbon monoxide and hydrogen → 22
- Carbon monoxide, carbon dioxide, hydrogen and nitrogen → 43
- Carbon monoxide, hydrogen and nitrogen → 13
- Carbon dioxide, hydrogen and nitrogen → 3

19

34 Your answer is wrong. If a factor of four were to appear in front of each substance, the equation as a whole could be divided by four. A detailed discussion on this point is to be found in 45 .

---

35 Correct. At room temperature ammonia is a pungent *gas.* At temperatures below −33 °C it is liquid.

Industrially, ammonia is synthesized by the *Haber* Process. (It's worth remembering this name.)

In this process nitrogen and hydrogen are reacted together to form ammonia thus:

$$3\ H_2 + N_2 \rightarrow 2NH_3$$

Using Avogadro's Law it is now possible to draw certain conclusions from this equation concerning the volumes of the gases involved in the reaction.

Which of the three statements below is correct?

- 6 volumes of hydrogen and 2 volumes of nitrogen yield 6 volumes of ammonia → [47]
- 1 volume of hydrogen and 1 volume of nitrogen yield 1 volume of ammonia → [59]
- 3 volumes of hydrogen and 1 volume of nitrogen yield 2 volumes of ammonia → [56]

---

36 You have forgotten one of the valences. Hint: consider the valence of the nitrogen in NO_2.

Now go back to [54] and select the correct answer

---

37 Your answer is wrong. Certain concepts need to be clarified. Please read through [4] carefully.

---

38 Your answer is wrong. Back in Program 18 we said that 1000 L of ammonia dissolve in 1 L of water.

Thus, 1000 m^3 of ammonia gas will dissolve in one cubic meter of water. → [10]

---

39 Your answer is wrong. Ammonia is a compound having the chemical formula NH_3. The nitrogen component of mixed gas is vital for the synthesis of ammonia.

Write out now the equation for the synthesis of ammonia by the Haber Process.

$$\ldots\ldots\ldots\ldots + \ldots\ldots\ldots\ldots \rightarrow NH_3$$

Check your answer at [44] .

---

40 The correct equation is:

$$CO + H_2O \rightarrow CO_2 + H_2$$

What is the chemical significance of this equation?

19

- It represents oxidation of the carbon monoxide → 52
- It represents reduction of the water → 29
- It represents oxidation of the carbon monoxide and simultaneous reduction of the water → 63

---

41 Your answer is wrong. Sodium hydrogen carbonate does not react at all with carbon dioxide.

If CO_2 is bubbled into an aqueous solution of sodium carbonate, the product formed is sodium hydrogen carbonate:

$$Na_2CO_3 + H_2O + CO_2 \rightarrow 2\ NaHCO_3$$

Any further amounts of CO_2 bubbled in will not change this product.

How may sodium hydrogen carbonate be converted into sodium carbonate in a commercially viable way?

- By heating it up → 15
- By reacting it with sodium hydroxide → 5

42

Haber
hydrogen
nitrogen
500 °C
pressure

19

Even under these conditions the reaction is still *incomplete.* Unreacted starting materials left over are *recovered* and used again in the process.

The preparation of the starting materials for the Haber Process rests upon two important technical processes:

1. The process for the manufacture of *producer gas.*

Air is passed over red-hot coke. The oxygen in the air reacts with the carbon to form carbon monoxide:

$$2\ C + O_2 \rightarrow 2\ CO$$

Because air is being used here, the producer gas formed must contain .... in addition to carbon monoxide. → 33

---

43 No carbon dioxide is formed, either during the production of producer gas or in the manufacture of water gas. It is therefore not possible for it to be contained in mixed gas.

Read through [42] once again.

---

44 $N_2 + 3\ H_2 \rightarrow 2\ NH_3$

Now answer the question in [13] .

---

45 The following equation is to be balanced:

$$NH_3 + O_2 \rightarrow NO + H_2O$$

The NH_3 on the left contains 3 H atoms whereas the H_2O on the right contains only 2. We should therefore start out with 2 NH_3:

$$2\ NH_3 + O_2 \rightarrow 2\ NO + 3\ H_2O$$

The number of N and H atoms is now the same on both sides of the equation, although the oxygen is not yet balanced! On the right we have 5 oxygen atoms but only 2 in the O_2 molecule on the left. If we were to put a factor of 2 in front of the O_2 we would have one oxygen atom too few and if we were to put a 3 in front we should have one oxygen atom too many.

It thus becomes necessary to double all the factors in our equation thus:

$$4\ NH_3 + O_2 \rightarrow 4\ NO + 6\ H_2O$$

What factor should be inserted in front of the O_2? Write out now the correct version of this equation. → [21]

19

---

46 Your answer is wrong. Back in Program 18 we said that 1000 L of ammonia dissolve in 1 L of water.

Thus, 1000 m^3 of ammonia gas will dissolve in one cubic meter of water. → [10]

---

47 Your answer reveals that you have not completely understood Avogadro's Law.

A detailed explanation of this is to be found in [11] .

48 No, in the first stage of the Solvay Process for the manufacture of sodium carbonate, gaseous carbon dioxide is pumped into an aqueous solution of ammonia until the saturation point is reached. The product formed is ammonium hydrogen carbonate.

Write out the relevant equation for this reaction. → 20

49 The structural formula of NO is:

$N{=}O$

The nitrogen in NO has a valence of *two*.

The NO obtained from the catalytic combustion of NH_3 is reacted with more oxygen in the second stage of the process. This takes place at room temperature and the product formed is nitrogen dioxide, NO_2.

$$NO + O_2 \rightarrow NO_2$$

Write down the balanced version of this equation. → 61

50

Valence of 2: NO
Valence of 3: NH_3
Valence of 4: NO_2
Valence of 5: HNO_3

For revision purposes answer now the question below.

Which of the following processes is employed for the production of ammonia?

- The Contact Process → 17
- The Lead Chamber Process → 67
- The Water Gas Process → 60
- The Haber Process → 26
- The Producer Gas Process → 6

51 Contact gas consists of:

N_2 (nitrogen)

H_2 (hydrogen)
and CO_2 (carbon dioxide)

Let's take a closer look at where these components come from:

The nitrogen is in *producer gas*

The hydrogen is in *water gas,* and also comes from the reaction of carbon monoxide and steam in the Contact Gas Process

The carbon dioxide comes from the Contact Gas Process

Because of its great solubility in water, carbon dioxide under pressure can readily be washed out of contact gas.

Nitrogen and hydrogen are then the starting materials left behind for the synthesis of ammonia.

What raw materials are used in the synthesis of ammonia?

- Air, water and coal → 25
- Water gas and producer gas → 37
- Hydrogen and nitrogen → 14

---

52 Your answer is correct but incomplete.

Let's have another look at the relevant equation:

$$CO + H_2O \rightarrow CO_2 + H_2$$

Certainly an oxidation of the carbon monoxide to carbon dioxide has taken place, since the CO_2 contains twice as much oxygen as the CO. However, a reduction of the water has also taken place, since the .... has been removed from the water. → 63

---

53 Your answer is wrong.

A detailed explanation of this is given in 45 .

---

54 $4\ NO_2 + 2\ H_2O + O_2 \rightarrow 4\ HNO_3$

Nitrogen dioxide Water Oxygen Nitric acid

Here once again are all three stages in the manufacture of nitric acid:

Stage I: $4\ NH_3 + 5\ O_2 \rightarrow 4\ NO + 6\ H_2O$

Stage II: $2\ NO + O_2 \rightarrow 2\ NO_2$

Stage III: $4\ NO_2 + 2\ H_2O + O_2 \rightarrow 4\ HNO_3$

In discussing the manufacture of nitric acid from ammonia, we mentioned most of the important valences nitrogen can have in its compounds.

What valences does nitrogen have in its compounds?

- 2, 3 and 5 → 36
- Only 3 → 9
- Only 5 → 19
- 2, 3, 4, 5 and 6 → 28
- 3 and 5 → 23
- 2, 3, 4, and 5 → 66

---

55 The *Haber Process.*

The right hand side of the overall equation for the Solvay Process gives the *end products* which are formed:

Na_2CO_3 sodium carbonate

NH_4Cl ammonium chloride

Sodium carbonate is further used in numerous chemical processes, whereas ammonium chloride finds application as a fertilizer.

We have already mentioned other ammonium salts which can serve as fertilizers. These are the ammonium salts of sulfuric and carbonic acids.

Write down the names and formulas of both of these salts. → 8

---

56 Your answer is quite right.

Three volumes of hydrogen react with one volume of nitrogen to form two volumes of ammonia:

$$3\ H_2 + N_2 \rightarrow 2\ NH_3$$

Although this equation may look simple, the reaction is actually very difficult to carry out in practice. Elevated temperatures of around 500 °C are necessary, as well as very high pressure of several hundred atmospheres. In addition to this, special catalysts have to be employed. Even then the hydrogen and the nitrogen do not react to completion. The residual starting materials have to be recovered and used again in the process. The ammonia is extracted from the final reaction mixture by dissolving it out with water.

Write out now completed versions of the sentences below:

Ammonia is produced by the .... Process from the elements .... and .... These two gases are reacted together at a high temperature (...... °C) and a high .... using a special catalyst. → 42

57

1. *Proteins consist of complicated, nitrogen-containing organic compounds.*

2. *Human beings and animals get sufficient amounts of the nitrogen essential for their nourishment from plant and animals proteins.*

3. *Plants get the nitrogen necessary for their production of proteins from nitrogen compounds present in the soil.*

4. *Justus von Liebig first introduced the idea of artificial fertilization of the soil.*

Ammonia is used not only in the Solvay Process and in the manufacture of fertilizers; it is also used as a starting material for the manufacture of nitric acid.

Write out the chemical formula for nitric acid. → 64

---

58 Your answer is quite correct.

Here once again are the three stages of the Solvay Process:

Stage I: $NH_3 + H_2O + CO_2 \rightarrow NH_4HCO_3$

Stage II: $NH_4HCO_3 + NaCl \rightarrow NaHCO_3 + NH_4Cl$

Stage III: $2\ NaHCO_3 \rightarrow Na_2CO_3 + H_2O + CO_2$

The carbon dioxide obtained from the third stage can be recovered and used again in the first stage.

Write out now the equation for III to complete your listing of the three stages of the Solvay Process.

At this point you should double equations I and II by writing a 2 in front of each compound which appears in them. Now add together separately the left and right hand sides of all three equations to obtain the overall equation for the process. Check at 65 to see if your overall equation for the Solvay Process is correct.

---

59 Your answer reveals that you have not yet fully understood Avogadro's Law.

A detailed explanation of it is given in 11 .

---

60 Your answer is wrong. In the Water Gas Process steam is passed over red-hot coke. The products formed are H_2 and CO.

Go back now to 50 .

---

61 $2\ NO + O_2 \rightarrow 2\ NO_2$

Write down the structural formula for NO_2. What is the valence of the nitrogen in NO_2? → 12

19

62 1. The *Contact Process:*

The manufacture of *sulfuric acid* by the oxidation of sulfur dioxide with oxygen in the presence of catalysts.

2. The *Water Gas Process:*

The production of *hydrogen* and *carbon monoxide* by the reaction of coke and steam.

3. The *Producer Gas Process:*

The production of a mixture of *carbon monoxide* and *nitrogen* by the reaction of coke and air.

End of Program 19.

63

Your answer is correct. In addition to the oxidation of the carbon monoxide, a reduction of the water also takes place. The *oxygen* is removed from the water.

This process is carried out industrially by passing mixed gas and steam over suitable catalysts, i.e. contacts, at an elevated temperature. The gas which is formed is referred to as *contact gas*.

Now list the components of contact gas. → 51

64

Nitric acid, HNO_3.

The nitrogen in nitric acid has a valence of five.

The process for the manufacture of nitric acid proceeds via several stages again.

In the first stage, a mixture of ammonia and air at a high temperature is passed over a catalyst. The ammonia gets "burnt up" at this stage:

$$NH_3 + O_2 \rightarrow NO + H_2O$$

See if you can balance this equation. What factors need to be inserted into it?

The missing factors are:

In front of NH_3	In front of O_2	In front of NO	In front of H_2O	
• 2	4	2	3	→ 53
• 4	4	4	4	→ 34
• 4	5	4	6	→ 21

- I obtained different factors from those above → 45

19

65

$$2\ NH_3 + 2\ H_2O + 2\ CO_2 + 2\ NH_4HCO_3 + 2\ NaCl + 2\ NaHCO_3 \rightarrow$$

$$2\ NH_4HCO_3 + 2\ NaHCO_3 + 2\ NH_4Cl + Na_2CO_3 + H_2O + CO_2$$

After cancelling out the compounds and factors which appear on both the left and right hand sides of this overall equation, we obtain the simplified equation:

$$2\ NH_3 + H_2O + CO_2 + 2\ NaCl \rightarrow Na_2CO_3 + 2\ NH_4Cl$$

This simplified version has all the *starting materials* for the Solvay Process on the left hand side, namely:

NaCl (rock salt), which is obtained from salt mines.

CO_2, which is obtained by the complete combustion of carbon.

$NH_3 + H_2O$ is simply an aqueous solution of ammonia.

Jot down the name of the process used to manufacture ammonia. → 55

---

66 Your answer is quite right.

In its various compounds nitrogen can have valences of 2, 3, 4, or 5. The most important of its valences are, however, 3 and 5.

Write down now nitrogen-containing compounds in which the nitrogen has each of the valences listed above. → 50

19

---

67 Your answer is wrong. The *Lead Chamber Process* is used for the manufacture of *sulfuric acid.* The sulfur dioxide is oxidized to sulfur trioxide by means of nitrogen oxides which act as catalysts.

Go back now to 50 .

Program 20

Nitrogen (III)

1 This is the last of the Programs in which we shall be discussing nitrogen.

As you know, nitrogen combines with oxygen to form a series of compounds usually referred to as *oxides of nitrogen* or, more simply, as *nitrogen oxides*. The two most important oxides of nitrogen are nitrogen monoxide and nitrogen dioxide.

Draw out the structural formulas for these two oxides of nitrogen. → 13

2 The manufacture of nitric acid:

1. $4\ NO_2 + 2\ H_2O + O_2 \rightarrow 4\ HNO_3$

2. $NaNO_3 + H_2SO_4 \rightarrow HNO_3 + NaHSO_4$

When sodium nitrate reacts with sulfuric acid the product sodium hydrogen sulfate is formed.

It is also possible to carry out the reaction in such a way that the product formed is sodium sulfate, Na_2SO_4.

Write out the relevant equation for this latter reaction. → 42

3 Your answer is wrong. Remember that in any equation the same number of atoms of each kind must appear on both the left and right hand sides of that equation. After checking this in your equation, you should arrive at the correct answer. → 14

4 Your answer is wrong.

An explanation of this is given in 23 .

5 Quite right! When gunpowder burns the sulfur is oxidized to sulfur dioxide and the carbon to carbon dioxide.

Other important examples of nitrates include *ammonium nitrate* and *silver nitrate.*

Ammonium nitrate, NH_4NO_3, is produced on a large scale industrially by the neutralization of ammonia with nitric acid. It is an extremely valuable fertilizer as it has a very high nitrogen content.

Silver nitrate, $AgNO_3$, is used as a reagent to test for chloride ions. (Note that Ag is the symbol for silver.) In solutions containing chloride ions a white, flocculent precipitate of silver chloride is formed when silver nitrate solution is added.

An example of this is provided by the reaction of sodium chloride with silver nitrate:

$$NaCl + AgNO_3 \rightarrow AgCl + NaNO_3$$

Write down now the equation for the reaction of $AgNO_3$ with HCl. → 21

---

6 *KNO_2 is potassium nitrite.*

The oxygen released from potassium nitrate when it is heated can, of course, react with other substances.

$$2\ KNO_3 \xrightarrow{\text{heat}} 2\ KNO_2 + O_2$$

Nitrates can therefore be classified as:

- Reducing agents → 39
- Oxidizing agents → 17
- Catalysts → 28

---

7 Your answer is wrong. Although carbon dioxide is formed, *no* sulfur trioxide is formed.

An explanation of this is given in 20 .

---

8 NO_2 is a brown gas which is very poisonous.

It is easy to recognize by its characteristic brown color.

On the contrary, nitrogen monoxide, NO, is colorless. It is, however, immediately recognizable whenever it flows out of a vessel or apparatus.

In what way can the NO be recognized when it flows out?

Jot down your answer, and then turn to 45 .

9

The complete equations are as follows:

Stage I: $4\ NH_3 + 5\ O_2 \rightarrow 4\ NO + 6\ H_2O$

Stage II: $4\ NO + 2\ O_2 \rightarrow 4\ NO_2$

Stage III: $4\ NO_2 + O_2 + 2\ H_2O \rightarrow 4\ HNO_3$

By adding together separately the left and right hand sides of these equations we obtain a new equation. If we then cancel out those substances which appear on both sides of this equation, we arrive at the overall equation for this important synthesis.

Write out the overall equation which is finally obtained. → 18

10

Sodium hydrogen sulfate is correct.

Use keywords to describe the two methods we have discussed for the manufacture of nitric acid. Be sure to mention the names of all the substances involved.

Then check your answer at 16 .

11

Nitrites are formed when nitrates are heated.

Only very dilute solutions of nitrous acid can be prepared. This acid is very unstable and readily breaks down into oxides of nitrogen and water. Its formula is HNO_2.

Great care is necessary when working with nitrites!

If nitrites come into contact with acids, extremely poisonous nitrogen dioxide gas is formed. Be on your guard!

This appearance of NO_2 when acids are added to nitrites is in fact used as a test for nitrites.

What is the color of nitrogen dioxide?

- Brown → 44

- Yellow → 24
- Colorless → 35

---

12 $HNO_3 \rightarrow H^{\oplus} + NO_3^{\ominus}$

With increasing dilution nitric acid loses its ability to act as an oxidizing agent. The dilute acid in fact behaves just like any other strong acid.

What salt is formed when dilute nitric acid reacts with zinc? Hint: zinc, Zn, forms ions bearing a doubly positive charge.

Write out the relevant equation, and give the name of the salt which is formed. → 43

---

13 N=O O=N=O

The production of nitrogen monoxide by the catalytic combustion of ammonia has already been discussed.

Write out the relevant equation for this reaction. → 32

---

14 $2\ NO + O_2 \rightarrow 2\ NO_2$

When nitrogen dioxide reacts with water in the presence of oxygen the product formed is nitric acid. The unbalanced equation for this reaction is:

$$NO_2 + H_2O + O_2 \rightarrow 4\ HNO_3$$

20

The factors are missing on the *left hand* side of the equation, i.e. the equation *needs to be balanced.*

What factors should be put in front of the NO_2, the H_2O and the O_2?

In front of NO_2	In front of H_2O	In front of O_2	
● 2	2	none	→ 25
● 4	2	none	→ 46
● 4	4	2	→ 36
● 2	4	2	→ 3

15 Your answer is wrong.

An explanation of this is to be found in 23 .

16 The processes for the manufacture of nitric acid are based on the reactions of:

1. Nitrogen dioxide with water and oxygen.
2. Sodium saltpeter (Chile saltpeter) with sulfuric acid. The by-product is sodium hydrogen sulfate.

Write out now the relevant equations for these two processes once again, and then continue your reading at 2 .

17 Nitrates are oxidizing agents. The oxidizing power of nitrates was exploited in historical times for the manufacture of gunpowder and explosives. Nowadays these products contain organic nitro-compounds.

The historical gunpowder, invented several hundred years ago, is a mixture of *potassium nitrate*, *sulfur* and *coal powder*. When it burns, the sulfur and the coal react with the oxygen given off by the potassium nitrate. The reaction products are the combustion gases formed, which expand enormously because of the heat generated by the reaction. Gunpowder thus explodes when it is confined within a bullet and ignited. On the other hand, if a small amount of gunpowder is ignited in an open dish, a harmless puff of gas is formed since then the combustion products can escape freely.

What combustion products are formed from gunpowder?

- Carbon dioxide and sulfur trioxide → 7
- Carbon dioxide and sulfur dioxide → 5
- Carbon monoxide and sulfur trioxide → 41
- Carbon monoxide and sulfur dioxide → 29

18

$$4\ NH_2 + 8\ O_2 + 4\ NO + 4\ NO_2 + 2\ H_2O \rightarrow$$

$$4\ NO + 6\ H_2O + 4\ NO_2 + 4\ HNO_3$$

The overall equation is:

$$4\ NH_3 + 8\ O_2 \rightarrow 4\ HNO_3 + 4\ H_2O$$

or, after dividing it throughout by 4:

$$NH_3 + 2\,O_2 \rightarrow HNO_3 + H_2O$$

Concentrated nitric acid is roughly a 60% solution of HNO_3 in water. It is a powerful oxidizing agent. It becomes an even more powerful oxidizing agent when it is mixed with concentrated hydrochloric acid. This mixture is referred to as *aqua regia* which means "royal water". It is given this name because it is capable of dissolving the "king of metals" – gold.

Great care is necessary whenever concentrated nitric acid or aqua regia are handled as both are extremely corrosive.

Use keywords to jot down now the most important properties of concentrated nitric acid and aqua regia. Then continue your reading at [30] .

---

[19] Nitrogen monoxide, NO, is colorless, gaseous and reacts very readily with oxygen.

$$2\,NO + O_2 \rightarrow 2\,NO_2$$

Nitrogen dioxide is a brown and extremely poisonous gas. It is particularly insidious as large amounts of it can be breathed in without any apparent ill effects. It is only after several hours that the effects become evident, and it is often then too late to do anything.

Accordingly, you should always take the greatest possible care when working with nitrogen dioxide!

Make a short summary now of the major properties of NO_2. → [8]

---

[20] It should be borne in mind that gunpowder is a mixture of potassium nitrate, sulfur and coal powder.

When gunpowder is ignited, the potassium nitrate gives up some of its oxygen and the sulfur and coal react with this oxygen. Both substances are, of course, oxidized.

Sulfur can be oxidized to either sulfur dioxide or sulfur trioxide. Its oxidation to sulfur trioxide is, however, difficult and is possible only in the presence of catalysts.

When gunpowder burns no suitable catalyst is present, of course, and so the sulfur is oxidized only to sulfur dioxide.

Carbon can be oxidized to either carbon monoxide or carbon dioxide. Carbon monoxide is formed only when there is incomplete combustion of the carbon. As there is ample oxygen available in gunpowder from the potassium nitrate, the carbon is actually oxidized to carbon dioxide.

When gunpowder burns, the sulfur is oxidized to .... and the carbon to .... → 5

---

21 Now check your equation:

$HCl + AgNO_3 \rightarrow AgCl + HNO_3$

It is important to remember that any solution to be tested for chloride ions should be acidified with dilute nitric acid before the silver nitrate solution is added. This is necessary because there are several ions which form a white precipitate with silver ions in neutral or weakly alkaline solutions. The acid thus prevents the formation of any precipitate which is not due to chloride ions.

Now write down the equation for the reaction of $CaCl_2$ with $AgNO_3$. → 33

---

22 The test solution is mixed with iron(II) sulfate solution and concentrated sulfuric acid is then poured in slowly to form a lower layer. A brown ring which forms between the two layers indicates the presence of a nitrate or nitric acid.

To return now to nitrous acid, we said that the salts of nitrous acid are called nitrites. These salts are formed when nitrates are .... Oxygen gas is given off during the reaction. → 11

---

23 When the formulas for sodium nitrate and nitric acid are placed alongside one another thus:

$NaNO_3 \quad HNO_3$

it is easy to see that both contain the nitrate ion and thus the same number of oxygen atoms.

In an *oxidation* an element or a compound combines with oxygen whereas in a *reduction* oxygen is removed from a compound. Neither by oxidation nor by reduction is it possible to produce nitric acid from sodium nitrate.

Remember here that sodium nitrate is a salt of nitric acid. We are trying to prepare the acid from this salt.

Now make another attempt at the question in 46 .

---

24 Your answer is wrong. The color of nitrogen dioxide is darker. Nitrogen dioxide is *brown* whereas nitrogen monoxide is .... → 44

25 Your answer is wrong. Remember that in any equation the same number of atoms of each kind must appear on both the left and right hand sides of that equation.

This condition is satisfied only in the case of the hydrogen in the equation you have set up. You should now check the factors in front of the NO_2 and O_2 once again. → 14

26 Your answer is quite correct.

The non-volatile sulfuric acid drives nitric acid from its salt.

The relevant equation is as follows:

$$NaNO_3 + H_2SO_4 \rightarrow HNO_3 + NaHSO_4$$

What is the name of the salt formed with the nitric acid?

- Sodium hydrogen sulfate → 10
- Sodium sulfate → 37

27 Now check your answers:

1. NO, NO_2
2. a) from Chile saltpeter and sulfuric acid:
   $NaNO_3 + H_2SO_4 \rightarrow HNO_3 + NaHSO_4$
   b) from nitrogen dioxide, oxygen and water:
   $4\ NO_2 + O_2 + 2\ H_2O \rightarrow 4\ HNO_3$
3. $4\ NH_3 + 5\ O_2 \rightarrow 4\ NO + 6\ H_2O$
4. $N_2 + O_2 \rightarrow 2\ NO$
5. The oxides of nitrogen are extremely poisonous.
6. Concentrated nitric acid is very corrosive.

7. When nitrites are mixed together with acids the very poisonous oxides of nitrogen are produced.

8. Nitrous acid: nitrites
   Nitric acid: nitrates

Here are some additional questions:

9. How can one test for a nitrate?

10. Name some examples of important fertilizers.

11. How can one test for a nitrite?

12. Write down the equations for the:
    a) Producer Gas Process
    b) Water Gas Process.

13. What are the component gases in:
    a) producer gas
    b) water gas
    c) mixed gas?

14. What process is used for the industrial synthesis of ammonia? Give the name of the process, the equation and the conditions under which the reaction is carried out.

15. How can one test for the ammonium ion in ammonium compounds? Give the relevant equation, taking as your example ammonium sulfate.

16. What is the valence of the nitrogen in:
    a) nitric acid
    b) ammonia
    c) sodium nitrite
    d) nitrogen monoxide?

17. Complete the following equations:
    a) $HNO_3 + NH_4OH \rightarrow$
    b) $KNO_3 \xrightarrow{\text{heat}}$
    c) $NH_4Cl + \qquad\qquad \rightarrow NH_3 + H_2O + NaCl$

Check all your answers at **40** .

---

28 Your answer is wrong. *A catalyst is a substance which speeds up a chemical reaction, but which at the end of the reaction, is present in the same form as at the beginning.*

Write out this definition now.

The very first catalyst we came across was manganese dioxide. It speeds up the production of oxygen when potassium chlorate is heated, without itself being changed in any way.

KNO_3 certainly gets changed, however, when it is heated!

Read through [6] once again.

---

29 Your answer is wrong. Sulfur dioxide is formed, but *no* carbon monoxide.

An explanation of this is given in [20] .

---

30 Concentrated nitric acid is a powerful oxidizing agent. **Caution:** it is also very corrosive!

Aqua regia is a mixture of conc. nitric acid and conc. hydrochloric acid. It can dissolve gold. **Caution:** it is very corrosive!

In water, nitric acid dissociates into singly charged positive hydrogen ions and singly charged negative nitrate ions.

Write down the equation for this dissociation. → [12]

31

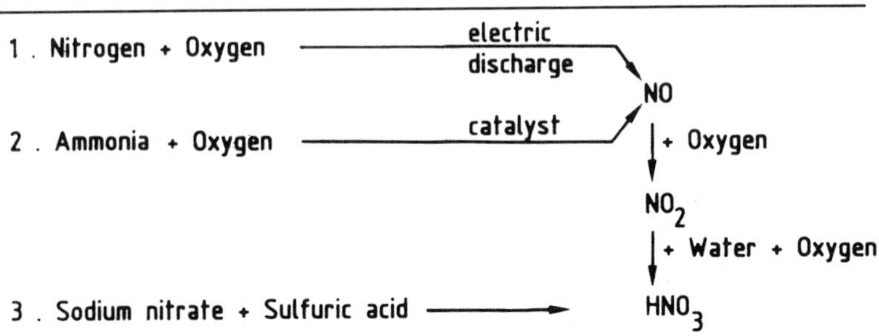

The most important process for the manufacture of nitric acid is the second one here which starts out from ammonia. The incomplete equations for the three reactions involved are as follows:

I. $NH_3 + 5\ O_2 \rightarrow 4\ NO +$

II. $4\ NO + O_2 \rightarrow NO_2$

III. $NO_2 + 2\ H_2O +$ $\rightarrow 4\ HNO_3$

Copy down and complete each of these equations. → 9

---

32 $4\ NH_3 + 5\ O_2 \rightarrow 4\ NO + 6\ H_2O$

If you got the equation wrong, have another look at the correct version and copy it down. This reaction takes place at a high temperature and in the presence of a catalyst.

The production of NO from the elements nitrogen and oxygen is possible only under certain special conditions owing to the fact that nitrogen is so unreactive. Appropriate conditions can be created by passing electric discharges through the mixture. The heat generated in this way is sufficient to enable the reaction to take place.

Nitrogen and oxygen combine together to form nitrogen monoxide according to the equation:

$$N_2 + O_2 \rightarrow 2\ NO$$

Nitrogen monoxide is a *colorless* gas which reacts with oxygen very readily.

Using keywords jot down the properties of nitrogen monoxide. Then write out the equation for the reaction of nitrogen monoxide with oxygen. → 19

---

33 $CaCl_2 + 2\ AgNO_3 \rightarrow 2\ AgCl + Ca(NO_3)_2$

After describing the test for a chloride with silver nitrate, we now move on to the test for nitric acid and its salts, i.e. the nitrates.

The test solution is mixed together with a solution of iron(II) sulfate, $FeSO_4$. A few cm^3 of pure, concentrated sulfuric acid are then carefully poured in and form a lower layer. Note that the test tube is held at an angle of about 45° and that the acid is allowed to trickle down the side of it and settle at the bottom. The conc. sulfuric acid will not mix with the test solution if this operation is carried out carefully as it has a greater density than that of the solution. Two layers are thus formed: a lower one of conc. sulfuric acid and an upper one of the test solution. If either nitric acid or a nitrate is present, a brown ring will form between these two layers.

Using keywords, indicate how you would test for nitric acid or a nitrate. → 22

34 Only very *dilute* solutions of nitrous acid can be prepared. This acid is highly *unstable* and decomposes *readily.*

Its formula is HNO_2; its salts are called *nitrites.*

To test your knowledge of the material we have covered in all three of the Programs dealing with nitrogen, you should now write out answers to the following questions. You can check whether your answers are correct at 27. Before checking, however, you should read all your answers carefully to eliminate possible errors.

1. Which oxides of nitrogen have we discussed?
2. In what ways can nitric acid be prepared? Give the relevant equations.
3. Give the equation for the combustion of ammonia in the presence of a catalyst.
4. Give the equation for the reaction of N_2 with O_2 in an electric discharge.
5. Why is caution necessary when working with the oxides of nitrogen?
6. Why is caution necessary when working with concentrated nitric acid?
7. Why is caution necessary when nitrites come into contact with acids?
8. What do we call the salts of nitrous and nitric acids?

Now check your answers at 27 .

35 Your answer is wrong. Nitrogen dioxide is *brown* whereas nitrogen monoxide is .... → 44

36 Your answer is wrong. Remember that in any equation the same number of atoms of each kind must appear on both the left and right hand sides of that equation.

This is satisfied only in the case of the nitrogen in the equation you have set up. You should now check the factors in front of the H_2O and O_2 once again. → 14

37 Your answer is wrong.

The compound

$NaHSO_4$

still contains some hydrogen.

The compound involved is therefore.............. → 10

---

38 Your answer is wrong. We have already discussed a number of electrolytic processes. In such processes the substances electrolyzed have always been split up into their component parts. Thus, the electrolysis of hydrochloric acid yields hydrogen and chlorine, and the electrolysis of water yields hydrogen and oxygen.

Remember here that sodium nitrate is a salt of nitric acid. It is our aim to liberate the acid from this salt. How can this be done?

- By reacting the salt with sulfuric acid → 26
- By means of oxidation → 4
- By means of reduction → 15

---

39 Your answer is wrong. An oxidation is a chemical process in which a substance *combines with oxygen*. Substances which have the necessary oxygen for this chemical process to take place are termed oxidants.

Read through 6 once again.

---

40 Now check your answers:

9. Test solution + iron(II) sulfate solution. Form lower layer of conc. sulfuric acid. Brown ring forms at the contact interface.

10. $NaNO_3$, KNO_3, NH_4NO_3, $(NH_4)_2SO_4$.

11. When the test substance is acidified, oxides of nitrogen are given off.

12. a) $2\ C + O_2 \rightarrow 2\ CO$
    b) $C + H_2O \rightarrow CO + H_2$

13. a) CO, N_2
    b) CO, H_2
    c) CO, N_2, H_2

14. The Haber Process:
    $N_2 + 3\ H_2 \rightarrow 2\ NH_3$
    500 °C and over 200 atm.

15. $(NH_4)_2SO_4 + 2\ NaOH \rightarrow Na_2SO_4 + 2\ NH_3 + 2\ H_2O$
    NH_3 is evolved; smell, litmus paper

16. a) HNO_3 pentavalent
    b) NH_3 tervalent
    c) $NaNO_2$ tervalent
    d) NO bivalent

17. a) $HNO_3 + NH_3 + H_2O \rightarrow NH_4NO_3 + H_2O$
    b) $2\ KNO_3 \xrightarrow{\text{heat}} 2\ KNO_2 + O_2$
    c) $NH_2Cl + NaOH \rightarrow NH_3 + H_2O + NaCl$

If you were unable to answer more than two of these questions or made mistakes in more than three of them, you are advised to re-read all the Programs on nitrogen again.

End of Program 20.

---

41 Your answer is wrong. Neither carbon monoxide nor sulfur trioxide is formed.

An explanation of this is given in 20 .

---

42 $2\ NaNO_3 + H_2SO_4 \rightarrow 2\ HNO_3 + Na_2SO_4$

Most of the nitric acid which is manufactured is not in fact made from sodium nitrate, but rather from nitrogen dioxide. The NO required for the production of NO_2 is produced in countries having relatively cheap electric power simply by combining nitrogen and oxygen together in an electric discharge.

In the United States, however, the process involving the catalytic combustion of ammonia is more economic:

$$4\ NH_3 + 5\ O_2 \rightarrow 4\ NO + 6\ H_2O$$

Copy down now the flow chart which represents the various ways of manufacturing nitric acid. Insert the missing substances where appropriate.

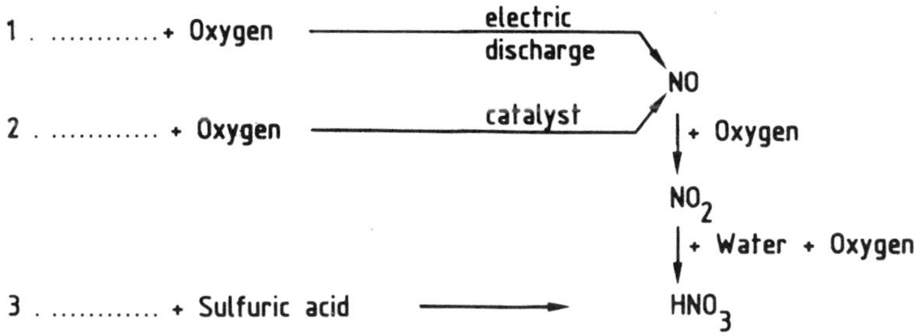

The complete version of this flow chart is given in 31 .

43

$$Zn + 2\ HNO_3 \rightarrow \underset{\text{zinc nitrate}}{Zn(NO_3)_2} + H_2$$

When discussing sulfur we introduced:

Sulfuric acid	H_2SO_4	(a stable acid)
Sulfurous acid	H_2SO_3	(an unstable acid)
Hydrogen sulfide	H_2S	(a weak, stable acid)

A similar situation obtains for nitrogen:

Nitric acid	HNO_3	(a stable acid)
Nitrous acid	HNO_2	(an unstable acid)
Ammonia	NH_3 or H_3N	(a base)

In the case of sulfur:

The salts of sulfuric acid are called: sulfates.
The salts of sulfurous acid are called: sulfites.

In the case of nitrogen:

The salts of nitric acid are called: nitrates.
The salts of nitrous acid are called: nitrites.

The sodium nitrate mentioned earlier occurs naturally as sodium saltpeter or Chile saltpeter. Like potassium nitrate (potassium saltpeter) it is used as a fertilizer.

When both sodium and potassium nitrates are heated up, some of the nitrogen in the molecule is given off. In the case of potassium nitrate the relevant equation is:

$$2\ KNO_3 \rightarrow 2\ KNO_2 + O_2$$

Write down the name of the substance KNO_2. → 6

---

44 Correct, nitrogen dioxide, NO_2, is a *brown* gas, whereas nitrogen monoxide is *colorless*.

Write down the completed version of the sentence below:

Only very .... (dilute/concentrated) solutions of nitrous acid can be prepared. This acid is highly .... (stable/unstable) and decomposes .... (readily/with difficulty).

Its formula is .... (HNO_2/HNO_3); its salts are called .... (nitrates/nitrites). → 34

---

45 Your answer is correct if you remembered that the NO streaming out immediately combines with the oxygen in the air to form NO_2. Thus, a brown gas is formed.

Write out the relevant equation, and then check it at 14 .

---

46 Your answer is quite right.

The complete equation is as follows:

20

$$4\ NO_2 + 2\ H_2O + O_2 \rightarrow 4\ HNO_3$$

There is a second process to consider. Here one starts out from naturally occurring saltpeter from Chile. Sodium saltpeter, or Chile saltpeter, is sodium nitrate.

How can nitric acid be produced from sodium nitrate?

- By means of electrolysis → 38
- By reacting it with sulfuric acid → 26
- By means of oxidation → 4
- By means of reduction → 15

Program 21

Chlorine and its Compounds

1 Chlorine is an element that has been mentioned in previous Programs. We shall start here by briefly reviewing what was said about chlorine earlier on. Do you remember the physical properties associated with chlorine? If so, the following question should be easy for you to answer:

What is chlorine?

- A metal → 5
- A non-metal → 8
- Neither of these → 11

---

2 Your answer is almost correct.

The compound formed when chlorine and hydrogen combine, HCl, is, however, called hydrogen chloride. It is the aqueous solution of hydrogen chloride which is referred to as hydrochloric acid.

Continue your reading at 23 .

---

3 Correct, chlorine is obtained by electrolysis of a sodium chloride solution. However, if sufficiently large quantities of hydrogen chloride are available, one can also electrolyze hydrochloric acid.

We have already discussed the electrolysis of sodium chloride solutions. What products are formed in the process, apart from chlorine?

- Caustic soda → 7
- Caustic soda and hydrogen → 10
- Sodium and hydrogen → 18

---

4 Your answer is wrong.

Continue your reading at 19 .

---

5 Your answer is wrong.

Continue your reading at 14 .

6 Your answer is wrong.

Copy down now the following sentences:

To test for hydrochloric acid and chlorides one acidifies with nitric acid and then adds silver nitrate solution. A white, flocculent precipitate of silver chloride forms if chloride ions are present.

End of Program 21.

7 Your answer is incomplete.

When an aqueous sodium chloride solution is electrolyzed, hydrogen is formed in addition to chlorine and caustic soda. Now copy down the following sentence:

The electrolysis of an aqueous sodium chloride solution yields ...., ... and ....

Insert the missing words, and then check your answers at 10 .

8 Quite right!

Chlorine is a *non-metal.* In fact it is a pungent, greenish gas.

Chlorine is very poisonous.

Chlorine is denser than air.

Chlorine is a very reactive element which readily combines with other elements, including

hydrogen
sodium and
potassium.

Write down now the equations for the reaction of chlorine with hydrogen, sodium and potassium.

What substances are formed?

- Hydrochloric acid, sodium chloride, potassium chloride → 2
- Hydrogen chloride, sodium oxide, potassium oxide → 13
- Hydrogen chloride, sodium chloride, potassium chloride → 23

---

9 Your answer is not quite right.

Although chlorine can be obtained by the electrolysis of hydrochloric acid, such a process is viable only when very large amounts of hydrogen chloride are produced. Usually, however, an aqueous solution of sodium chloride is electrolyzed.

Sodium chloride is, of course, available in nature in very large quantities.

Continue your reading at 3 .

---

10 The electrolysis of an aqueous sodium chloride solution yields *chlorine, caustic soda* and *hydrogen.*

So far we have discussed only one chlorine acid.

Write down the name and formula for this acid. What sort of acid is it?

- A monobasic acid → 17
- A dibasic acid → 20
- A tribasic acid → 24

---

11 Your answer is wrong. Every chemical element is either a *metal* or a *non-metal.*

Now continue your reading at 14 .

---

12 Only the valence of the chlorine in hypochlorous acid is not correct. Draw out now the structural formula for hypochlorous acid (HOCl).

All that you will then need to do is simply to count up the valence bonds emanating from the chlorine atom.

Go through [53] once again.

---

13 Your answer is wrong. We asked you for the reaction of chlorine with hydrogen, sodium and potassium. Since oxygen is not involved in any of these reactions, no oxide can possibly be formed.

Read through [8] once again.

---

14 *Metals* are elements which conduct both heat and electric current well. Metals have a lustrous appearance and are solid under normal conditions. (The one exception to this rule is mercury which is liquid.)

Non-metals conduct heat and electric current poorly. Under normal conditions non-metals can be gaseous, liquid or solid. (The only non-metal which is liquid is bromine.) Non-metals are frequently colored.

Chlorine is a greenish, pungent and *poisonous* gas.

Now complete the sentence below:

Chlorine is a .... (metal/non-metal). → [8]

---

15 Your answer is wrong. If we heat a sodium chloride solution, the initially cold solution becomes a hot one. Continued heating will cause the water to evaporate. Solid sodium chloride will be left behind. There is thus no *chemical* process taking place at all, but rather a *physical* process.

Make another attempt at the question in [23] .

---

16 Your answer is wrong. Only the valence of the chlorine in HCl is correct. You have just drawn out the structural formulas for chloric and perchloric acids. You should now draw out the structural formula for hypochlorous acid (HOCl) too.

All that you then need do is simply to count up the valence bonds emanating from each of the chlorine atoms.

Now read through [53] once again.

---

17 Hydrogen chloride, HCl, is a *monobasic acid.*

Hydrogen chloride may be obtained from the elements chlorine and hydrogen. To do this, however, great caution is necessary, for chlorine and hydrogen react together very violently.

Another process for the manufacture of hydrogen chloride is based on the reaction of sulfuric acid with sodium chloride. Depending on the relative amounts of sulfuric acid and sodium chloride used, this reaction can take place according to either of the two equations below:

1. $NaCl + H_2SO_4 \rightarrow NaHSO_4 + HCl$
2. $2\ NaCl + H_2SO_4 \rightarrow Na_2SO_4 + 2\ HCl$

How many moles of sulfuric acid react with one mole of sodium chloride in each of these equations?

	In equation 1	In equation 2	
•	1 mol	2 mol	→ 25
•	1 mol	1 mol	→ 28
•	1 mol	1/2 mol	→ 30

18 Your answer is only partially correct. The hydrogen ions are discharged at the cathode and hydrogen is formed. The sodium ions, however, are not discharged.

The electrolysis of an aqueous sodium chloride solution yields ...., ... and ....

Jot down the missing words, and then check your answers at 10 .

19 In water, acids split up (dissociate) to form hydrogen ions and acid radicals.

Hydrogen chloride dissociates in water according to the following equation:

$HCl \rightarrow H^{\oplus} + Cl^{\ominus}$

Upon neutralization, 1 mol of hydrogen chloride reacts with 1 mol of caustic soda:

$HCl + NaOH \rightarrow NaCl + H_2O$

Hydrogen chloride is a *monobasic* acid.

Sulfuric acid dissociates in water according to the equation:

$$H_2SO_4 \rightarrow 2\ H^{\oplus} + SO_4^{2\ominus}$$

Sulfuric acid is a *dibasic* acid.

Jot down now the missing word in the sentence below:

Hydrogen chloride, HCl, is a .... basic acid. → 17

---

20 Your answer is wrong.

The chlorine acid we have discussed so far is hydrogen chloride. Its aqueous solution is called hydrochloric acid. Hydrogen chloride has the formula HCl. What is the basicity of this acid?

- It is monobasic → 17
- It is dibasic → 32
- It is tribasic → 4

---

21 $Zn + 2\ HCl \rightarrow ZnCl_2 + H_2$

When hydrochloric acid reacts with zinc the products zinc chloride and hydrogen are formed.

The salts of hydrochloric acid are called chlorides.

The chlorides of sodium, potassium, ammonium and calcium are of especial importance.

Copy down the names of these salts and give their chemical formulas.

→ 50

---

22 Your answer is wrong. Take another good look at the structural formulas in 33 and count up the valence bonds emanating from the chlorine atom in each case. You should then have no difficulty reaching the correct answer.

---

23 Your answer is quite right.

Check once again the equations you have written. The correct versions are as follows:

$$Cl_2 + H_2 \rightarrow 2\ HCl$$

$$Cl_2 + 2\ Na \rightarrow 2\ NaCl$$

$$Cl_2 + 2\ K \rightarrow 2\ KCl$$

You have already learnt one method for the production of chlorine.

How is chlorine produced industrially?

- By the electrolysis of a sodium chloride solution → 3
- By the electrolysis of hydrochloric acid → 9
- By heating a sodium chloride solution → 15

24 Your answer is wrong.

The chlorine acid we have discussed so far is hydrogen chloride. Its aqueous solution is called hydrochloric acid. Hydrogen chloride has the formula HCl. What is the basicity of this acid?

- It is monobasic → 17
- It is dibasic → 32
- It is tribasic → 4

25 Your answer is wrong. You should remember here that:

In equations NaCl means 1 mol of sodium chloride.
In equations H_2SO_4 means 1 mol of sulfuric acid.

Have another careful look at the equations in 17, and then make one more attempt at this question.

26 $HOCl \rightarrow H^{\oplus} + OCl^{\ominus}$

$$HClO_3 \rightarrow H^{\oplus} + ClO_3^{\ominus}$$

The salts of hypochlorous acid are called *hypochlorites.*

The salts of chloric acid are called *chlorates.*

Important examples of salts of these two acids include:

Sodium hypochlorite and potassium hypochlorite;
Sodium chlorate and potassium chlorate.

What is the valence of the chlorine in each of these salts?

	In sodium hypochlorite	In potassium hypochlorite	In sodium chlorate	In potassium chlorate	
●	2	2	5	5	→ 47
●	1	2	5	6	→ 42
●	1	1	5	5	→ 38
●	5	5	1	1	→ 51

27 Your answer is wrong.

An explanation of this is given in 46 .

28 Your answer is wrong. Your should remember here that:

21

In equations, NaCl means 1 mol of sodium chloride.
In equations, H_2SO_4 means 1 mol of sulfuric acid.

Have another careful look at the equations in 17 , and then make one more attempt at this question.

29 Now check your equation:

$$2\ KClO_3 \rightarrow 2\ KCl + 3\ O_2$$

Was your equation correct? If so, you are working very well.

The fact that potassium chlorate readily gives off its oxygen when heated shows that it is a strong oxidizing agent.

Accordingly, great care is necessary whenever handling potassium chlorate. Combustible substances can react with potassium chlorate very vigorously, at times even explosively!

You may recall that we added a catalyst to facilitate the reaction of potassium chlorate upon heating. The catalyst concerned was manganese dioxide.

Now write out your answer to the following question:

What is a catalyst? → 53

---

30 Your answer is quite right.

In equation 1 we have 1 mol of NaCl reacting with 1 mol of H_2SO_4, whereas in equation 2 we have 1 mol of NaCl reacting with 1/2 mol of H_2SO_4:

1. $NaCl + H_2SO_4 \rightarrow NaHSO_4 + HCl$

2. $2\ NaCl + H_2SO_4 \rightarrow Na_2SO_4 + 2\ HCl$

Hydrogen chloride is a colorless gas which instantly forms a cloud of hydrochloric acid in the presence of damp air. The hydrogen chloride is said to "fume" in the air. Hydrogen chloride is extremely soluble in water; its aqueous solution is referred to as hydrochloric acid. Hydrochloric acid is a very strong acid.

Hydrochloric acid is corrosive and is particularly dangerous to the eyes. Protective goggles must therefore always be worn when working with hydrochloric acid!

Many metals are readily dissolved by hydrochloric acid. One example we have already discussed is that of zinc. Write down now the equation for the dissolution of zinc by hydrochloric acid. Hint: zinc, Zn, is divalent. → 21

---

31 Your answer is wrong.

An explanation of this is to be found in 46 .

---

32 Your answer is wrong.

Continue your reading at 19 .

33

$HCl + AgNO_3 \rightarrow HNO_3 + AgCl$

$KCl + AgNO_3 \rightarrow KNO_3 + AgCl$

$CaCl_2 + 2\ AgNO_3 \rightarrow Ca(NO_3)_2 + 2\ AgCl$

There are also several oxygen-containing chlorine acids. The two most important are:

Hypochlorous acid, $HOCl$, and
Chloric acid, $HClO_3$.

The structural formula for each of these acids is given below to illustrate the various valences which chlorine can adopt. However, you should not forget that such structural formulas do not reflect the actual bonding arrangement in a molecule. Structural formulas are no more than aids which enable us to depict in a straightforward way the valences of the atoms in a molecule.

Hypochlorous acid Chloric acid

Copy down both of these names and structural formulas.

What are the respective valences of the chlorine in hypochlorous acid and chloric acid?

- 1 and 3 → 22
- 1 and 5 → 44
- 1 and 7 → 48

34

Your answer is quite right.

You have now thoroughly reviewed and updated your knowledge of chlorine and its compounds.

End of Program 21.

35

Only the valence of the chlorine in perchloric acid is incorrect.

You have just drawn the structural formulas for chloric and perchloric acids.

All you need do now is simply to count up the valence bonds emanating from the chlorine atom.

Then you should re-read 53 .

---

36 Here are the relevant equations:

$H_2 + Cl_2 \rightarrow 2\ HCl$

$NaCl + H_2SO_4 \rightarrow NaHSO_4 + HCl$ or

$2\ NaCl + H_2SO_4 \rightarrow Na_2SO_4 + 2\ HCl$

The most important oxygen-containing chlorine acids are hypochlorous acid, HOCl, chloric acid, $HClO_3$, and perchloric acid, $HClO_4$.

What are the salts of these acids called? → 49

---

37 Your answer is quite right.

You should now even be able to formulate the reaction for the decomposition of potassium chlorate by heating. Potassium chlorate breaks down into potassium chloride and oxygen.

Write out the relevant equation now, and then turn to 29 .

---

38 Quite right.

When chlorine is bubbled into water some of the chlorine dissolves. The product formed is called *chlorine water.*

Part of the chlorine actually undergoes a chemical reaction with the water as follows:

$$Cl_2 + H_2O \rightarrow HCl + HOCl$$

Copy down this equation.

What are the names of the sodium salts of the two acids on the right hand side of this equation?

- Sodium chloride and sodium chlorate → 31

- Sodium hypochlorite and sodium chlorate → 27
- Sodium chloride and sodium hypochlorite → 52
- I don't know → 46

---

39 Your answer is wrong. Potassium chlorate is not used in any tests.

Have another shot at the question in 52 .

---

40 Your answer is quite correct.

For review purposes, you should now copy down the completed version of the sentence below:

Chlorine is obtained industrially by the .... of an aqueous sodium chloride solution. The by-products formed are .... and ....

Continue your reading at 54 .

---

41 Your answer is wrong.

Now copy down the sentence below:

To test for hydrochloric acid and chlorides, one acidifies with nitric acid and then adds silver nitrate solution. A *white*, flocculent precipitate of silver chloride forms if chloride ions are present.

End of Program 21.

21

---

42 No. Another careful inspection of the structural formulas for hypochlorous and chloric acids in 33 , which you have just copied down, should enable you to determine the correct valences. The valence of chlorine in a salt is precisely the same as its valence in the corresponding acid.

Now continue your reading at 33 .

---

43 You have just copied down the structural formulas for chloric and perchloric acids and given the correct valence of the chlorine in each of them. However, the other valences you specified were wrong. Draw out now structural formulas for hydrogen chloride (HCl) and hypochlorous acid (HOCl).

All you need do then is to count up the valence bonds emanating from the chlorine atom in each case.

Read through 53 once again.

---

44 Yes, the valence of the chlorine in hypochlorous acid is one and its valence in chloric acid is five.

Both of these acids dissociate in water to form hydrogen ions, which bear a single positive charge, and negatively charged acid radical ions.

Write down now the equations for the dissociation of these two acids.

$HOCl \rightarrow$

$HClO_3 \rightarrow$

→ 26

---

45 Your answer is wrong.

A catalyst is a substance which speeds up a chemical reaction, but which at the end of the reaction is present in the same form as at the beginning.

Make another attempt at the question posed in 52 .

---

46 Copy down and memorize the following structural formulas:

Hydrogen chloride	HCl	H-Cl	Salts: Chlorides
Hypochlorous acid	HOCl	H-O-Cl	Salts: Hypochlorites
Chloric acid	$HClO_3$	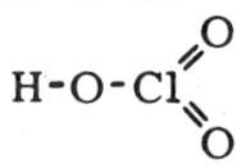	Salts: Chlorates

Now continue your reading at 52 .

---

47 Your answer is wrong. Another careful inspection of the structural formulas for hypochlorous and chloric acids in 33 , which you have just copied down, should enable you to determine the correct valences. The valence of chlorine in a salt is precisely the same as its valence in the corresponding acid.

Now continue your reading at 33 .

48

Your answer is wrong. Have another careful look at the structural formulas in 33 and count up the valence bonds emanating from each of the chlorine atoms. It should then be possible for you to arrive at the correct answer.

49

The salts of hypochlorous acid are called hypochlorites, the salts of chloric acid chlorates and the salts of perchloric acid perchlorates.

The oxygen-containing chlorine acids and their salts are all powerful oxidizing agents since they are able to give off oxygen.

To test for a chloride, one acidifies with nitric acid and then adds silver nitrate solution. What product is formed?

- A white, powdery precipitate → 6
- A white, flocculent precipitate → 34
- A black precipitate → 41

50

Now check your answers:

Sodium chloride	$NaCl$
Potassium chloride	KCl
Ammonium chloride	NH_4Cl
Calcium chloride	$CaCl_2$

Both chlorides and HCl dissociate completely in water to form ions. Since chloride ions in the presence of silver ions yield an insoluble precipitate of silver chloride, a silver nitrate solution can be used to test for chlorides and hydrochloric acid. One starts by acidifying the test solution with nitric acid. Then silver nitrate solution is added and, if chloride ions are present, a white, flocculent precipitate of silver chloride, AgCl, is immediately formed.

The reaction of silver nitrate with sodium chloride may be represented by the following ionic equation:

$$Na^{\oplus} + Cl^{\ominus} + Ag^{\oplus} + NO_3^{\ominus} \rightarrow AgCl + Na^{\oplus} + NO_3^{\ominus}$$

or, more conventionally, by the equation:

$$NaCl + AgNO_3 \rightarrow AgCl + NaNO_3$$

Write down now the equations for the reactions of silver nitrate with HCl, KCl and $CaCl_2$. Do not write ionic equations. → 33

51 Your answer is wrong. Another careful inspection of the structural formulas for hypochlorous and chloric acids in 33, which you have just copied down, should enable you to determine the correct valences. The valence of chlorine in a salt is precisely the same as its valence in the corresponding acid.

Now continue your reading at 33 .

52 Correct! When chlorine is bubbled into water, both hydrochloric and hypochlorous acids are formed:

$$Cl_2 + H_2O \rightarrow HCl + HOCl$$

The sodium salts of these two acids are called sodium chloride and sodium hypochlorite.

Chlorine water is a powerful oxidizing agent. It has this effect because of the hypochlorous acid which readily decomposes into hydrochloric acid and oxygen:

$$2\ HOCl \rightarrow 2\ HCl + O_2$$

To come now to chloric acid, one salt of this acid – potassium chlorate – has already been introduced.

For what purpose was potassium chlorate used?

- As a test solution → 39
- As a catalyst → 45
- For the production of oxygen in the laboratory → 37

53 A catalyst is a substance which speeds up a chemical reaction, but which at the end of the reaction is present in the same form as at the beginning.

There is one oxygen-containing chlorine acid which contains even more oxygen than chloric acid. For this reason it is referred to as *perchloric acid* and its salts are known as *perchlorates.*

To enable you to determine some valences, we now present the structural formulas for chloric and perchloric acids.

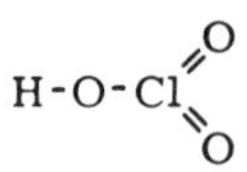

Choric acid

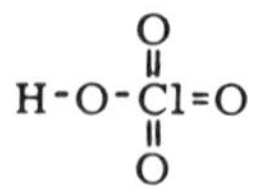

Perchloric acid

Copy down these two structural formulas and then attempt the following question:

What is the valence of the chlorine in HCl, HOCl, $HClO_3$ and $HClO_4$?

HCl	HOCl	$HClO_3$	$HClO_4$	
● 1	3	5	7	→ [12]
● 1	1	5	7	→ [40]
● 2	2	5	7	→ [43]
● 1	2	3	5	→ [16]
● 1	1	5	5	→ [35]

[54] Now check your answer:

Chlorine is obtained industrially by the *electrolysis* of aqueous sodium chloride solutions. The by-products formed are *caustic soda* and *hydrogen.*

Hydrogen chloride may be obtained either from its elements chlorine and hydrogen or via the reaction of sodium chloride with sulfuric acid.

Write out the relevant equations for these two reactions, and then continue your reading at [36] .

Program 22

Phosphorus and Silicon

1 Phosphorus has the symbol P. Phosphorus can occur in several different forms (allotropic modifications). The most important of these are *yellow* phosphorus and *red* phosphorus.

Yellow phosphorus is very poisonous and so highly flammable that it ignites spontaneously if left lying about in the air. For this reason it should always be stored under water and treated with the greatest respect.

Yellow phosphorus glows in the dark.

Red phosphorus is not poisonous and can be ignited only at a fairly high temperature. Red phosphorus is used in the manufacture of matches.

Using keywords, write out now the properties of the two phosphorus allotropes.

We have already introduced several elements which exist in the form of two or three different allotropes. In which of the rows below do all of the elements listed occur in the form of allotropes?

- Chlorine, nitrogen, sulfur → 9
- Sulfur, carbon, hydrogen → 20
- Nitrogen, carbon, phosphorus → 14
- Sulfur, carbon, phosphorus → 17
- Chlorine, sulfur, carbon → 22

2 Here again is our definition:

A compound which reacts with water to form an acid is called an acid anhydride.

We have already discussed a number of acid anhydrides. Carbon dioxide is the anhydride of carbonic acid; sulfur dioxide is the anhydride of sulfurous acid; and sulfur trioxide is the anhydride of sulfuric acid. The equations for the reaction of each of these anhydrides with water to form the corresponding acid are as follows:

$$CO_2 + H_2O \rightarrow H_2CO_3$$

$$SO_2 + H_2O \rightarrow H_2SO_3$$

$$SO_3 + H_2O \rightarrow H_2SO_4$$

An analogous equation is that representing the reaction of phosphorus(V) oxide (phosphorus pentoxide) with water:

$$P_2O_5 + 3\ H_2O \rightarrow 2\ H_3PO_4$$

Accordingly, phosphorus(V) oxide is the .... of phosphoric acid.

Copy down and complete this last sentence, and then check your version at 21 .

---

3 The chemical formula of phosphorus(V) oxide (phosphorus pentoxide or, more precisely, diphosphorus pentoxide) is P_2O_5.

See if you can also draw out the structural formula for this compound.

Check your formula at 13 .

---

4 Copy down the sentence:

Quartz consists of silicon dioxide, SiO_2.

Now continue your reading at 66 .

---

5 Phosphoric acid has the structural formula:

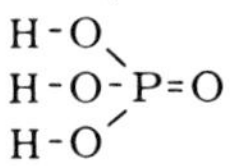

Its chemical formula is thus H_3PO_4.

In water, phosphoric acid dissociates into three positively charged hydrogen ions and one triply negative phosphate radical.

$$H_3PO_4 \rightarrow 3\ H^{\oplus} + PO_4^{3\ominus}$$

Phosphoric acid is a ...... basic acid.

Copy down and complete this sentence, and then check your answer in 26 .

---

6 Stage 3: $HPO_4^{2\ominus} \rightleftarrows H^{\oplus} + PO_4^{3\ominus}$

The overall equation is:

$$H_3PO_4 \rightleftarrows 3\ H^{\oplus} + PO_4^{3\ominus}$$

As a *tri*basic acid, phosphoric acid forms *three* different series of salts. Let's consider here the simple case of the sodium salts. The chemical formulas of the three sodium salts of phosphoric acid are:

NaH_2PO_4 Na_2HPO_4 Na_3PO_4

See if you can correctly assign the following three names to these three formulas: trisodium phosphate, sodium dihydrogen phosphate and disodium hydrogen phosphate.
Write out your answer as follows:
NaH_2PO_4 is called ....
Na_3PO_4 is called ....
Na_2HPO_4 is called .... → 15

If you need any help here, you should turn first to 23 .

7 The special type of glass used in laboratories is called *pyrex glass.*

It is necessary to use this special glass because the component substances of ordinary types of glass can be dissolved out - even with hot water. This occurrence can greatly interfere with experiments performed in the laboratory. What sort of reaction will these components have?

- I don't know → 52
- They will be neutral → 68
- They will be acidic → 62
- They will alkaline → 57

8 $P_2O_5 + 3\ H_2O \rightarrow 2\ H_3PO_4$

If necessary, you should correct the equation you wrote down. It should not now be difficult for you to decide what sort of acid phosphoric acid is.

- It is a monobasic acid → 16

22

- It is a dibasic acid → 24
- It is a tribasic acid → 26
- It is a tetrabasic acid → 35

---

9 Your answer is wrong. In the row you have chosen only one element is listed which exists in several allotropic forms, namely the sulfur. Sulfur can exist in brittle and plastic forms. The two remaining elements, chlorine and nitrogen, are gases under normal conditions. It is thus not possible for them to exist as differing allotropes.

Now select another answer to the question in 1 .

---

10 Your answer is wrong. Calcium sulfate is certainly a by-product in the reaction of tricalcium phosphate and sulfuric acid. But if three calcium ions and two SO_4^{2-} ions appear on the left hand side of the equation, the same number of each must also appear on the right hand side.

Work your way through the equation in 28 once again.

---

11 Here once again is the equation for the production of superphosphate:

$$Ca_3(PO_4)_2 + 2\ H_2SO_4 \rightarrow Ca(H_2PO_4)_2 + 2\ CaSO_4$$

Tricalcium phosphate + Sulfuric acid → Calcium dihydrogen phosphate + Calcium sulfate

At this point we turn our attention to the element *silicon.*

Silicon has the chemical symbol Si.

22

As regards its properties, it is fairly similar to carbon.

Now complete the sentence below:

Si is a ...... -metal and has a valence of ....

Continue your reading at 64 .

---

12 The most important naturally occurring phosphate is *calcium phosphate* which has the formula $Ca_3(PO_4)_2$, i.e. tricalcium diphosphate.

The phosphates of calcium form an essential ingredient of bones.

Phosphorus compounds are frequently found as impurities in iron ores. In producing the pig iron formed from such ores, a phosphorus-containing by-product known as "Thomas slag" is separated off. In a finely ground form this slag is designated "Thomas meal" and is used as a phosphorus-rich fertilizer.

Other important phosphorus compounds include *phosphorus (III) chloride (phosphorus trichloride)* and *phosphorus (V) chloride (phosphorus pentachloride.)*

Jot down the names of these two compounds and write out their chemical formulas.

Then continue your reading at [38] .

---

[13] Here is the structural formula of phosphorus(V) oxide:

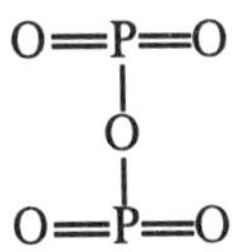

Phosphorus(V) oxide reacts with water to form phosphoric acid.

The relevant equation is:

$$P_2O_5 + 3\ H_2O \rightarrow 2\ H_3PO_4$$

Accordingly, we may designate phosphorus(V) oxide as:

- An acid anhydride → [21]
- A phosphorus and oxygen compound → [25]
- I don't know → [2]

---

[14] Your answer is wrong. The row you have chosen lists one element which is gaseous under normal conditions, namely nitrogen. But gases are not able to exist in the form of different allotropes. Carbon and phosphorus do, however, exist in several allotropic forms.

Carbon: soot, graphite, diamond

Phosphorus: yellow and red phosphorus

Now make another attempt at the question in [1] .

---

15 Now check your answers:

NaH_2PO_4 is called sodium dihydrogen phosphate.

Na_3PO_4 is called trisodium phosphate

Na_2HPO_4 is called disodium hydrogen phosphate.

Trisodium phosphate is frequently designated more briefly as sodium phosphate. However, the phosphates of calcium are considerably more important than those of sodium.

Write down the chemical formula for calcium phosphate. Note here that:

The calcium ion = $Ca^{2\oplus}$

The phosphate radical = $PO_4^{3\ominus}$ → [28]

---

16 Your answer is wrong.

An explanation of this is to be found in [5] .

---

17 Your answer is quite right.

22

The gaseous elements chlorine, nitrogen and hydrogen cannot occur in the form of different allotropes. However, we know that sulfur can exist in two forms (brittle and plastic sulfur), carbon in three forms (diamond, soot and graphite), and phosphorus in two forms (yellow and red phosphorus).

We can prove that yellow and red phosphorus are only differing forms of the same element in the following way:

Upon combustion, both allotropes yield the same combustion products.

The combustion product in question is phosphorus(V) oxide or, as it is commonly known, phosphorus pentoxide. In this compound the phosphorus has a valence of five. It should now be easy for you to write down the chemical formula for this compound.

After writing it down, continue reading at [3] .

---

[18] Your answer is wrong. Calcium sulfate is certainly a by-product of the reaction of tricalcium phosphate with sulfuric acid. But when you have three calcium ions and two SO_4 ions on the left hand side, the same number of each must also appear on the right hand side.

Take another careful look at the equation in [28] .

---

[19] According to your answer, we would have three calcium atoms on the left hand side of the equation as against four on the right hand side. This is obviously incorrect.

Take another careful look at the equation in [27] .

---

[20] Your answer is wrong. The row you have chosen lists one element which is gaseous under normal conditions, namely hydrogen. But gases are not able to exist in the form of different allotropes. Sulfur and carbon, however, do exist in several allotropic forms.

Sulfur: brittle and plastic forms

Carbon: soot, graphite, diamond.

Make another attempt at the question in [1] .

---

[21] Correct. Phosphorus(V) oxide (phosphorus pentoxide) is the *anhydride* of phosphoric acid.

See if you can set up the equation for the formation of phosphoric acid, H_3PO_4, from phosphorus(V) oxide, P_2O_5, and water.

Check your equation at [8] .

---

[22] Your answer is wrong. The row you have chosen lists one element which is gaseous under normal conditions, namely chlorine. But gases are not able to exist in the form of different allotropes. Carbon and sulfur, however, do exist in several different forms.

Carbon: soot, graphite, diamond

Sulfur: brittle and plastic forms

Make another attempt at the question in 1 .

---

23 The following material, which is designed to help you, should be read through very carefully.

The names of chemical compounds frequently have prefixes attached, such as mono = one, di = two, tri = three, tetra = four or penta = five. These prefixes are always placed *in front* of the atom to which they refer. In chemical formulas, however, the number is placed *behind* the relevant atom.

Here are some examples:

Sulfur trioxide, SO_3, because there are three (= tri) *oxygen* atoms present.

Sulfur dioxide, SO_2, because there are two (= di) *oxygen* atoms present.

Phosphorus pentoxide, P_2O_5, because there are five (= penta) *oxygen* atoms present.

Read through 6 once again.

---

24 Your answer is wrong.

An explanation of this is given in 5 .

---

25 Your answer is wrong. Phosphorus(V) oxide is certainly a compound of phosphorus and oxygen, though it is more correctly designated as an acid anhydride.

The concept of an acid anhydride is outlined again in some detail in 2 . Please read through this frame now.

22

---

26 Phosphoric acid is a *tri*basic acid. As is the case with other polybasic acids, its dissociation takes place in several stages.

The first stage involves the dissociation:

$$H_3PO_4 \rightleftarrows H^{\oplus} + H_2PO_4^{\ominus}$$

A positive hydrogen ion and an acid radical bearing a single negative charge are formed. $H_2PO_4^{\ominus}$ is the dihydrogen phosphate ion.

The second stage involves the dissociation of the dihydrogen phosphate ion:

$$H_2PO_4^{\ominus} \rightleftarrows H^{\oplus} + HPO_4^{2\ominus}$$

Write down now the third stage in the dissociation of phosphoric acid. Check your equation at [6] .

27 Now check your answer:

On the left hand side of the equation: tricalcium phosphate (the designation "calcium phosphate" is also correct).

On the right hand side of the equation: calcium dihydrogen phosphate.

The mixture of calcium dihydrogen phosphate and calcium sulfate, $Ca(H_2PO_4)_2$ and 2 $CaSO_4$, is given the name "superphosphate". Superphosphate is an important fertilizer.

If tricalcium phosphate and sulfuric acid are *heated* together, the course of the reaction is different. The equation then takes the form:

$$Ca_3(PO_4)_2 + 3\ H_2SO_4 \rightarrow \ldots\ldots\ldots\ldots + 3\ CaSO_4$$

What is the missing entry in this equation?

- $CaHPO_4$ → [19]
- 2 H_3PO_4 → [42]
- H_3PO_4 → [49]

28 $Ca_3(PO_4)_2$ is correct.

The phosphate radical bears a triply negative charge and the calcium ion a doubly positive charge. We know that the same number of positive and negative charges must be present in calcium phosphate. This can be achieved by taking two phosphate radicals and three calcium ions thus:

$$3 \times Ca^{2\oplus} = 6^{\oplus}$$

$$2 \times [PO_4]^{3\ominus} = 6^{\ominus}$$

$Ca_3(PO_4)_2$ is a mineral which occurs widely in nature. It is known as tricalcium phosphate.

The phosphates of calcium are important fertilizers. Tricalcium phosphate is insoluble in water. To convert it into a *soluble* form and *thereby render it useful as a fertilizer,* it is reacted with sulfuric acid in the cold.

The relevant equation is:

$$Ca_3(PO_4)_2 + 2\ H_2SO_4 \rightarrow Ca(H_2PO_4)_2 + \ldots\ldots\ldots\ldots$$

What is the missing entry on the right?

- 2 $CaSO_3$ → 40
- $CaSO_4$ → 18
- 3 $CaSO_4$ → 10

---

29 From the chemical viewpoint glass consists of *sodium, potassium* and *calcium silicates.*

In the various types of glass known, several different silicates normally coexist together. The properties of a given type of glass depend upon the composition of the glass. Cheaper types of glass, e.g. that used for window panes, can have small amounts of alkaline substances washed out of them-even by warm water. Such a type of glass would obviously be quite unsuitable for the manufacture of laboratory apparatus. A specially developed stable glass, known as *pyrex glass,* is however suitable for the manufacture of laboratory glassware. It can be subjected to thermal shocks to a much greater degree than ordinary glass, i.e. sudden temperature changes will not cause it to crack. For strongly corrosive chemicals which are to be used at high temperatures though, one employs quartz apparatus instead of glass.

You have probably seen objects such as vases or ash trays fashioned from heavy *lead crystal glass.* This especially valuable glass contains among other things *lead silicates.*

What is quartz made of?

- A variety of silicates → 60
- Silicon dioxide → 66
- I don't know → 4

30

Now check your answers:

tricalcium phosphate

$Ca_3(PO_4)_2$

reduction

fertilizers

We have already come across the particularly valuable fertilizers known as superphosphate and Thomas meal.

How is superphosphate produced?

- By reacting tricalcium phosphate with sulfuric acid → 11
- Superphosphate is mined as it is found in nature → 39
- By purifying phosphorus-containing pig iron → 43
- I don't know → 54

31

The industrial production of phosphorus is based on the reactions:

$$2\ H_3PO_4 \rightarrow P_2O_5 + 3\ H_2O$$

$$P_2O_5 + 5\ C \rightarrow 2\ P + 5\ CO$$

Phosphoric acid is also used in the manufacture of fertilizers, e.g. the ammonium phosphates.

Two of these of especial importance are:

1. Triammonium phosphate and
2. Diammonium hydrogen phosphate.

Write down now chemical formulas for these two salts. Check the correct pair below.

- 1. $(NH_4)H_2PO_4$
  2. $(NH_4)_2HPO_4$ → 37
- 1. $(NH_4)_3PO_4$
  2. $(NH_4)\ H_2PO_4$ → 45

- 1. $(NH_4)_3PO_4$
  2. $(NH_4)_2HPO_4$ → 50
- I need some help → 56

---

32 Now check your answer:

SiO_2

Silicon dioxide is very abundant in nature. Sand, gravel and sandstone all consist mainly of silicon dioxide. It should now be clear why silicon is the second most common element on earth – after oxygen. Silicon is always found in the form of compounds.

Sand, gravel and sandstone are very important raw materials in the construction industry:

Mortar is a mixture of slaked lime, water and *sand.*

Concrete is a mixture of cement, water and *gravel.*

Sandstone can be used directly as a building material.

Write out an answer to the following question:

In what form is silicon dioxide found in nature and for what purpose is it used?

Continue your reading at 53 .

---

33 Your answer is quite correct.

The acid anhydride (SiO_2) reacts with sodium hydroxide to form the salt sodium silicate (Na_2SiO_3) and water thus:

$$SiO_2 + 2\ NaOH \rightarrow Na_2SiO_3 + H_2O$$

From the chemical standpoint, this equation represents a *neutralization.* Let's now compare this equation with the analogous equation for the neutralization of carbon dioxide:

$$CO_2 + 2\ NaOH \rightarrow Na_2CO_3 + H_2O$$

Just as carbon dioxide is the anhydride of carbonic acid, silicon dioxide is the anhydride of *silicic acid.*

Write down the chemical formula for silicic acid, and then continue your reading at 61 .

---

34 Using keywords, write out the three major advantages which quartz apparatus has over glassware. If you can do this easily, you have been working well.

You should then continue your reading at 63 .

If this seems difficult, however, you should in your own interest read 53 once again.

---

35 Your answer is wrong.

An explanation of this is given in 5 .

---

36 Now check your answers and correct them where necessary:

Na_3PO_4

Trisodium phosphate

Na_2HPO_4

Disodium hydrogen phosphate

NaH_2PO_4

Sodium dihydrogen phosphate

We have already discussed the occurrence of the phosphates. The most important naturally occurring phosphate is ...., which has the formula ....

Jot down the relevant name and formula, and then go on to 12 .

---

37 Your answer is wrong.

An explanation on this is given in 56 .

---

38 Now check your answers:

Phosphorus(III) chloride: PCl_3

Phosphorus(V) chloride: PCl_5

Both of these compounds are used to carry out syntheses, especially in organic chemistry.

PCl_3 and PCl_5 are extremely sensitive to water. In fact, they react quite violently with water. Both of these substances should be handled only in a fume hood.

For review purposes, you should now complete the sentences below:

Phosphorus is solid. Two allotropes are known, namely .... and .... phosphorus. When they burn both forms of phosphorus yield ...., which has the formula .... This substance reacts with water to form ...., which has the formula .... The salts of this acid are called ....

Continue your reading at 46 .

---

39 No, superphosphate has to be produced by a chemical process.

Read through 30 once again.

---

40 Correct! Here once again is the complete equation:

$$Ca_3(PO_4)_2 + 2\ H_2SO_4 \rightarrow Ca(H_2PO_4)_2 + 2\ CasO_4$$

Jot down the names of the two phosphates in this equation.

Then continue your reading at 27 .

---

41 Now check your answers:

Mortar is a mixture of *sand* and slaked *lime.*

Concrete is a mixture of *gravel* and cement.

The chemical behavior of silicon is similar to that of carbon.

We shall now compare certain carbon compounds with their silicon counterparts. Copy down and complete the following table:

Carbon Compounds		Silicon Compounds	
Carbon dioxide	....	Silicon	....
....	H_2CO_3	.... acid	....
Sodium carbonate	....	Sodium	....

Continue your reading at 65 .

42 Your answer is quite right. The complete equation is:

$$Ca_3(PO_4)_2 + 3\ H_2SO_4 \rightarrow 2\ H_3PO_4 + 3\ CaSO_4$$

Phosphoric acid is manufactured industrially according to this equation. From phosphoric acid one can make phosphoric acid anhydride, i.e. phosphorus pentoxide. Reduction of the latter with coke yields the element phosphorus:

$$2\ H_3PO_4 \rightarrow \ldots\ldots + 3\ H_2O$$

$$P_2O_5 + 5\ C \rightarrow \ldots\ldots\ldots\ldots + 5\ CO$$

Write out completed versions of the two equations, and then continue your reading at 31 .

43 No, in the purification of phosphorus-containing pig iron *Thomas slag* is separated off. The latter contains the phosphorus and is used in a finely ground form as a fertilizer (known as Thomas meal).

Read through 30 once again.

22

44 Your answer is wrong. A catalyst is a substance which speeds up a chemical reaction but which at the end of the reaction is present in the same form as at the beginning. All the substances which appear in this equation are however changed!

You should therefore read through 63 once again.

45 Your answer is wrong.

An explanation of this is given in 56 .

46 Now check your answers:

yellow phosphorus
red phosphorus
phosphorus(V) oxide (phosphorus pentoxide), P_2O_5
phosphoric acid, H_3PO_4
phosphates

Jot down the missing words in the sentences below:

Yellow phosphorus is extremely poisonous and ignites spontaneously. Accordingly, it should always be stored under .... Great caution is necessary when working with yellow phosphorus!

Two important compounds of phosphorus and chlorine are ...., which has the formula ......, and ...., which has the formula ......

These two substances are also very poisonous and react violently with water.

Continue your reading at 55 .

47 Closer inspection of the chemical process involved will reveal that what is taking place is the reaction of an acid anhydride (SiO_2) with a base (NaOH). Bearing this in mind, you should now make another attempt at the question in 63 .

48 Sodium silicate, Na_2SiO_3

Sodium silicate dissolves in *hot* water.

We have converted insoluble silicon dioxide into a soluble salt by melting it down with sodium hydroxide. Chemical processes in which insoluble compounds are rendered soluble by melting them down with appropriate substances are referred to as *fusion* processes.

Write out now the equation for the fusion reaction of silicon dioxide mentioned above.

Then continue your reading at 69 .

49 According to your solution, we would have two phosphate radicals on the left hand side but only one on the right hand side. Obviously, something must be wrong here!

Take another careful look at the equation in 27 .

50 Your answer is quite right.

Write down now the chemical formulas for trisodium phosphate, disodium hydrogen phosphate and sodium dihydrogen phosphate.

Continue your reading at 36 .

51 Your answer is wrong. Although it is true that melting is involved in this process, such a statement gives us no information about the chemical process which is taking place.

Read through 63 once again.

52 It is just as well that you did not guess here. Read through 29 once again.

53 Sand, gravel and sandstone consist of silicon dioxide and are used in the construction industry as raw materials.

Silicon dioxide occurs naturally in an especially pure form known as *quartz.* Good crystalline specimens of quartz are sometimes referred to as *rock crystal.*

Quartz is used in the manufacture of laboratory apparatus such as crucibles, dishes, test tubes, flasks and so on. Although equipment of this type looks like glass, it has the following advantages over glass:

1. It can be used at much higher temperatures, since quartz has a very high melting point.

2. It is much more resistant to attack by chemicals.

3. Because quartz expands very little when heated, quartz apparatus is relatively insensitive to temperature variations compared to glass. For instance, a red-hot quartz vessel can be plunged into cold water without cracking.

You should now memorize these three advantages of quartz.

Continue your reading at 34 .

---

54 It is just as well that you did not guess here.

In the purification of phosphorus-containing pig iron, *Thomas slag* is separated off. The latter contains the phosphorus and is used in a finely ground form as a fertilizer (known as *Thomas meal).*

In the manufacture of *superphosphate,* one starts out from tricalcium phosphate which occurs naturally. This is treated with sulfuric acid at ambient temperature thus:

$$Ca_3(PO_4)_2 + H_2SO_4 \rightarrow Ca(H_2PO_4)_2 + 2\ CaSO_4$$

Copy down this equation and write out the name of each substance present underneath its formula.

Then continue your reading at 11 .

---

55 Now check your answers:

under water
phosphorus(III) chloride (phosphorus trichloride), PCl_3
phosphorus(V) chloride (phosphorus pentachloride), PCl_5

Jot down the missing words in the sentences below:

22

Phosphorus is not found in the free state in nature. It occurs rather in combined form as phosphates. For example, it is found in bones in the form of ...., which has the formula ...... and in several iron ores. Phosphorus is produced by the .... of phosphorus(V) oxide (phosphorus pentoxide) with coke. Phosphorus compounds are frequently used as agricultural .... → 30

---

56 Phosphoric acid, H_3PO_4, is a tribasic acid. It can thus form three different acid radicals when it dissociates:

$H_2PO_4^{\ominus}$ the dihydrogen phosphate ion

$HPO_4^{2\ominus}$ the hydrogen phosphate ion

$PO_4^{3\ominus}$ the phosphate ion

The ammonium ion, $NH_4^{\oplus}$, bears a single positive charge like sodium and potassium ions. → 31

---

57 Correct. Hot water dissolves alkaline substances from the cheaper types of glass.

For review purposes, you should now write out answers to the following three questions:

1. What is the most commonly occurring silicon compound in nature?

2. In what form is this compound found?

3. What is this compound used for?

Continue your reading at 67 .

---

58 Your answer is wrong. Although heat is applied in the reaction of silicon dioxide with sodium hydroxide, this fact tells us nothing about the actual chemical process taking place.

Read through 63 once again.

---

59 Now check your equations:

The chemical processes occurring during the manufacture of glass are as follows:

$$Na_2CO_3 + SiO_2 \rightarrow Na_2SiO_3 + CO_2$$

$$K_2CO_3 + SiO_2 \rightarrow K_2SiO_3 + CO_2$$

$$CaCO_3 + SiO_2 \rightarrow CaSiO_3 + CO_2$$

Silicon dioxide (in the form of sand) is thus melted down with sodium, potassium and calcium carbonates. The CO_2 which forms is given off and the resulting molten mass is made into glass apparatus, window panes, glass tubes and other glassware.

After inspecting the above three equations, it should be easy to complete the sentence below:

Chemically, glass consists of ...., ... and ....

Now continue your reading at [29] .

60 Your answer is wrong. You are too superficial in your approach. You are therefore advised to copy down the two sentences below in italics:

Quartz consists of silicon dioxide.

Glass consists of various different silicates.

Write down now three component silicates in glass.

Then continue your reading at [29] .

61 Silicic acid: H_2SiO_3

The salts of silicic acid are called *silicates.*

Write down now the formula for sodium silicate, and then continue your reading at [48] .

62 Your answer is wrong. You should now turn to [29] , where some assistance is given.

63 The advantages of quartz equipment are:

1. Improved durability at high working temperatures.

2. Greater resistance to chemical attack.

3. Very high stability to temperature variations.

There are only very few substances which will attack SiO_2, e.g. sodium hydroxide.

When silicon dioxide and sodium hydroxide are heated together, the following reaction takes place:

$$SiO_2 + 2\ NaOH \rightarrow Na_2SiO_3 + H_2O$$

What is actually taking place in this chemical process?

- A catalysis reaction → 44
- A melting process → 51
- A heating process → 58
- A neutralization reaction → 33
- I don't know → 47

---

64 Si is a *non*-metal; it has a valence of four.

Silicon is not found in the free state in nature. The most important silicon compound is silicon dioxide.

Write down the formula for silicon dioxide, and then continue your reading at 32 .

---

65 Now check your answers:

Carbon Compounds		Silicon Compounds	
Carbon dioxide	CO_2	Silicon dioxide	SiO_2
Carbonic acid	H_2CO_3	Silicic acid	H_2SiO_3
Sodium carbonate	Na_2CO_3	Sodium silicate	Na_2SiO_3

End of Program 22.

---

66 Correct, quartz consists of silicon dioxide.

Do you still remember what the special type of glass often used in the laboratory is called?

Jot down its name, and then continue your reading at 7 .

67 Now check your answers:

1. Silicon dioxide, SiO_2.
2. In the form of sand, gravel, sandstone, quartz and rock crystal.
3. For the manufacture of mortar and concrete, and as a starting material for the manufacture of glass.

Complete the two statements below:

Mortar is a mixture of .... and slaked ....

Concrete is a mixture of .... and cement.

Continue your reading at 41 .

68 Your answer is wrong. You are advised to turn to 29 where some assistance is given.

69 The fusion of silicon dioxide by melting it down with sodium hydroxide entails the reaction:

$$SiO_2 + 2\ NaOH \rightarrow Na_2SiO_3 + H_2O$$

An aqueous solution of sodium silicate is referred to as *"water glass"*. If the water is evaporated off, a glassy mass of Na_2SiO_3 remains behind. Among its various uses, water glass is used to seal cracks.

Silicon dioxide is not only of the greatest importance in the construction industry – in the form of sand, gravel, sandstone, mortar and cement – but is also vital to the *glass industry.*

The manufacture of glass is quite a complicated business. We are primarily interested here in the chemical processes involved. These may be expressed in a simplified way by means of the following equations:

$$Na_2CO_3 + SiO_2 \rightarrow Na_2SiO_3 + CO_2$$

$$K_2CO_3 + SiO_2 \rightarrow$$

$$CaCO_3 + SiO_2 \rightarrow$$

Write out the completed version of each of these equations, and then continue your reading at 59 .

Program 23

Iron (I)

1 Iron and steel are among the most important construction materials available to us. Without iron and steel there would be no automobiles, no railroads, no aircraft manufacture and no steel and concrete buildings. In addition, the chemical, mining and machine engineering industries as well as agriculture are all heavily dependent upon iron and steel. In fact, it is fair to say that our whole civilisation would be unthinkable without iron and steel. In this Program we shall be discussing in some detail the commercial production of iron and the various technical processes used in the manufacture of steel.

Iron has the chemical symbol Fe; it has valences of two and three.

Write down now formulas for the two possible iron chlorides. Then continue your reading at 23 .

2 When pyrites is roasted sulfur dioxide is formed. This gas is then used in the production of sulfuric acid. The sulfur dioxide must, of course, first be oxidized to sulfur trioxide. This is carried out in either the Lead Chamber Process or the Contact Process.

You should now make a note to re-read as soon as possible details of both of these processes used for the manufacture of sulfuric acid!

Now continue your reading at 22 .

3 Your answers should read:

Hematite: Fe_2O_3

Limonite: $Fe_2O_3 \cdot 2\ H_2O$

Another important ore of iron is known as magnetite and it has the formula Fe_3O_4. Magnetite can be regarded as a mixed iron oxide. Instead of writing it as Fe_3O_4, we could also write is as FeO, Fe_2O_3.

Magnetite is a very valuable iron ore since it contains around 70% of iron.

What is the valence of the iron atoms in magnetite?

- Three → 35

- Two → 29
- Two and three → 12

4

Iron ores are in fact reduced in a blast furnace. They cannot therefore be reducing agents! A reducing agent is a substance which brings about a reduction.

Read through 43 once again.

5

In the form of *coke.*

At this point we shall briefly repeat the chemical processes which take place in a blast furnace. You should copy down the following equations and complete them on the right hand side:

1. Coke and oxygen in the air combine together to form carbon monoxide.

$$2\,C + O_2 \rightarrow$$

2. The carbon monoxide reduces the iron oxide ores.

$$Fe_2O_3 + CO \rightarrow$$

$$Fe_3O_4 + CO \rightarrow$$

3. The iron(II) oxide which forms is reduced by carbon to metallic iron.

$$FeO + C \rightarrow$$

Now continue your reading at 27 .

6

Your equations should read:

$$Fe_3O_4 + CO \rightarrow 3\,FeO + CO_2$$

23

$$Fe_2O_3 + CO \rightarrow 2\,FeO + CO_2$$

The upper region of a blast furnace is used to preheat the initially cold ores which are fed in. Ores which contain water lose their water at this stage. For example, limonite loses its water according to the equation:

$$Fe_2O_3 \cdot 2\,H_2O \rightarrow Fe_2O_3 + 2\,H_2O$$

In the three equations given here three different iron ores appear on the left hand sides. State what these ores are in the sequence given above.

The three ores in the equations are:

- Magnetite, hematite and limonite → 28
- Magnetite, limonite and hematite → 36
- Hematite, limonite and magnetite → 39
- Pyrites, hematite and limonite → 41

---

7 The equation describing the reduction of magnetite by carbon monoxide to form iron(II) oxide is as follows:

$$Fe_3O_4 + CO \rightarrow 3\ FeO + CO_2$$

See if this equation is balanced by checking the number of each kind of atom on both the left and right hand sides of the equation. You will discover that the equation is in fact balanced. If we were to start out with iron(III) oxide, Fe_2O_3, instead of Fe_3O_4, the equation would be quite similar. Write down the relevant equation, and then attempt the following question:

What products appear on the right hand side of the equation you have just written down?

- Fe and CO_2 → 30
- 3 Fe and CO_2 → 21
- 2 FeO and CO_2 → 19

---

8 Now check your answer and make any corrections which are necessary.

$$4\ FeS_2 + 11\ O_2 \rightarrow 2\ Fe_2O_3 + 8\ SO_2$$

Sulfide iron ores have to be converted into oxide iron ores before they can be treated to yield metallic iron. Oxide iron ores on the other hand can be used directly.

In what way are oxide iron ores treated to obtain iron from them?

- They are reduced → 43
- They are oxidized → 16
- I don't know → 31

9 *slag*

For an *acid* gangue one employs a *basic* flux, e.g.

$$SiO_2 + CaO \rightarrow CaSiO_3$$

acid gangue + basic flux → slag

Instead of calcium oxide, one can also use magnesium oxide as the basic flux.

Write out the relevant equation for this latter case, and then continue your reading at 32 .

10 Your answer is quite correct. Now we shall consider in some detail the construction of a blast furnace.

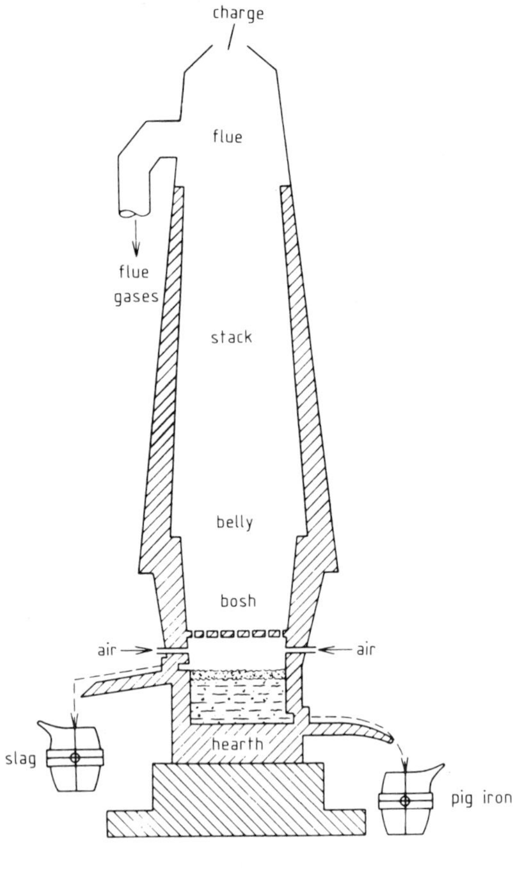

The lower part of a blast furnace is known as the hearth. Above this are located the bosh and the stack. At the top of the furnace is the flue. The widest part of the furnace – where the bosh and stack meet – is called the belly. In the hearth there are so-called tuyères (jets) through which hot air can be blown into the furnace.

You will probably find it easier to memorize the details of the construction of this furnace by copying down this diagram. When drawing the furnace, be sure to insert the names of the various parts of the furnace.

We now outline – in somewhat simplified form – the chemical processes which take place in a blast furnace.

The air blown in from the bottom of the furnace converts the coke – coke is almost pure carbon – into carbon monoxide.

Write down the relevant equation for this reaction, and then continue your reading at 34 .

---

11 All of the products formed during this reaction appear in the equation:

$FeS_2 + O_2 \rightarrow Fe_2O_3 + SO_2$

You should now make one more attempt to determine the correct factors.

After writing out the balanced equation, continue your reading at 8 .

---

12 Your answer is quite correct.

In magnetite, Fe_3O_4, the iron has valences of *two* and *three*. Thus, we may regard magnetite as iron(II, III) oxide.

Now, all of the iron ores we have come across so far contain *oxygen*. Such ores are referred to as *oxide ores*. These may be used *directly* in the production of iron.

But, in addition to oxide iron ores, there are also some sulfur-containing iron ores. These are referred to as *sulfide iron ores.*

Sulfide iron ores *cannot be used directly* for the production of iron. Such ores first have to be "roasted", i.e. they have to be *oxidized in air.*

An important example of a sulfide iron ore is pyrites, which has the chemical formula FeS_2. This ore should be known to you already from our discussion on the manufacture of sulfuric acid.

What product is formed when pyrites is roasted?

- I don't know → 2
- Sulfur dioxide → 22
- Sulfur trioxide → 37

---

13 The important ores of iron are:

Oxide Iron Ores

Hematite Fe_2O_3

Limonite $Fe_2O_3 \cdot 2\ H_2O$

Magnetite Fe_3O_4

Sulfide Iron Ores

Pyrites FeS_2

To conclude this Program, we take another brief look at the diagram of the blast furnace.

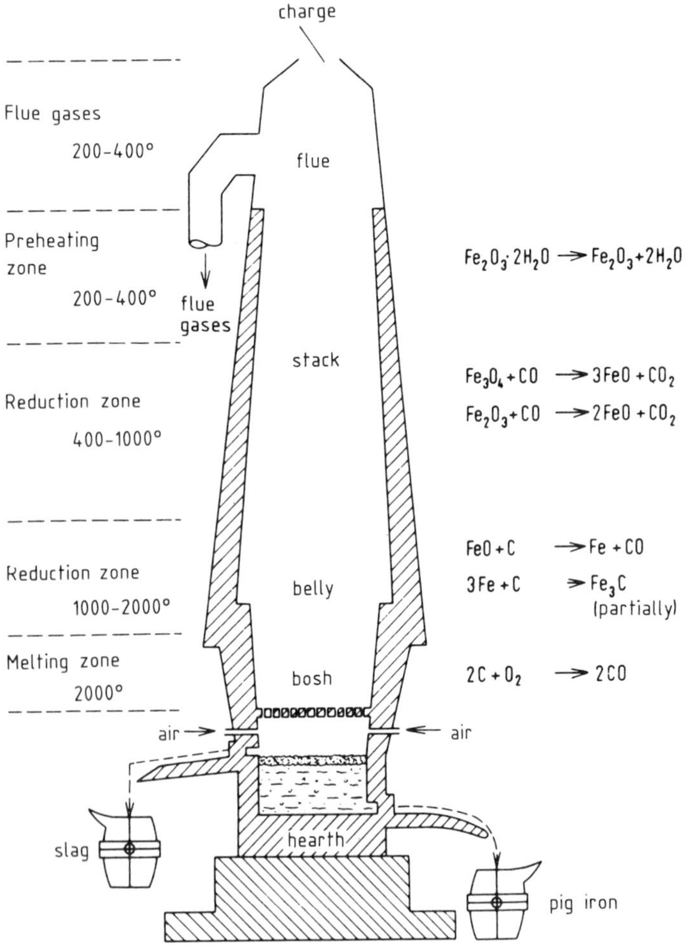

End of Program 23.

---

14 Your answer is wrong.

An explanation of this is to be found in [7] .

---

15 Now check your equations:

1. $FeO + C \rightarrow Fe + CO$

2. $3\ Fe + C \rightarrow Fe_3C$

Roughly half way up the furnace, temperatures lying between 400 to 1000 °C are attained. It is in this region that the various oxides of iron are reduced to iron(II) oxide.

Write out the respective equations for the reduction of Fe_3O_4 and Fe_2O_3, and then continue reading at [6] .

---

16 Your answer is wrong.

Now continue your reading at [31] .

---

17 $SiO_2 + CaCO_3 \rightarrow CaSiO_3 + CO_2$

$MgCO_3 + SiO_2 \rightarrow MgSiO_3 + CO_2$

We shall now discuss one further important reaction which occurs in the blast furnace.

At the very high temperatures prevailing in a blast furnace some of the iron formed combines with the carbon present.

The product produced in this way is *iron carbide,* Fe_3C:

$3\ Fe + C \rightarrow Fe_3C$

The presence of the iron carbide has the effect of lowering the melting point of the iron and so the iron becomes liquid. Together with the slag, the iron trickles down to the bottom of the furnace and collects in the hearth. The less dense slag floats as a liquid on the surface of the iron and thereby prevents the iron from being oxidized by the air which is being blown into the furnace.

For what reaction is air used in a blast furnace? Write out the relevant equation. → 40

---

18 *slag*

There are two different varieties of gangue, namely:

1. "Acid" gangue, such as quartz. Quartz is silicon dioxide, SiO_2.

2. "Basic" gangue, such as magnesium carbonate, $MgCO_3$.

The designation of gangue as either acid or basic should not be understood to mean that we are dealing here with acids and bases in the strict sense. The terms do not have the precise chemical significance they had in our earlier discussion on acids and bases. They are merely convenient labels coined by workers in the foundry business.

A gangue is said to be "acid" because it can react with a "basic" flux to produce a salt. In a similar way, "basic" gangue can react with an "acid" flux.

The salts produced by the reaction of gangue with flux form low melting .... → 9

---

19 Your answer is quite right.

Here once again are the equations for the reduction of magnetite and iron(III) oxide with carbon monoxide in the blast furnace process:

$$Fe_3O_4 + CO \rightarrow 3\ FeO + CO_2$$

$$Fe_2O_3 + CO \rightarrow 2\ FeO + CO_2$$

The further reduction of the iron(II) oxide to metallic iron is carried out in the *belly* of the furnace. (Which part of the furnace is this? Check with your own diagram!) Here the iron(II) oxide reacts with the carbon to form metallic iron and carbon monoxide thus:

$$FeO + C \rightarrow Fe + CO$$

In what form is the carbon introduced into the blast furnace?

Note down your answer and then turn to 5 .

20 Flue gas is the gas which escapes from the top of a blast furnace. As it does not come into contact with the iron ores again, this gas cannot be the reducing agent in question.

Read through 43 once again.

21 You are not concentrating enough on your work.

Read through 34 once again.

22 Clearly, when pyrites is roasted, the elements present in the pyrites will be oxidized. The products formed will be sulfur dioxide and iron(III) oxide. The iron(III) oxide can then be used for the production of iron, while the sulfur dioxide can be used in the manufacture of sulfuric acid.

Write out now the equation for the roasting of pyrites.

When you have completed this equation – and only then! – continue your reading at 8 . If you run into difficulties here, turn instead to 11 .

23 Now check your answers:

Iron(II) chloride $FeCl_2$

Iron(III) chloride $FeCl_3$

Iron is not normally found as the free metal. Iron usually occurs in the form of iron ores. Two examples of important iron ores are:

Hematite, an iron(III) oxide and
Limonite, an iron(III) oxide with two molecules of water.

Jot down the names of these two iron ores, and write alongside each the corresponding formula.

Then continue your reading at 3 .

24 A reducing agent brings about a reduction, i.e. it brings about the removal of oxygen. It is, of course, not possible to bring about a reduction by the use of air. On account of its oxygen content, air will act as an oxidizing agent.

Read through 43 once again.

25 Now check your equation:

$$CaCO_3 + SiO_2 \rightarrow CaSiO_3 + CO_2$$

basic gangue + acid gangue → slag

See if you can answer the following question:

In which row below are compounds listed which constitute a basic gangue in ores?

- Calcium carbonate, silicon dioxide → 33
- Silicon dioxide, magnesium carbonate → 42
- Calcium carbonate, magnesium carbonate → 38

26 Your answer is wrong.

An explanation of this is given in 7 .

27 1. $2\ C + O_2 \rightarrow 2\ CO$

2. $Fe_2O_3 + CO \rightarrow 2\ FeO + CO_2$

$Fe_3O_4 + CO \rightarrow 3\ FeO + CO_2$

3. $FeO + C \rightarrow Fe + CO$

At this point it is necessary to go into somewhat greater depth on the process used for the production of iron.

Iron ores are never found in an entirely pure state in nature. They always contain significant amounts of impurities which are referred to as *gangue*. These impurities have to be removed; this is accomplished by converting the gangue into a low melting *slag*. For this purpose a substance known as a *flux* has to be added to the iron ore.

In summary we have:

Gangue	An impurity present in an iron ore
Flux	A substance added during the blast furnace process which converts the gangue into low melting slag

Gangue and flux together yield ....

Jot down the missing word, and then continue reading at 18 .

---

28 Your answer is quite correct.

Now write down again the names of the most important ores of iron and give the chemical composition of each:

Hematite
Limonite
Magnetite
Pyrites

Check your formulas at 13 .

---

29 Have another careful look now at the formula for magnetite, Fe_3O_4 or FeO, Fe_2O_3. From the second of these formulas it is clear that the iron is combined with differing amounts of oxygen. The iron atoms must therefore have two different valences.

Jot down the two valences of iron in magnetite. → 12

---

30 The initial reduction in a blast furnace of iron oxide ores by carbon monoxide gas goes only as far as FeO, iron(II) oxide. The reduction does not result in the formation of elemental iron. You should already be aware of this fact. In future you are advised to work more carefully. Start now by reading through 34 once again.

---

31 Oxide iron ores are compounds of iron and oxygen.

If we wish to produce iron from them, it is obviously necessary to remove the oxygen contained in them. The removal of oxygen from any chemical compound constitutes a reduction. The iron ores will therefore have to be .... to obtain iron from them.

Jot down the missing word. → 43

---

32 Now check your equations:

SiO_2	+	MgO	⟶	$MgSiO_3$
acid gangue	+	basic flux	⟶	slag

When there is a *basic* gangue present, use is made of an *acid* flux, e.g.

$MgCO_3$	+	SiO_2	⟶	$MgSiO_3 + CO_2$
basic gangue	+	acid flux	⟶	slag

If the gangue consists of calcium carbonate one can similarly use quartz as the acid flux. The equation is, of course, almost identical. Write down this latter equation, and then turn to 25 .

---

33 Calcium carbonate acts as a basic gangue in ores, although silicon dioxide does not.

In order to ensure that you have fully understood this topic, you are now advised to re-work carefully the sections dealing with gangue.

Start by re-reading 18 .

---

34 Now check your equation:

$$2\ C + O_2 \rightarrow 2\ CO$$

23

Carbon monoxide acts as a reducing agent for the oxide iron ores. These ores are initially reduced only as far as iron(II) oxide thus:

$$Fe_2O_3 + CO \rightarrow 2\ FeO + CO_2$$

If the oxide iron ore used is magnetite, Fe_3O_4, the equation for the reduction of this ore will be similar. Write out this latter equation. It should then be easy for you to answer the question:

What products are formed on the right hand side of the equation?

- 2 FeO and CO_2 → 14
- 3 FeO and CO_2 → 19
- 3 Fe and 4 CO_2 → 26

35 Have another careful look now at the formula for magnetite, Fe_3O_4 or FeO, Fe_2O_3. From the second of these formulas it is clear that the iron is combined with differing amounts of oxygen. The iron atoms must therefore have two different valences.

Jot down the two valences of iron in magnetite. → 12

36 Your answer is only partially correct.

Fe_3O_4 is magnetite, a mixed iron oxide or Fe(II,III) oxide.
Fe_2O_3 is hematite; the Fe here is trivalent.
$Fe_2O_3 \cdot 2\ H_2O$ is limonite; here too the Fe is trivalent.
Limonite, of course, also contains some water of crystallization.

Now continue your reading at 13 .

37 Chemically speaking, the roasting of pyrites represents an oxidation. The sulfur contained in pyrites is oxidized *only as far as sulfur dioxide.* The further oxidation of sulfur dioxide to sulfur trioxide is not at all easy. To do this special catalysts have to be employed. We recall here the production of sulfur trioxide in the Lead Chamber and Contact Processes for the manufacture of sulfuric acid.

Note: you should re-read sections on the manufacture of sulfuric acid at this point!

Continue your reading at 22 .

38 Your answer is quite correct.

Now copy down the equations in italics, and then complete the two specimen equations in both words and symbols.

Acid gangue	+	*Basic flux*	$\longrightarrow$	*Slag* +	*Carbon dioxide*
Quartz	+	Calcium carbonate	$\longrightarrow$		
Basic gangue	+	*Acid flux*	$\longrightarrow$	*Slag* +	*Carbon dioxide*
Magnesium carbonate	+	Quartz	$\longrightarrow$		

Blast furnace slag is an important by-product of the production of iron. It is used for instance in the manufacture of sidewalk paving stones.

Now continue your reading at 17 .

---

39 Your answer is only partially correct.

Fe_3O_4 is magnetite, a mixed iron oxide or Fe(II, III) oxide.
Fe_2O_3 is hematite; the Fe here is trivalent.
$Fe_2O_3 \cdot 3\,H_2O$ is limonite; here too the Fe is trivalent.
Limonite, of course, also contains some water of crystallization.

Now continue your reading at 13 .

---

40 $2\,C + O_2 \rightarrow 2\,CO$

Let's now consider the various temperatures which exist in the different zones of a blast furnace.

Starting from the bottom, the temperature in the bosh is around 2000 °C and is the highest temperature in the furnace. In this zone the iron and slag melt and trickle down to the hearth.

Above this zone the temperature ranges from 1000 °C to 2000 °C, In this zone the slag is formed and two other chemical processes take place, namely:

23

1. The reduction of iron oxide to metallic iron, and

2. The formation of iron carbide.

Write out the two relevant equations for these processes, and then continue reading at 15 .

---

41 Your answer is only partially correct.

Fe_3O_4 is magnetite, a mixed iron oxide or Fe(II, III) oxide.
Fe_2O_3 is hematite; the Fe here is trivalent.
$Fe_2O_3 \cdot 2\ H_2O$ is limonite; here too the Fe is trivalent.
Limonite, of course, also contains some water of crystallization.

Now continue your reading at 13 .

---

42 Magnesium carbonate acts as a basic gangue in ores, although silicon dioxide does not.

In order to ensure that you have fully understood this topic, you are now advised to re-work carefully the sections dealing with gangue.

Start by re-reading 18 .

---

43 Yes, the oxide iron ores are *reduced* in a blast furnace.

A blast furnace consists basically of a stack some 25 meters high into which iron ore and coke are fed from the top. Air is blown in through jets at the bottom.

A series of chemical reactions take place in a blast furnance, all of them at fairly elevated temperatures. The oxide iron ores are reduced to metallic iron and the metal is drawn off from the furnace. Slag is also formed in addition to the iron. Flue gas escapes from the upper part of the furnace.

What is used as the reducing agent in a blast furnace?

- Flue gas → 20
- Iron ores → 4
- Coke → 10
- Air → 24

Program 24

Iron (II)

1 The blast furnace process leads to the production of *pig iron.* Pig iron melts at around 1100 °C whereas pure iron melts at over 1500 °C. The lower melting point of the pig iron is caused by the various impurities it contains.

One of these impurities is a compound of iron and carbon which we have already mentioned.

Write down the name and chemical formula of this compound, and then continue reading at 13 .

2 It is worth memorizing the *correct* answer, namely that:

Gray pig iron contains silicon as an impurity in additon to graphite. Its gray color derives from the graphite it contains.

White pig iron contains .... and .... as its impurities.

Now continue your reading at 9 .

3 *acidic* lining

A special procedure has to be adopted in the production of steel whenever phosphorus-bearing iron ores have been used to manufacture the pig iron. Phosphorus-bearing iron ores always yield pig iron which contains phosphorus. For a long time it was not possible to produce steel from such material in a Bessemer Converter. By an oxidation process, however, the phosphorus can be transformed into a product whose name and formula you should know.

Write down the name and formula of this product, and then continue your reading at 25 .

4 Your answer is quite right.

Iron can have valences of *two* and *three.* Examples of important iron compounds include:

$FeCl_2$, $FeCl_3$, Fe_2O_3 and FeO.

In which of these compounds is the iron

a) divalent
b) trivalent?

Jot down the appropriate valence for each of the compounds, and then check your answers at 48 .

---

5 Now check your answers:

a) Gravimetric analysis: balance, filtration apparatus, weights, funnel, filter paper.

b) Volumetric analysis: buret, pipet, measuring cylinder, indicator.

Let's now summarize what we have covered so far.

Iron seldom occurs as the free metal. The most important ores of iron – listed according to their iron content – are:

Magnetite	Iron(II, III) oxide
Hematite	Iron(III) oxide
Limonite	Iron(III) oxide with two water molecules
Pyrites	Iron sulfide, FeS_2

How many of these ores can be used directly in the production of pig iron?

- All of them → 74
- Three of them → 49
- Two of them → 12
- One of them → 43
- I don't know → 57

---

6 Your answer is wrong.

An explanation of this is given in 65 .

---

7 Your answer is quite right.

You should now turn to 15 for a more detailed coverage.

8 Iron can have valences of *two* and *three*. Consider, for instance, the following compounds:

$FeCl_2$, $FeCl_3$, Fe_2O_3 and FeO.

In which of these compounds is the iron

a) divalent
b) trivalent?

Jot down the appropriate valence for each of the compounds, and then check your answers at 48 .

9 Your answer is quite correct.

The main impurities in gray pig iron are silicon and carbon in the form of graphite.

White pig iron contains *manganese* and *iron carbide* as its impurities.

At what temperature does pig iron melt?

- At about 500 °C → 30
- At about 1100 °C → 21
- At about 1500 °C → 16

10 Your answer is wrong.

Now continue your reading at 13 .

24

11 Your answer is wrong because you are not concentrating on your work. If you read through section 41 carefully, you should be able to arrive at the correct answer.

Turn now to 41 .

12 Your answer is wrong.

An explanation of this is given in 57 .

13 Iron carbide = Fe_3C

A distinction is made between *two different kinds of pig iron,* namely:

1. White pig iron which contains as its main impurities *manganese* and *iron carbide (Fe_3C).*

2. *Gray pig iron* which contains as its main impurities *silicon* and carbon in the form of *graphite* (from which the gray color derives).

Which will have the lower melting point: the two kinds of pig iron or pure iron?

- Pure iron → 23
- White and gray pig iron → 27
- I don't know → 1

14 Iron can have valences of *two* and *three.* Consider, for instance, the following compounds:

$FeCl_2$, $FeCl_3$, Fe_2O_3 and FeO.

In which of these compounds is the iron

a) divalent
b) trivalent?

Jot down the appropriate valence for each of the compounds, and then check your answer at 48 .

24

15 In our discussion of the blast furnace process we said that:

To remove impurities present in iron ores – the so-called gangue – various fluxes are added to the ore.

Acid gangue, such as quartz (SiO_2), combines with *basic* fluxes, such as calcium oxide (CaO), to form slag.

Basic gangue, such as magnesium carbonate ($MgCO_3$), similarly combines with *acidic* fluxes, such as quartz (SiO_2), to form slag.

An analogous reaction takes place in the Bessemer Process:

Oxidation of the manganese contained in pig iron yields *basic* manganese dioxide. This then combines with the .... lining in the Bessemer Converter.

Now continue your reading at [3] .

---

16 The melting point of *pure* iron is over 1500 °C. Because pig iron contains various impurities it melts at a lower temperature.

Please note that:

Pig iron melts at around 1100 °C.

Continue your reading at [21] .

---

17 Your answer is quite correct.

If a sample of white pig iron is dissolved in dilute hydrochloric acid, there will be *no* residue left behind. This is because the impurities in pig iron, namely iron carbide and manganese, are soluble in hydrochloric acid.

If, however, a sample of gray pig iron is dissolved in dilute hydrochloric acid, a black residue of insoluble graphite will be left behind.

What other impurity does gray pig iron contain?

- Manganese → [34]
- Silicon → [9]
- Iron carbide → [2]

---

18 Now check your equation:

$$2\ Fe(OH)_3 \xrightarrow{\text{roasting}} Fe_2O_3 + 3\ H_2O$$

The iron (III) oxide is weighed after it has cooled down. From its mass, it is then easy to calculate the amount of iron in the original solution.

The determination of iron content by such a procedure is based on the use of an analytical balance. A quantitative analytical procedure of this type is referred to as *gravimetric* analysis.

Now write out your answer to the question below.

Which of the equipment listed below is used in

a) gravimetric analysis
b) volumetric analysis?

Balance, buret, pipet, filtration apparatus, weights, funnel, measuring cylinder, filter paper, indicator.

Continue your reading at [5] .

---

[19] Your answer is wrong.

Continue reading now at [13] .

---

[20] Your answer is wrong.

An explanation on this is to be found in [15] .

---

[21] Your answer is quite right.

Gray pig iron is used in the form of *cast iron* or is *converted into steel.* White pig iron is too hard and brittle to be used as cast iron. It is therefore *converted into steel.*

Iron which is not cast but which is either forged or rolled is referred to as *wrought iron.* Accordingly, we may assert that:

Steel is wrought iron.

To make steel out of pig iron, it is first necessary to remove the impurities the latter contains.

The impurities in pig iron are removed by means of oxidation. The oxidation products formed in the process are as follows:

From the carbon and iron carbide: carbon dioxide
From the manganese: manganese dioxide
From the silicon: silicon dioxide

Write down now both the names and formulas for these oxidation products, remembering here that manganese, Mn, is tetravalent and that its oxide is designated manganese(IV) oxide.

Continue your reading at [37] .

---

[22] Your answer is wrong.

Read through [66] once again.

---

[23] Your answer is wrong.

Read through [1] once again.

---

[24] An alloy is obtained when two or more metals are melted down together.

We now turn to a discussion of the chemical and physical properties of iron.

Iron is a heavy metal having a density of 7.8 g · cm^{-3}. Iron is attracted by magnets because it becomes magnetized in the presence of magnets. It loses its magnetism, however, when the magnet is removed.

The chemical properties of iron have already been covered elsewhere. You should by now know, for instance, quite a lot about the valences of iron.

What are the possible valences which iron can have?

- Two → [8]
- Three → [14]
- Two and three → [4]
- Two, three and five → [35]
- I have forgotten → [39]

---

[25] When phosphorus-containing pig iron is oxidized, the phosphorus is converted into phosphorus(V) oxide (also known as phosphorus pentoxide), P_2O_5, the acid anhydride of phosphoric acid.

You can probably now guess how pig iron containing phosphorus is converted into steel.

What sort of lining has to be used in a Bessemer Converter when pig iron containing phosphorus is converted into steel?

- A basic lining → 46
- An acidic lining → 36
- I don't know → 41

---

26 Your answer is wrong.

Read through 66 once again.

---

27 Both white and gray pig iron contain impurities which have the effect of lowering their melting points with respect to that of pure iron. These two types of pig iron will therefore melt at a lower temperature than pure iron.

What are the major impurities in

a) white pig iron
b) gray pig iron?

- a) Manganese and iron carbide
  b) Silicon → 19
- a) Manganese
  b) Silicon and carbon (graphite) → 33
- a) Manganese and carbon (graphite)
  b) Silicon and iron carbide → 10
- a) Manganese and iron carbide
  b) Silicon and carbon (graphite) → 17

---

28 Potassium ferricyanide is used to test for iron(II) salts. In fact, it is only in the presence of iron(II) salts that potassium ferricyanide will yield a blue precipitate.

To test for iron(III) salts one has to use potassium ferrocyanide.

Now read through [66] once again.

---

[29] In the Bessemer Converter air is blown through liquid pig iron which is glowing hot. Some of the iron becomes oxidized by this air blast, a fact which is attested to by the brown clouds of finely-divided iron oxide fumes seen over steel foundries. In the main, however, it is the *impurities* present in the pig iron which are oxidized. The oxidation products either escape in the form of gases, e.g. as carbon dioxide, or are absorbed by the lining of the Bessemer Converter.

As an example, we mention that the oxidation of manganese leads to the formation of basic manganese(IV) oxide (also called manganese dioxide).

What type of lining must the Bessemer Converter have to absorb these basic impurities?

- I don't know → [15]
- A basic lining → [20]
- An acidic linig → [7]

---

[30] Please note that:

Pig iron melts at around 1100 °C.

Now continue your reading at [21] .

---

[31] Iron and iron compounds are certainly not fertilizers.

Read through [46] once again.

---

[32] *Nickel, chromium, manganese*

Now write out an answer to the question: what is an alloy?

If you find this difficult, read through [44] once again.

If you are able to answer the question, do so and then turn to [24] .

---

33 Your answer is wrong.

Continue your reading now at 13 .

34 You should memorize the *correct* answer, which is that:

Gray pig iron contains silicon as an impurity in addition to graphite (from which its gray color derives!).

White pig iron contains .... and .... as its impurities.

Now continue your reading at 9 .

35 Iron can have valences of *two* and *three*. Consider, for instance, the following compounds:

$FeCl_2$, $FeCl_3$, Fe_2O_3 and FeO.

In which of these compounds is the iron

a) divalent
b) trivalent?

Jot down the appropriate valence for each of the compounds, and then check your answers at 48 .

36 Your answer is wrong, though it is fair to point out that the question was not an easy one.

Continue your reading now at 41 .

37 Carbon dioxide, CO_2; manganese dioxide, MnO_2; silicon dioxide, SiO_2.

The oxidation of the impurities present is brought about by air. The process is carried out in a *Bessemer Converter.* A Bessemer Converter consists of a pear-shaped, steel container which is coated with a layer of refractory material. This layer is referred to as the *lining* of the furnace.

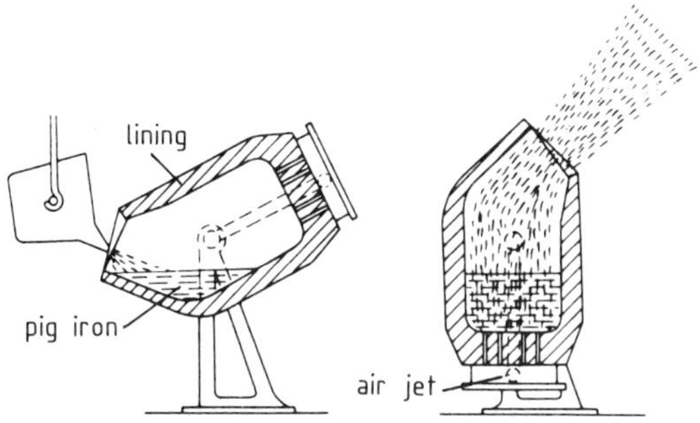

Copy down the diagram of the Bessemer Converter on the right. Then continue your reading at [29] .

---

[38] Your answer is quite right.

Here once again are the two equations:

$$Fe + 2\ HCl \rightarrow FeCl_2 + H_2$$

$$Fe + H_2SO_4 \rightarrow FeSO_4 + H_2$$

The iron(II) sulfate crystallizes with seven molecules of water of crystallization. Accordingly, it has the chemical formula

$$FeSO_4 \cdot 7\ H_2O$$

What happens when concentrated sulfuric acid acts on iron?

- Iron(II) sulfate is formed → [64]
- Iron(III) sulfate is formed → [52]
- There is no reaction → [71]

---

[39] Iron can have valences of *two* and *three*. Consider, for instance, the following compounds:

$FeCl_2$, $FeCl_3$, Fe_2O_3 and FeO.

In which of these compounds is the iron

a) divalent
b) trivalent?

Jot down the appropriate valence for each of these compounds, and then check your answers at 48 .

---

40 You have obviously mixed things up here.

Read through 48 once again.

---

41 *Basic* products can be absorbed by an *acidic* lining.

Acidic products can be absorbed by a *basic* lining.

The oxidation of various sorts of pig iron which contain phosphorus will yield an *acidic* product, namely phosphorus(V) oxide (phosphorus pentoxide).

What type of lining should therefore be used when phosphorus-bearing pig iron is to be converted into steel?

- A basic lining → 46
- An acidic lining → 11

---

42 Your answer is quite right.

Do you still remember what reagent one uses to test for iron(II) salts?

- It is potassium thiocyanate → 73
- It is potassium ferricyanide → 59
- It is potassium ferrocyanide → 26

---

43 Your answer is wrong.

An explanation of this is to be found at 57 .

---

44 The principal areas of steel production in the United States are Chicago and Pennsylvania state.

The steel produced by any of the processes we have outlined here can be improved in a variety of ways by the addition of other metals. Examples of metals which may be added include nickel, chromium and manganese. The steel is simply melted down with one or more of these metals. A product formed in this way by melting together two or more metals is referred to as an *alloy*.

Alloys are often possessed of special properties and so can be used for various specific applications. One example is provided by steel which has been alloyed with the element *chromium*. This particular alloy does not rust and can therefore be used as a so-called *stainless steel.*

Jot down now the three metals with which steel is commonly alloyed. Then continue your reading at 32 .

---

45 No, an important fertilizer is a by-product of the Thomas Process.

In the *Thomas* Process one starts off with various sorts of iron containing phosphorus as an impurity. The Thomas meal formed in the course of the process can be used as a fertilizer because of its high phosphorus content.

Now continue your reading at 63 .

---

46 Your answer is quite correct.

The process used for converting phosphorus-bearing iron into steel is known as the Thomas Process.

So far we have outlined two processes which are used for the production of steel:

1. The *Bessemer Process:* this involves oxidation of the impurities present in the iron by air. The major impurity is manganese which is oxidized to basic manganese(IV) oxide (manganese dioxide) and then absorbed by an acidic furnace lining.

2. The *Thomas Process:* this involves oxidation of the impurities present in the iron by air. The major impurity is phosphorus which is oxidized to acidic phosphorus (V) oxide (phosphorus pentoxide) and then absorbed by a basic furnace lining.

The linings in both the Bessemer and Thomas Converters consist of refractory bricks. Such material is necessary as the operating temperature in each case is around 1400 °C. These linings have to be replaced from time to time. The old linings from Thomas Converters are ground down into a fine powder and then sold as a fertilizer known as Thomas meal.

Which component of Thomas meal is effective as a fertilizer?

- The iron → [31]
- The potassium → [62]
- The nitrogen → [54]
- The phosphorus → [58]

---

[47] Your answer is quite correct.

In addition to *volumetric analysis,* there is another important quantitative analytical procedure known as *gravimetric analysis.* In this latter procedure substances to be analysed are converted into a form in which they can be weighed. Consider the following example:

If an aqueous solution of ammonia is added to an iron(III) chloride solution, a brown precipitate of iron hydroxide will be formed. The relevant equation is:

$$FeCl_3 + 3\ NH_3 + 3\ H_2O \rightarrow Fe(OH)_3 + 3\ NH_4Cl$$

To make a quantitative determination of the amount of iron present in the solution, the precipitated iron hydroxide is collected on a filter paper, rinsed and then dried together with the filter paper. Iron(III) hydroxide, however, is not a suitable compound for gravimetric analysis as it always retains some water. Accordingly, the filter paper with the precipitate is placed in an evaporating dish and heated to glowing heat. In the process the filter paper burns away and the iron(III) hydroxide is converted into iron(III) oxide by splitting off water.

Write out now the equation for the splitting off process of water from iron(III) hydroxide. Then continue your reading at [18] .

---

[48] The iron has a valence of two in $FeCl_2$ and FeO.
The iron has a valence of three in $FeCl_3$ and Fe_2O_3.

Iron is attacked by dilute acids, especially by dilute hydrochloric and sulfuric acids. On the contrary, however, iron is not attacked by concentrated sulfuric acid. For this reason concentrated sulfuric acid can be stored and treated in iron containers. When left in air iron is attacked by the oxygen and forms rust. Rust consists of a variety of iron oxides and iron hydroxides.

Now write down equations for the reaction of iron with:

a) dilute hydrochloric acid and
b) dilute sulfuric acid.

Calculate the molecular mass of each of the iron(II) salts formed. The relevant atomic masses are: Fe=56, Cl=35.5 and the radical SO_4 = 96.

What molecular masses do you obtain?

- a) 127 b) 152 → 38
- a) 162.5 b) 152 → 60
- a) 152 b) 127 → 40
- My results are different → 68

---

49 Your answer is quite right.

Only the ore pyrites has to be roasted before it can be converted into pig iron. The other three iron ores listed are all oxide ores and so can be used directly.

The production of iron is carried out in a blast furnace. The furnace is loaded with iron ores, coke and various fluxes. Hot air blown in from the bottom of the furnace combines with the carbon to form .... This substance then reduces the iron ores to iron(II) oxide. The reduction of the iron(II) oxide to metallic iron is brought about by the .... present. Some of the iron actually reacts with the carbon to form ....

Jot down the missing words, and then continue your reading at 61 .

---

50 Your answer is quite right.

End of Program 24.

---

51 No, an important fertilizer is a by-product of the Thomas Process.

In the *Thomas* Process one starts off with various sorts of iron containing phosphorus as an impurity. The Thomas meal formed in the course of the process can be used as a fertilizer because of its high phosphorus content.

Now continue your reading at 63 .

---

52 You seem to have overlooked an important point here.

The correct answer is to be found by reading through [48] very carefully!

53 Now check your answers:

Potassium ferricyanide, for *iron(II) salts*
Potassium ferrocyanide, for *iron(III) salts*

Another test for iron(III) salts – which is very sensitive – is the reaction of these salts with potassium thiocyanate. If a potassium thiocyanate solution is added to a solution of an iron(III) salt, a blood red coloration results.

What is observed when iron(III) salts are reacted with:

a) Potassium ferricyanide and
b) Potassium thiocyanate?

- a) A blue precipitate is formed
  b) A blood red coloration is seen → [28]
- In both cases a blue precipitate is formed → [22]
- a) No precipitate is formed
  b) A blood red coloration is seen → [42]

54 Although there are fertilizers which consist of nitrogen compounds, no such compounds are present in Thomas meal.

Read through [46] once again.

55 The Siemens-*Martin* Process is carried out in large open hearth furnaces. To oxidize the impurities present in pig iron use is made of the bonded oxygen in *iron oxide.*

The iron oxides are introduced in the form of rusted *scrap iron.*

The technical importance of iron is enormous. In fact, the industrial performance of a country can be measured by its output of iron and steel. The world production of steel in 1980 amounted to some 750 million tons, and that of the United States in 1980 was around 125 million tons.

The annual output of iron and steel in the United States weighs about 20 times that of all other manufactured metals together.

Which areas of the United States are especially noted for the production of iron and steel?

Write out your answer and then turn to [44] .

---

[56] Your answer is almost right.

According to you, the complete equation reads as follows:

$$1\ Fe_2O_3 + 6\ HCl \rightarrow 2\ FeCl_3 + 3\ H_2O$$

However, when a 1 appears in front of a chemical substance, the 1 is not written but is omitted. The correct version of this equation should thus read:

$$Fe_2O_3 + 6\ HCl \rightarrow 2\ FeCl_3 + 3\ H_2O.$$

Now continue your reading at [66] .

---

[57] A distinction is made between two differing sorts of iron ore, namely:

1. Oxide iron ores, e.g. hematite, limonite and magnetite

2. Sulfide iron ores, e.g. pyrites.

The oxide ores can be used directly for the production of iron in a blast furnace. The sulfide ores on the other hand first have to be roasted, i.e. they are heated to glowing heat in the presence of air. In the process oxides of iron and sulfur dioxide are formed. A typical equation for such a reaction is the following:

$$4\ FeS_2 + 11\ O_2 \rightarrow 2\ Fe_2O_3 + 8\ SO_2$$

Now continue your reading at [49] .

---

[58] Your answer is quite right.

When Thomas meal is used as a fertilizer, the effective component is the phosphorus it contains.

Now jot down the missing words in the sentences below:

The two processes used for the production of steel are the .... and the .... Processes. Both of these processes are carried out in steel converters. In order to .... the impurities present in the pig iron, ... is blown through the molten pig iron.

Continue your reading at [69] .

---

59 Your answer is quite right.

We shall now take up the question how trivalent iron compounds may be determined, not only qualitatively but also quantitatively.

One procedure for performing a quantitative analysis has already been described. What was this procedure called?

- Volumetric analysis → [47]
- Electrolysis → [72]
- Catalysis → [67]

---

60 When hydrochloric acid acts on iron, the salt formed is iron(II) chloride and *not* iron(III) chloride.

Read through [48] once again.

---

61 Now check your answers:

carbon monoxide, carbon, iron carbide

In any steel-making process it is first necessary to remove the impurities present in pig iron. There are three ways of doing this, namely:

1. Oxidation of the impurities by air. When the major impurity is manganese, the oxidation process will yield basic manganese(IV) oxide (manganese dioxide). The latter substance can be absorbed by an acidic furnace lining. The whole process is carried out in a converter.

2. Oxidation of the impurities by air. When the major impurity is phosphorus, the oxidation process will yield acidic phosphorus(V) oxide (phosphorus pentoxide). The latter substance can be absorbed by a basic furnace lining. The whole process is carried out in a converter.

3. Oxidation of the impurities by oxygen, which is supplied in the form of iron oxides. The whole process is carried out in a low open hearth furnace.

Write down now the name of each of these processes in the sequence given, and then continue your reading at 75 .

---

62 Although there are potassium fertilizers, having as their effective component salts of potassium, the latter are not present in Thomas meal.

Read through 46 once again.

---

63 Your answer is quite right.

Iron is a heavy metal. When it comes into contact with magnets or is placed in a magnetic field, it becomes magnetic itself. Iron is attacked by moist air; it then undergoes the process of rusting. With dilute acids it forms divalent iron compounds. Iron does not react with concentrated sulfuric acid.

To test for iron compounds the following reagents may be used:

a) For divalent iron compounds: potassium ferricyanide
b) For trivalent iron compounds: potassium ferrocyanide
c) For trivalent iron compounds: potassium thiocyanate.

What is observed in each of these three test reactions?

- a) A blue precipitate is formed
  b) A blue precipitate is formed
  c) A red precipitate is formed → 70

- a) A blue coloration is observed
  b) A blue coloration is observed
  c) A red coloration is observed → 6

- a) A blue precipitate is formed
  b) A blue precipitate is formed
  c) A red coloration is observed → 50

---

64 You seem to have overlooked an important point here.

The correct answer is to be found by reading through 48 very carefully!

---

65 Although you managed to get the colors right for the three test reactions for iron, it seems you are uncertain whether a precipitate is formed or not during each of these reactions. The correct answers should read:

a) Divalent iron compound + potassium ferricyanide → blue precipitate

b) Trivalent iron compound + potassium ferrocyanide → blue precipitate

c) Trivalent iron compound + potassium thiocyanate → red coloration

End of Program 24.

---

66 Your answer is quite right.

When testing for iron compounds, a distinction has to be made between the divalent and trivalent compounds of iron.

The reagents used to test for iron salts are rather complicated inorganic compounds. All that you will need to know about them are their names.

To test for iron(II) salts use is made of *potassium ferricyanide.*

To test for iron(III) salts use is made of *potassium ferrocyanide.*

In the actual test for an iron(II) salt, the salt solution is mixed with some potassium ferricyanide solution. If a deep blue precipitate forms, the presence of an iron(II) salt is indicated.

Similarly, in the test for an iron(III) salt, the salt solution is mixed with some potassium ferrocyanide solution. If a deep blue precipitate forms, the presence of an iron(III) salt is indicated.

Write down now the names of these two reagents and then indicate alongside each the type of iron compound which can be tested for.

Continue your reading at 53 .

---

67 No, catalysis is something quite different.

A catalyst is a substance which speeds up a chemical reaction, but which at the end of the reaction is present in the same form as at the beginning.

A example of a catalytic reaction is the following:

When potassium chlorate is heated, its oxygen is given off much more easily and smoothly if some manganese dioxide has been added to the chlorate. The manganese dioxide (manganese(IV) oxide) acts as a catalyst for this particular reaction.

Now read through 59 once again.

---

68 If you obtained figures different from those given, you must have made an error somewhere.

When hydrochloric acid acts on iron, the compound iron(II) chloride is formed; when dilute sulfuric acid acts on iron, the compound iron(II) sulfate is formed.

In addition to this hydrogen gas is also given off.

Are the above statements in accord with your equations? If necessary, you should now correct your equations and check through your calculations.

→ 48

---

69 Now check your sentences:

The two processes used for the production of steel are the *Bessemer* and *Thomas* Processes. Both of these processes are carried out in steel converters. In order to *oxidize* the impurities present in the pig iron, *air* is blown through the molten pig iron.

If you were uncertain about the answer or got it wrong, you should in your own interest now read through 37 again.

There is also a third process for the production of steel from pig iron. This is known as the *Siemens-Martin Process*.

As in the other two processes, the impurities present in pig iron are oxidized. To do this, use is made of the bonded oxygen in the various iron oxides.

The Siemens-Martin Process is carried out in large flat pans (open hearth furnaces). As before, the oxidized impurities from the pig iron are absorbed in a suitable lining.

The iron oxides required for the Siemens-Martin Process are added to the molten pig iron in the form of rusted scrap iron. Rusted scrap iron contains a fair amount of various iron oxides.

Now jot down the missing words in the sentences below:

The Siemens- .... Process is carried out in large open hearth furnaces. To oxidize the impurities present in pig iron use is made of the bonded oxygen in .... The iron oxides are introduced in the form of ....

Now continue your reading at [55] .

---

70 Your answer is wrong.

An explanation of this is given in [65] .

---

71 Correct! Iron is *not* attacked by concentrated sulfuric acid, although iron does react with dilute sulfuric acid to form iron(II) sulfate. With hydrochloric acid iron forms iron(II) chloride.

One way of making iron(III) chloride is to dissolve iron(III) oxide in hydrochloric acid. The incomplete equation for this reaction is as follows:

$$Fe_2O_3 + HCl \rightarrow FeCl_3 + H_2O$$

The factors are missing from this equation. Can you say what the correct factors should be?

	In front of Fe_2O_3	In front of HCl	In front of $FeCl_3$	In front of H_2O	
•	–	6	–	3	→ [76]
•	–	6	2	3	→ [66]
•	1	6	2	3	→ [56]

---

72 By electrolysis is meant the decomposition of a chemical compound by means of direct electric current, e.g. the decomposition of water into hydrogen and oxygen.

24

Read through [59] once again.

---

73 Your answer is wrong.

Read through [66] once again.

74 Your answer is wrong.

An explanation of this is given in 57 .

---

75 The three processes are:

1. The Bessemer Process
2. The Thomas Process
3. The Siemens-Martin Process.

In which of these three processes is one of the by-products an important fertilizer?

- In the Bessemer Process → 45
- In the Thomas Process → 63
- In the Siemens-Martin Process → 51

---

76 You have forgotten the factor which goes in front of the $FeCl_3$.

By taking another careful look at the equation in 71 , you should be able to arrive at the correct version.

Turn back now to 71 .

Program 25

Copper, Lead, Zinc and Tin

1 In this Program we shall be discussing four metals, all of which have manifold industrial uses. We shall start with *copper* which is a reddish and lustrous heavy metal. It is known to conduct heat and electric current very well.

Copper is stable in dry air, provided that the temperature is not too high. Under these circumstances it retains its reddish luster. After a few weeks or months in moist air, however, it becomes coated with a greenish layer consisting mainly of copper hydroxide and copper carbonate. This layer, which is frequently seen on old copper rooves and domes, is referred to as patina or verdigris.

Using keywords, write out answers now to the following questions:

What are the main properties of copper?
How is patina formed?
What is patina?

Continue your reading at 24 .

---

2 Your answer is correct only for the $CuFeS_2$. In the case of the oxygen, you appear to have forgotten that this will appear in the equation as O_2.

Make another attempt at the question in 12 .

---

3 The formulas you have written down are probably wrong.

Below we give the correct structural formulas for:

Copper (I) oxide and Copper (II) oxide

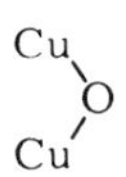

Cu=O

Take a careful look at these two formulas and count up the valence bonds present in each. In this way you will be able to determine the valences of both the copper and the oxygen. Then write down the chemical formula for each of the two copper oxides.

Now continue your reading at 11 .

4 Your answer is quite correct. The complete equation is as follows:

$$2\ CuS + 3\ O_2 \rightarrow 2\ CuO + 2\ SO_2$$

You should also have no difficulty in finding the missing factors for the equation below. This equation describes the roasting of copper pyrites. Copper pyrites consists of the sulfides of copper and iron; its chemical formula is $CuFeS_2$. When it is roasted the products formed are copper oxide, iron oxide and sulfur dioxide thus:

$$CuFeS_2 + O_2 \rightarrow CuO + Fe_2O_3 + SO_2$$

What are the appropriate factors for this equation?

	In front of $CuFeS_2$	In front of O_2	In front of CuO	In front of Fe_2O_3	In front of SO_2	
•	2	13	2	2	4	→ 21
•	4	13	4	2	8	→ 25
•	I don't know					→ 12
•	2	10	6	4	4	→ 28
•	4	6	2	2	4	→ 32

5 Now copy down the statement below:

The test for a copper compound using an excess of aqueous ammonia solution leads to the formation of a deep blue coloration.

Continue reading at 84 .

6 You have not written down the required equation correctly.

Turn now to 27 where some assistance on this is given.

25

7 Now check your equations:

1. $CuO + H_2 \rightarrow Cu + H_2O$

2. $CuO + C \rightarrow Cu + CO$

In the industrial context coke is used for the reduction of copper oxide. The carbon in the coke represents a cheaper reducing agent than hydrogen.

Copper in the form of its sulfide ore cannot be directly reduced to metallic copper. In fact, the same procedure is adopted in this case as that for the sulfide ores of other metals.

Jot down the missing words in the sentences below:

Sulfide ores have to be .... before they can be converted into metals. By roasting is meant the heating of the sulfide ore in the presence of .... The metal sulfides then yield the corresponding ....
and .... → 22

---

8 You have mixed up the two molecular masses!

Read through 24 once again.

---

9 Your answer is quite right.

You should also note here that these blue ammoniacal copper salt solutions act as important solvents for cellulose. By using such solvents it is possible to produce artifical silk – so-called cuprammonium silk.

End of Program 25.

---

10 Copy down the following:

To test for lead compounds one uses hydrogen sulfide. A black precipitate of lead sulfide forms in each case, e.g.

$$PbCl_2 + H_2S \rightarrow PbS + 2\,HCl$$
$$Pb(NO_3)_2 + H_2S \rightarrow PbS + 2\,HNO_3$$

Now continue your reading at 84 .

---

11 Your answer is quite correct.

The chemical formulas for the two copper oxides are:

Copper (I) oxide	Copper (II) oxide
Cu_2O	CuO

Both of these oxides are formed when copper is heated in air.

Write out the two relevant equations for these oxidation reactions, and then continue your reading at 39.

---

12 The correct factors for this equation are indeed rather difficult to determine. The task will be made easier for you, however, by considering the ratios of copper to iron and sulfur in the molecule $CuFeS_2$. The ratios are clearly 1:1:2. Accordingly, we should start by writing on the right hand side the products:

$$\rightarrow 2\ CuO + Fe_2O_3 + 4\ SO_2$$

This automatically leads us to write 2 $CuFeS_2$ on the left hand side:

$$2\ CuFeS_2 + O_2 \rightarrow 2\ CuO + Fe_2O_3 + 4\ SO_2$$

At this stage consider the number of oxygen atoms on the right. There are 13 such atoms in all, an *uneven* number. But as there can only ever be an even number of oxygen atoms on the left, the right hand side of the equation must be multiplied by two:

$$2\ CuFeS_2 + O_2 \rightarrow 4\ CuO + 2\ Fe_2O_3 + 8\ SO_2$$

This equation is still not balanced as it stands. What factors should be placed in front of the $CuFeS_2$ and the O_2 on the left hand side?

In front of $CuFeS_2$	In front of O_2		
• 2	13	→	37
• 4	26	→	2
• 4	13	→	25
• 2	26	→	34
• I need some help		→	23

---

13 When copper sulfide is roasted *copper oxide* and *sulfur dioxide* are formed. The unbalanced equation for this reaction is as follows:

$$CuS + O_2 \rightarrow CuO + SO_2$$

You may have some difficulty in finding the correct factors for this equation.

What do you think the missing factors are?

	In front of CuS	In front of O_2	In front of CuO	In front of SO_2	
•	2	2	2	2	→ 6
•	3	2	2	2	→ 29
•	2	3	2	2	→ 4
•	3	3	2	3	→ 20
•	I need some help				→ 27

14 Continue your reading now at 3 .

15 We started out from the following unbalanced equation:

$$CuS + O_2 \rightarrow 2\ CuO + 2\ SO_2$$

On the right hand side of this equation there are two atoms each of copper and sulfur. The same number of each must, of course, also appear on the left hand side. To allow for this we can write initially:

$$2\ CuS + O_2 \rightarrow 2\ CuO + 2\ SO_2$$

Now we have too little oxygen on the left of our equation. On the right there are six oxygen atoms or 3 O_2, and so there must be this same number on the left. The final version of the equation thus takes the form:

$$2\ CuS + 3\ O_2 \rightarrow 2\ CuO + 2\ SO_2$$

Copy down this equation now, and then continue your reading at 4 .

16 You are unfortunately becoming slack in your work.

Read through 26 once again.

17 Now check your formulas:

Copper(I) chloride	CuCl
Copper(II) oxide	CuO
Copper(II) sulfide	CuS
Copper(II) carbonate	$CuCO_3$

Write down the names of the following copper compounds:

Cu_2O
$CuCl_2$
CuCl
$Cu(NO_3)_2$

Continue your reading at 43 .

18 No, it should be borne in mind that these ions are discharged at the oppositely charged electrode.

The anode is the positive electrode.
The cathode is the negative electrode.

Now make another attempt at the question in 55 .

19 Sulfide ores are in fact metal *sulfides*. When such ores are roasted they are heated in the presence of air. How then could carbon dioxide be formed?

Read through 72 once again.

20 You have not written down the required equation correctly.

Turn now to 27 where some assistance on this is given.

21 Your answer is wrong.

Some help on this question is to be found in 12 .

22 Now check your answers:

Sulfide ores have to be *roasted* before they can be converted into metals. By roasting is meant the heating of the sulfide ore in the presence of *oxygen (air)*. The metal sulfides then yield the corresponding *metal oxides* and *sulfur dioxide*.

Write down the products which are formed when copper sulfide is roasted.

Then continue your reading at [13] .

23 We started out from the following unbalanced equation:

$$2\ CuFeS_2 + O_2 \rightarrow 4\ CuO + 2\ Fe_2O_3 + 8\ SO_2$$

To arrive at the correct factor to be placed in front of the $CuFeS_2$, it is necessary to determine how many copper atoms there are on the right hand side of the equation. You should also check the number of iron atoms on each side to see that these are balanced.

To find the appropriate factor for the O_2 on the left, you should count up the number of oxygen atoms on the right. To come up with the correct factor you will, of course, have to divide the number of oxygen atoms by two. This is because the oxygen appears on the left as O_2.

Now make another attempt at the question in [12] .

24 Copper is a reddish, lustrous heavy metal; it conducts heat and electric current very well; it does not react with dry air, but does become coated with patina in damp air. Patina consist mainly of copper hydroxide and copper carbonate.

Copper ist not attacked by dilute acids, though it does react with concentrated sulfuric and nitric acids.

Copper has a valence of one and two in its compounds. Accordingly, there are two different oxides of copper, namely copper (I) oxide and copper (II) oxide. Write out formulas for these two copper oxides and then calculate their molecular masses. The relevant atomic masses are: Cu = 64 and O = 16.

What is the molecular mass of:

a) Copper (I) oxide
b) Copper (II) oxide?

- a) 96 b) 80 → 3
- a) 80 b) 144 → 8
- a) 144 b) 80 → 11
- I obtained other answers → 14

---

25 Your answer is quite right.

Here once again is the complete equation:

$$4\ CuFeS_2 + 13\ O_2 \rightarrow 4\ CuO + 2\ Fe_2O_3 + 8\ SO_2$$

In the production of copper it should be borne in mind that: Copper is found in the form of the metal and as the oxide and sulfide ores. Oxide ores can be immediately reduced (coke is used for this purpose by industry), whereas sulfide ores first have to be converted into oxide ores by roasting them in air.

Copper can also be produced by electrolysing solutions of copper salts. The most commonly used salt in this procedure is copper sulfate.

Write down the ions into which copper sulfate dissociates in water. → 55

---

26 The formula of copper sulfate having 5 molecules of water of crystallization is $CuSO_4 \cdot 5\ H_2O$. This salt was discussed earlier on. Do you remember what happens when it is heated?

- The water is driven off → 30
- The water is driven off and the salt turns black → 36
- The water is driven off and the blue crystals turn into a white powder → 51
- Nothing → 16
- The blue crystals turn into a white powder → 44

---

27 In the equation:

$$CuS + O_2 \rightarrow CuO + SO_2$$

there are 3 oxygen atoms on the right but only 2 on the left.

On the right there is thus an *uneven* number of oxygen atoms whilst on the left there must always be an *even* number of oxygen atoms. Multiplication of the left hand side by some factor cannot alter this, e.g. 2 O_2 will give us four oxygen atoms, 3 O_2 will give us six oxygen atoms, and so on. The factors thus have to be chosen so that we finish up with an *even* number of oxygen atoms on the right. Accordingly, we multiply the right hand side of the equation by 2. This yields the equation:

$$CuS + O_2 \rightarrow 2\ CuO + 2\ SO_2$$

It should now be easier for you to answer the question:

What factors have to be placed in front of the CuS, the O_2, the CuO and the SO_2 to balance the equation?

	In front of CuS	In front of O_2	In front of CuO	In front of SO_2	
●	2	3	2	2	→ 4
●	2	2	2	2	→ 35
●	I don't know				→ 15

28 Your answer is wrong.

Some assistance on this is given in 12 .

29 You have not written down the required equation correctly.

Continue your reading at 27 , where some assistance is given.

30 Your answer is incomplete.

In addition to water being driven off, a color change also occurs.

Read through 26 once again.

31 Bronze is an alloy of the metals copper and tin.

In its compounds copper has a valence of one and two.

Write down now the name and chemical formula of each of the compounds listed below:

Copper(I) chloride, copper(II) oxide, copper(II) sulfide, copper(II) carbonate. → 17

32 Your answer is wrong.

Some assistance on this is to be found in 12 .

33 Correct, the copper ions are discharged at the cathode and copper metal is deposited on the cathode.

The electrolysis of copper salt solutions is actually an important method of purifying copper. The method consists of using some crude and impure copper as the anode (the positive electrode) and some pure copper plate as the cathode (the negative electrode). These two electrodes are immersed in a copper sulfate solution.

When direct electric current is passed through the solution, copper is deposited on the cathode. The cathode thus increases in size whereas the impure copper which functions as the anode gradually goes into solution. The $Cu^{2\oplus}$ ions migrate from the anode to the cathode. The whole process is illustrated in the diagram below:

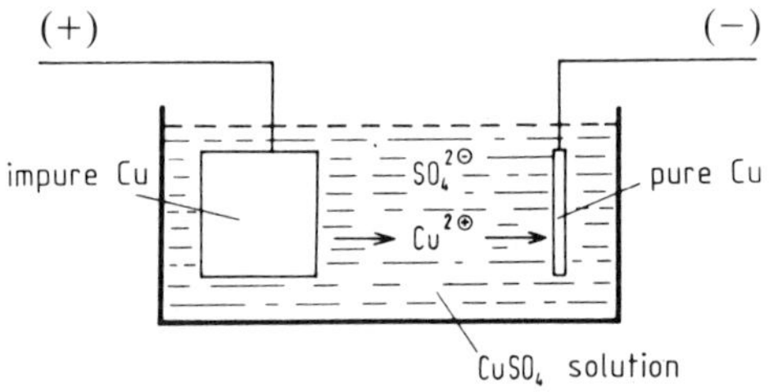

You are advised to copy down this diagram.

Now, it is not only by means of electrolysing copper salt solutions that copper can be prepared; there is one other important method in use.

When iron filings are dropped into a solution of copper sulfate, the iron dissolves and metallic copper separates out according to the equation:

$$Fe + CuSO_4 \rightarrow \ldots\ldots\ldots\ldots + Cu$$

Complete this equation, and then turn to [53] .

---

[34] None of these factors is correct.

You should now read through the further expanatory material given in [23] .

---

[35] Your answer is wrong.

Now continue your reading at [15] .

---

[36] Your answer is only partially correct.

Although water is clearly driven off when copper sulfate crystals containing water of crystallization are heated, it should not be forgotten that the blue crystals are converted into a *white* powder.

Now continue your reading at [51] .

---

[37] Your answer is correct only for the O_2. You appear to have forgotten that the factor in front of the $CuFeS_2$ also has to be changed.

Now make another attempt at the question in [12] .

---

[38] No, in the Lead Chamber Process the end product is *sulfuric acid.*

Now continue your reading at [42] .

---

[39] Now check your equations and make any necessary corrections:

$$4\ Cu + O_2 \rightarrow 2\ Cu_2O$$
$$2\ Cu + O_2 \rightarrow 2\ CuO$$

Copper is found in the form of copper metal, copper oxide, copper pyrites and copper sulfide. Copper oxide can readily be converted to metallic copper by means of reduction.

Write out the equation for the reduction of copper (II) oxide (CuO):

1. By means of hydrogen
2. By means of carbon.

Check your equations at 7 .

---

40 Copy down the following:

To test for lead compounds use is made of hydrogen sulfide. A black precipitate of lead sulfide is formed.

Here are two examples:

$$PbCl_2 + H_2S \rightarrow PbS + 2\,HCl$$
$$Pb(NO_3)_2 + H_2S \rightarrow PbS + 2\,HNO_3$$

Now continue your reading at 84 .

---

41 Copy down the following:

When testing for copper compounds using an excess of aqueous ammonia solution a deep blue coloration is observed.

Now continue your reading at 84 .

---

42 Yes, the Lead Chamber Process is used for the manufacture of sulfuric acid. In its compounds lead has valences of two and four.

Write down now the chemical formulas for the two oxides of lead, and calculate the molecular mass of each. The relevant atomic masses are: Pb = 207, O = 16.

	Lead(II) oxide	Lead(IV) oxide		
●	430	844	→	77
●	223	430	→	50
●	223	239	→	58
●	239	223	→	62
●	I need some help		→	66

25

43

Cu_2O	Copper(I) oxide
$CuCl_2$	Copper(II) chloride
$CuCl$	Copper(I) chloride
$Cu(NO_3)_2$	Copper(II) nitrate

Monovalent copper compounds are readily oxidized by air to divalent copper compounds. The divalent compounds of copper are more important and are more often encountered than those of monovalent copper. The salts of divalent copper are either blue or green. All copper salts are to some extent poisonous.

The most important salt of copper is copper sulfate. This crystallizes out with five molecules of water of crystallization. Write down its formula, and then continue your reading at 26 .

---

44 Your answer is incomplete. When crystalline copper sulfate is heated, the blue crystals turn into a white powder because a *loss of water* is taking place.

Continue your reading at 51 .

---

45 Your answer is wrong.

Zinc is used for the production of hydrogen.

In which context is this process used for the production of hydrogen?

- It is used in the laboratory → 63
- It is used in industry → 68

---

46 Your answer is quite right.

Now write out completed versions of the sentences below:

Zinc is used for the electroplating of ....
Brass is an alloy of .... and ....
Zinc is used for the production of .... in the laboratory.
Chemically, zinc white consists of ....
Lithopone is chemically a mixture of .... and ....

If you are able to complete all of these sentences, turn to 67 . If not, you should in your own interest go back to 54 .

25

47 When sulfur burns sulfur dioxide is formed. The further oxidation of the sulfur dioxide to sulfur trioxide is, however, not at all easy and can be brought about only in the presence of catalysts. (Recall here the manufacture of sulfuric acid from sulfur trioxide in the Lead Chamber and Contact Processes.)

Even when sulfide ores are roasted, the oxidation goes only as far as ....

Continue your reading now at 79 .

48 No, in the Lead Chamber Process the end product is sulfuric acid.

Continue your reading at 42 .

49

Having now completed our discussion on copper, we turn at this point to the element *lead.*

Lead is a heavy metal which has a silvery luster when freshly cut but which gradually darkens to a grayish-blue color thereafter. It is so soft and so easily abraded that it can be used to write on paper. We mention in passing that present day pencils no longer contain any lead as this has long since been replaced by graphite.

Lead and its compounds are poisonous!

In air lead loses its luster as it becomes coated with a layer of oxide. As a result lead is not attacked by sulfuric acid or by cold hydrochloric acid, although lead does dissolve in nitric acid.

Jot down now the colors of lead and copper, and then continue your reading at 70 .

50 Your answer is only half right. The chemical formula for lead (II) oxide is PbO. But your formula for lead (IV) oxide is wrong.

As a hint, we remind you here that the Roman numeral which appears in brackets after a metal indicates the valence of that metal in the compound in question.

Now attempt the question in 42 once again.

51 Your answer is quite correct.

The water of crystallization present in blue copper sulfate crystals is driven off when these crystals are heated and white, anhydrous copper sulfate is formed:

$$CuSO_4 \cdot 5\,H_2O \xrightarrow{\text{heat}} CuSO_4 + 5\,H_2O$$

If an excess of aqueous ammonia solution is added to a solution of a copper (II) salt, a very intense deep blue coloration is observed. This reaction is used as a test for copper salts.

What test is used for copper salts? → 80

52 Lead sulfate is the lead salt of sulfuric acid.

You should not have gotten this confused at your stage.

Now turn to 56 for some assistance.

53 $Fe + CuSO_4 \rightarrow FeSO_4 + Cu$

The positive charges originally on the copper ion are now to be found on the iron. A transfer of two electrons from the iron to the copper has taken place. In the process, the copper ion has been discharged and has been deposited as the metal whereas the iron has gone into solution in the form of ions. This sequence of reactions is described in greater detail under the general heading of **The Electrochemical Series,** Program 30.

Copper is used nowadays in the electrical engineering industry for the manufacture of cables, in the construction industries for the manufacture of radiators and heating coils, and as a general purpose metal for coating more reactive metals to protect them from attack.

In all of these various applications, the characteristic properties of copper play an important part, namely its high conductivity for both electric current and heat and its durability in the presence of dry air, water and dilute acids.

Use keywords to write down now the most important properties of copper and some of its uses which depend on these properties.

Then continue your reading at [86] .

---

[54] Correct! Now check your equations:

Heating of zinc carbonate: $ZnCO_3 \rightarrow ZnO + CO_2$
Roasting of limestone: $CaCO_3 \rightarrow CaO + CO_2$

Zinc may be obtained from its oxide by reducing the latter with coke:

$$ZnO + C \rightarrow Zn + CO$$

Zinc does not react with air. It is used to electroplate sheet iron and in the manufacture of metal alloys. *Brass,* for instance, is an alloy of copper and zinc. Using keywords, jot down now the various uses of this metal.

One additional use of zinc has already been discussed.

What use are we referring to here?

- To the production of oxygen → [45]
- To the production of hydrogen in the laboratory → [63]
- To the production of hydrogen in industry → [68]

---

[55] $CuSO_4 \rightarrow Cu^{2\oplus} + SO_4^{2\ominus}$

If direct electric current is passed through a solution of copper sulfate, metallic copper separates out.

At which of the electrodes will the copper be deposited?

- At the anode → [18]
- At the cathode → [33]

---

[56] For your convenience we repeat the table below:

Hydrogen sulfide:	H_2S	Salts: sulfides
Sulfurous acid:	H_2SO_3	Salts: sulfites
Sulfuric acid:	H_2SO_4	Salts: sulfates

Go back now to [75] and attempt the question there once again.

---

57 Now check your answers:

In order to make lead, it is first necessary to *roast* the lead sulfide. The products formed during this process are *lead oxide* and *sulfur dioxide.* The lead oxide is then reduced by coke to *lead.*

Lead has been used in the manufacture of water piping and of lead plates for accumulators. Because of its marked resistance to acids, it is also frequently used to line the insides of reaction vessels in chemical industry. Here we recall the Lead Chamber Process, where even the name of the process indicates that lead is being used for the reaction vessel. In addition to the above usage, lead is also a component in a variety of alloys.

What product is formed in the Lead Chamber Process?

- Nitric acid → [38]
- Sulfuric acid → [42]
- Hydrochloric acid → [48]

---

58 Your answer is quite right. The chemical formulas for the two oxides of lead are:

PbO	PbO_2
lead (II) oxide	lead (IV) oxide

PbO is yellow (although there is also a red form) whereas PbO_2 is blackish-brown.

When PbO is heated in the presence of air, a bright orange-red oxide of lead having the formula Pb_3O_4 is produced.

Pb_3O_4 is known as *red lead* or *minium.* It is used as a pigment in paints. Moreover, because it is able to protect iron against rusting, it is sometimes used as an undercoat when iron objects are painted.

Now complete the equation below for the production of red lead. Note that appropriate factors also have to be inserted.

$$\text{...........} + \text{....}O_2 \rightarrow \text{......}\ Pb_3O_4$$ → [75]

---

[59] *Zinc* is a heavy metal. (Metals having a density greater than 5 g/cm^3 are referred to as heavy metals.) It has the chemical symbol Zn. In its compounds it always has a valence of two. Zinc is not found in the free state in nature. Important examples of zinc-bearing minerals are *zinc sulfide* and *zinc carbonate.*

Write out now the names and formulas of these two zinc minerals. Then continue your reading at [69] .

---

[60] No, when sulfide ores are roasted they are always heated in the presence of air. The products formed are the corresponding metal oxide and sulfur dioxide, e.g.

$$2\ CuS + 3\ O_2 \rightarrow 2\ CuO + 2\ SO_2$$

Thus, it is the *first* of the equations in [73] which relates to the roasting of sulfide ores and *not* the *second.*

Now take another look at the two equations in [73] .

---

[61] Check your answers below:

Tin(II) chloride	$SnCl_2$
Tin(IV) chloride	$SnCl_4$
Tin(II) oxide	SnO
Tin(II) sulfate	$SnSO_4$

Now copy down the two formulas which follow and write alongside each its name.

SnO_2

$Sn(NO_3)_2$

Continue your reading at [72] .

---

[62] Although you have correctly calculated the molecular masses of the two lead oxides concerned, it would appear that you have mixed up their names.

Now continue your reading at 58 .

---

63 Your answer is quite right.

Write down the equation for the production of hydrogen using zinc in the laboratory.

Then continue your reading at 71 .

---

64 Lead sulfite is the lead salt of sulfurous acid. You should not be getting these salts mixed up at this stage.

Turn now to 56 where some additional help is given.

---

65 Your answer is quite correct.

Copper has a valence of either one or two in its compounds; lead has a valence of either two or four in its compounds; zinc has only a valence of two in its compounds; and tin has a valence of either two or four in its compounds.

The compounds of the four heavy metals we have discussed, especially those of copper and lead, are poisonous. Certain lead and zinc compounds are used as pigments in paints (red lead, zinc white and lithopone).

What reagent is used to test for lead compounds and what is observed in the test?

- Hydrogen sulfide is used to produce a white precipitate → 76
- Ammonia is used to produce a blue coloration → 40
- Hydrogen sulfide is used to produce a black precipitate → 84
- I am not really sure → 10

---

66 Lead has a valence of either two or four in its compounds.

The Roman numeral placed immediately after the metal when writing out the name of a metal oxide refers to the valence of that metal in the compound concerned.

For instance, in lead (II) oxide the valence of the lead is two.

Write down now the formulas for lead (II) oxide and lead (IV) oxide, and then turn to [42] to calculate their molecular masses again.

---

[67] Check your answers now and, if you detect any errors, make the necessary corrections.

Zinc is used to electroplate iron. Brass is an alloy of copper and zinc. Zinc is used for the production of hydrogen in the laboratory. Zinc white is chemically zinc oxide (ZnO). Lithopone is chemically a mixture of zinc sulfide and barium sulfate.

At this point we start our brief discussion on tin.

Tin has the chemical symbol Sn. Tin is also a heavy metal. It can readily be drawn out to form tin wire. Because of this characteristic it may be rolled into thin sheets, usually referred to as tin foil. Threads of this foil form the tinsel used to decorate Christmas trees.

Tin does not react with air, although it is attacked by acids. Tin is produced by the reduction of oxide ores. It is used primarily for the plating of sheet iron; by this means the surface of the iron is protected from rusting. Tin-plated iron is used, for example, in the manufacture of food and beer cans. It is referred to as tinplate. An important alloy of tin is known as bronze. Bronze consists of the metals tin and copper.

Tin has the same valences as those of lead. Jot down the valences which tin has in its compounds. → [81]

---

[68] The industrial production of hydrogen cannot be based on the reaction of zinc with dilute acids as this would prove to be far too expensive.

Hydrogen is produced commercially by the electrolysis of sodium chloride solutions during which it is formed as a by-product. Another method which can be used consists of passing steam over red-hot coke and then separating the resultant mixture of carbon monoxide and hydrogen.

Now continue your reading at [63] .

---

[69] *Zinc sulfide, ZnS* *Zinc carbonate,* $ZnCO_3$

To produce metallic zinc, it is first necessary to convert both of the zinc minerals into zinc oxide.

1. Zinc oxide is formed by roasting zinc sulfide (ZnS)

2. Zinc oxide is formed from zinc carbonate ($ZnCO_3$) simply by heating it. Carbon dioxide is given off in the process.

Write out the relevant equations for the above two processes, and then continue reading at 73 .

---

70 Copper is reddish in color whereas lead is silvery to bluish-grey, according to the age of the exposed surface.

Unlike copper, however, lead is not found as the free metal in nature.

The most important ore of lead is *lead sulfide*.

The production of lead from lead sulfide is based on an analogous process to that used for the production of other metals from their sulfide ores.

Now complete the following sentences:

In order to make lead, it is first necessary to .... the lead sulfide. The products formed during this process are .... and .... The lead oxide is then reduced by coke to ....

Continue your reading at 57 .

---

71 Now check your equation:

$$Zn + 2\ HCl \rightarrow ZnCl_2 + H_2$$

Certain zinc compounds are used as pigments in paints, e.g. zinc white (= zinc oxide, ZnO) and lithopone.

Lithopone is formed by reacting zinc sulfate with barium sulfide thus:

$$ZnSO_4 + BaS \rightarrow ZnS + BaSO_4$$

Both of the products formed on the right hand side of this equation exist as white powders which are insoluble in water.

What does lithopone consist of?

- A mixture of zinc sulfide and barium sulfate → 46

- A mixture of zinc sulfate and barium sulfate → 78
- Zinc oxide → 82

---

72 SnO_2, *tin(IV) oxide;* $Sn(NO_3)_2$, *tin(II) nitrate*

Let's now briefly summarize:

Copper, lead, zinc and tin are all heavy metals.

They are all produced by reduction of their oxide ores with coal. Whenever these metals occur in the form of sulfide ores, they first have to be converted into oxide ores by roasting before they can be reduced to the metal.

What by-product is formed during the roasting of sulfide ores?

- Carbon dioxide → 19
- Sulfur dioxide → 79
- Sulfur trioxide → 47

---

73 Now check your equations:

1. Roasting of zinc sulfide:

$$2\ ZnS + 3\ O_2 \rightarrow 2\ ZnO + 2\ SO_2$$

2. Heating of zinc carbonate:

$$ZnCO_3 \rightarrow ZnO + CO_2$$

If your equations were incorrect, you should now make the necessary corrections.

Which previously discussed reaction does the second of these equations remind you of?

- The calcining of limestone → 54
- The roasting of sulfide ores → 60

---

74 Your answer is incomplete.

Although copper, lead and tin can have the valences you have mentioned, they can also have other valences.

Continue your reading now at [65] .

---

75 Now check your equation:

$6\ PbO + O_2 \rightarrow 2\ Pb_3O_4$

Lead salts in aqueous solution may be tested for by bubbling in hydrogen sulfide gas or by adding an aqueous aolution of hydrogen sulfide to the solution. A black precipitate is then formed.

Here is a typical example:

$Pb(NO_3)_2 + H_2S \rightarrow PbS + 2\ HNO_3$

What is the name of the lead compound formed as the precipitate?

- Lead sulfide → [85]
- Lead sulfite → [64]
- Lead sulfate → [52]
- I don't know → [56]

---

76 Copy down the following:

To test for lead compounds one uses hydrogen sulfide. A *black* precipitate of lead sulfide forms in each case.

Here are two examples:

$PbCl_2 + H_2S \rightarrow PbS + 2\ HCl$

$Pb(NO_3)_2 + H_2S \rightarrow PbS + 2\ HNO_3$

Now continue your reading at [84] .

---

77 Your answer is wrong.

You are confusing matters here. The Roman numerals used in the names of oxides represent the valences of the metals.

Make another attempt now at the question in [42] .

---

78 You are not working carefully enough.

Read through 71 once again.

79 Your answer is quite right. Sulfur dioxide is formed.

Copper, lead, zinc and tin do not react with air. Moreover, lead is not at all attacked by acids (hence its use in the construction of chemical plant). Copper has a very high conductivity for both electric current and heat (hence its use in the electrical engineering industry). Zinc and tin are used for plating iron.

What are the valences of these four metals?

	Cu	Pb	Zn	Sn	
●	I	II	II	II	→ 74
●	II	II	II	II	→ 83
●	II	IV	II	IV	→ 87
●	I,II	II,IV	II	II,IV	→ 65

80 Solutions of copper salts turn deep blue in color when a excess of aqueous ammonia solution is added.

Go on to 49

81 Tin has valences of two and four in its compounds. Write down now the chemical formulas for the tin compounds listed below:

Tin(II) chloride, tin(IV) chloride, tin(II) oxide, tin(II) sulfate.

Continue your reading at 61 .

82 You are not working carefully enough.

Read through 71 once again.

83 Your answer is incomplete.

Although copper, lead and tin can have the valences you have mentioned, they can also have other valences.

Continue your reading now at 65 .

84 When lead compounds are tested for with hydrogen sulfide, a black precipitate of lead sulfide forms, e.g.

$$Pb(NO_3)_2 + H_2S \rightarrow PbS + 2\ HNO_3$$

We have also described a test for copper compounds. The reagent used was an excess of aqueous ammonia solution.

What was observed in this test?

- The formation of a blue precipitate → 41
- I am not very sure → 5
- The formation of a deep blue coloration → 9

85 Your answer is quite right. PbS is lead sulfide.

Just as lead salts can be tested for using hydrogen sulfide, the test can be applied the other way round to detect hydrogen sulfide or sulfides. One then starts out from a solution of a lead salt. The details of this test have already been outlined elsewhere.

This now brings us to the end of our discussion on lead. At this point we shall move on to a consideration of the latter two metals mentioned in the title of this Program: *zinc* and *tin*.

Zinc has a density of 7.1 g/cm^3. It is therefore a light or a heavy metal?

→ 59

86 Now check your answers:

Copper is possessed of:

a) Good electrical conductivity and so is used in the manufacture of cables in the electrical engineering industry;

b) Good heat conductivity and so is used in the manufacture of radiators and heating coils;

c) High resistance to attack by dilute acids and so is used as a protective coating for more reactive metals.

Copper has been known since the times of antiquity. It was used then together with tin for the production of bronze. Such mixtures of two different metals are referred to as *alloys*. Copper alloys are still being manufactured today; examples include bronze and brass.

From what metals is the alloy bronze made? → 31

---

87 Your answer is incomplete.

Although copper, lead and tin can have the valences you have mentioned, they can also have other valences.

Continue your reading now at 65 .

Program 26

Aluminum

1 In this Program we shall discuss one particular light metal, namely aluminum.

Aluminum, which has the chemical symbol Al, is a silvery *light metal* having a matt appearance.

Which of the following metals are light metals and which are heavy metals?

Iron, copper, zinc, lead, sodium, aluminum, silver, potassium, gold, mercury.

Jot them down now:

The light metals: ....

The heavy metals: .... → 7

2 In your chemical formula you have one aluminum atom and one oxygen atom.

But Al is trivalent whcreas O is only divalent. Something here is clearly wrong!

Go back to 15 again and see if you can write out the correct formula for aluminum oxide and then calculate its molecular mass. Alternatively, you could turn to 31 where some additional explanatory material is presented.

3 You have either miscalculated here or your formula was wrong.

Would you like to make another attempt at this?

- Yes → 67
- No, I prefer an explanation now → 56

4 The correct equation is:

$$2\ Al + 6\ HCl \rightarrow 2\ AlCl_3 + 3\ H_2$$

Was your equation correct? If not, make the necessary corrections.

Write out now the equations for the reaction of aluminum with nitric acid and with sulfuric acid. → 13

---

5 Let's work out the equation for the combustion of aluminum stepwise.

We start with the substances involved:

Aluminum reacts with oxygen to form aluminum oxide thus:

$$Al + O_2 \rightarrow Al_2O_3$$

Note that on the right hand side of this equation there is an uneven number of oxygen atoms. But, as there must be an O_2 on the left hand side, there will *always* be an *even number* of oxygen atoms on that side.

To reach an *even* number of oxygen atoms on the right, it is necessary to take 2 Al_2O_3:

$$Al + O_2 \rightarrow 2\ Al_2O_3$$

You should not now find it difficult to determine the correct factors for the left hand side of the equation.

Write out the complete version of the equation and then check it at 14 .

---

6 In your chemical formula you have one aluminum atom and two oxygen atoms.

But Al is trivalent whereas O is divalent. Something here is clearly wrong!

Go back to 15 to see if you can write out the correct formula for aluminum oxide, or, if you prefer, turn instead to 31 where some additional help is given.

---

7 Now check your answers:

The *light metals* are: sodium, aluminum and potassium.

The *heavy metals* are: iron, copper, zinc, lead, silver, gold and mercury.

Sodium and potassium react very readily with the oxygen in the air to form oxides. They react very violently with water to form hydroxides. For this reason the metals Na and K have to be stored under naphtha.

Aluminum also reacts with oxygen in the air and with water. In spite of this, however, saucepans and other cooking utensils are made of aluminum. This seeming contradiction is not a real one. In air aluminum quickly becomes coated with a layer of aluminum oxide. It is this layer which is able to protect the metal from further attack by oxygen in the air or by water.

The aluminum oxide forms a *protective layer* which adheres firmly to the underlying metal.

Now complete the sentence below:

In air aluminum quickly becomes coated with a .... consisting of .... → 15

---

8 Correct, an *aluminothermic process.*

Aluminothermic processes are also used to prepare metals from metal oxides which would be difficult to reduce in any other way.

For instance, *manganese* can be produced *by an aluminothermic process from manganese (IV) oxide (manganese dioxide, MnO_2).*

Write down now the equation for the reaction of MnO_2 with aluminum. The products formed are manganese and aluminum oxide.

What factors did you arrive at in your equation?

	In front of MnO_2	In front of Al	In front of Mn	In front of Al_2O_3	
•	2	3	2	4	→ 17
•	3	2	2	2	→ 29
•	3	4	3	2	→ 25

- I need some guidance here on how to determine the correct factors → 39

---

9 The factors in the equation you have written out are not correct.

You should now proceed as follows:

Write down on the left hand side the starting materials: $Al(OH)_3$ and H_2SO_4, and on the right hand side the end products: $Al_2(SO_4)_3$ and H_2O.

Then make the following comparisons:

How many Al atoms are there on the right? And on the left? How many SO_4 groups are there on the right? And on the left? Continue in this way.

See if you are now able to get the equation in [44] right.

---

[10] You should not be making such mistakes at this stage!

In the equation:

$$3\ MnO_2 + Al \rightarrow Mn + 2\ Al_2O_3$$

there are 3 Mn on the left hand side, and so there must be 3 Mn on the right hand side too. On the right hand side there are 4 Al and so there must be 4 Al on the left hand side.

Now write out the balanced equation and check your answer at [25] .

---

[11] The problem is easy if you know the formula for acetic acid. Its formula is:

CH_3COOH.

Now see if you can draw out its structural formula. You should remember that C is tetravalent, O is divalent and that H is monovalent.

Continue your reading at [77] .

---

[12] Na_2HPO_4 and $Al_2(SO_4)_3$ are *not* double salts.

Although phosphoric and sulfuric acids are polybasic acids, the definition of a double salt refers to something else, namely:

A double salt is a compound in which the negative charges on the acid radical of a polybasic acid are balanced by the positive charges of several different metal ions.

Are the two salts listed below double salts?

$(NH_4)MgPO_4$ $Ca(HCO_3)_2$

- Yes, both are double salts → [73]
- No, neither is a double salt → [37]

- Only $(NH_4)MgPO_4$ is a double salt → [58]
- Only $Ca(HCO_3)_2$ is a double salt → [61]

---

[13]

$$2\ Al + 6\ HNO_3 \rightarrow 2\ Al(NO_3)_3 + 3\ H_2$$

$$2\ Al + 3\ H_2SO_4 \rightarrow Al_2(SO_4)_3 + 3\ H_2$$

Aluminum is also attacked by alkalis because the protective layer of aluminum oxide gets dissolved away by alkalis. We shall discuss the chemistry of this later on.

Now complete the sentence:

Aluminum is attacked by .... and by strong .... → [24]

---

[14]

Here is the balanced equation:

$$4\ Al + 3\ O_2 \rightarrow 2\ Al_2O_3$$

The tendency of aluminum to combine with oxygen is so great that aluminum is capable of removing oxygen from oxygen-containing compounds. In such reactions a great deal of heat is generated.

One example is provided by the reaction of iron (III) oxide with aluminum:

$$Fe_2O_3 + 2\ Al \rightarrow 2\ Fe + Al_2O_3$$

Because of the great amount of heat which is generated, processes of this type are referred to as *aluminothermic processes.*

The name given to these processes is worth memorizing.

What is the role of the aluminum in an aluminothermic process?

- It acts as an oxidizing agent → [23]
- It acts as a reducing agent → [32]
- It provides the necessary heat → [42]
- It acts as a catalyst → [50]

---

15 In air, aluminum quickly becomes coated with a *firmly adhering, protective layer* of *aluminum oxide.*

Write out now the chemical formula for aluminum oxide and calculate its molecular mass. Al is trivalent and the relevant atomic masses are: Al =27 and O = 16.

What is the value you obtain?

- 43 → 2
- 59 → 6
- 75 → 22
- 102 → 26
- I will need some additional help → 31

---

16 The correct equation is as follows:

$$2\ Al(OH)_3 + 3\ H_2SO_4 \rightarrow Al_2(SO_4)_3 + 6\ H_2O$$

Note that there will be *no zero* in front of the $Al_2(SO_4)_3$ but rather a *one.* In this regard your answer was incorrect. The number 1 is, however, always omitted from the front of a chemical formula. As it stands, the $Al_2(SO_4)_3$ molecule appears *once* in the above equation, meaning that *one* molecule participates in the reaction. → 59

---

17 Either you are uncertain as to the valences of Al (trivalent) and Mn(tetravalent here) or you have miscalculated.

- Would you like to make one more attempt? → 8
- Would you prefer to have some guidance on setting up the desired equation? → 39

---

18 Not only is the $NaFe(SO_4)_2$ a double salt!

Make another attempt at answering the question in 69 after familiarizing yourself with the definition again.

19 You have either miscalculated here or your formula was wrong.

Would you like to make one more attempt?

- Yes → 67
- No, I should like some help → 56

20 You still seem to be uncertain about the bauxite purification procedures.

Repeat now from 27 .

Note particularly the sequence of the various steps involved.

21 $Fe_2O_3 + 2\ Al \rightarrow 2\ Fe + Al_2O_3$

In this reaction *a great deal of heat* is generated, so much in fact that the iron which is formed melts.

A mixture of aluminum and iron oxide is referred to as *thermite*.

Thermite is used for welding iron objects together. If, for instance, two railroad tracks are to be joined together, a thermite pack is placed between the two tracks at the spot to be welded. The pack is then ignited, and after the thermite mixture has finished burning the tracks will be welded together by the iron produced in the reaction.

This process, involving the welding of two iron objects by the *thermite* reaction, is an .... process. → 8

22 In your chemical formula you have 1 Al atom and 3 O atoms. But Al is trivalent whereas oxygen is only divalent. Something here is clearly wrong!

Go back to 15 again and see if you can write out the correct formula for aluminum oxide, or, if you prefer, go instead to 31 where some additional help is given.

23 *Oxidizing agents* are substances which are capable of *giving up oxygen.*

Reducing agents are substances which are capable of *removing oxygen* from other compounds.

Copy down these two definitions, and then see if you can work out the role of the aluminum in a thermite reaction. → [14]

---

24 Aluminum is attacked by *alkalis* and by strong *acids*.

The protective layer of aluminum oxide protects the metal from the oxygen in the air only at normal temperatures.

When heated, finely divided aluminum reacts very violently with oxygen. In the process the aluminum burns with a very intense flame to form aluminum oxide. Write down now the relevant equation for this reaction.

What factors will there be in front of the Al, the O_2 and the Al_2O_3?

	In front of Al	In front of O_2	In front of Al_2O_3	
●	4	3	1	→ [5]
●	4	3	2	→ [14]
●	2	3	2	→ [33]
●	I obtained other factors			→ [40]

---

25 Here is the balanced equation:

$$3\ MnO_2 + 4\ Al \rightarrow 3\ Mn + 2\ Al_2O_3$$

Did you manage to get the factors correct first time round? If so, you deserve to be congratulated!

Now write out the completed version of the sentences below:

Aluminum can remove the .... from compounds which contain oxygen. In the process a great deal of .... is generated. The mixture of iron (III) oxide and aluminum is known as .... → [45]

---

26 Your formula for aluminum oxide is correct: Al_2O_3

The protective layer of aluminum oxide can, of course, protect the underlying metal only from those substances which do not attack this layer. In fact, the oxide layer is able to protect the metal from attack by oxygen in the air, water and *weak* acids. It is not able to protect the metal from attack by alkalis and strong acids (even when the latter are *diluted).*

Aluminum is thus attacked by dilute strong acids such as dilute H_2SO_4 and dilute HCl.

When hydrochloric acid reacts with aluminum, it attacks first the thin oxide layer and then the metal. Aluminum and hydrochloric acid react together to form aluminum chloride and hydrogen.

Write out now the equation for the reaction of aluminum with hydrochloric acid with the name of each substance under its formula. Then continue reading at 4 .

---

27 In the production of aluminum, the following processes take place at the anode:

$$2\,C + O_2 \rightarrow 2\,CO$$

$$2\,CO + O_2 \rightarrow 2\,CO_2$$

Only *very pure* aluminum oxide can be used for the electrolysis. The manufacture of very pure aluminum oxide is actually quite difficult.

It is worth paying special attention to the various stages involved!

Aluminum is not found in the free state in nature, though it occurs widely in the form of different compounds. The most important ore of aluminum is *bauxite.*

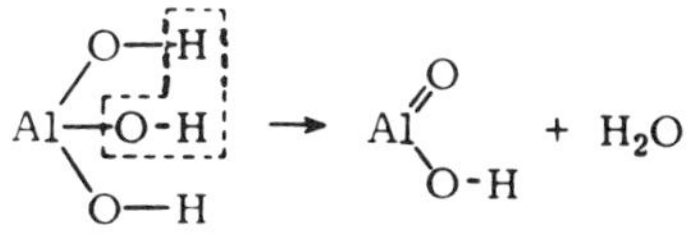

Bauxite

Bauxite is the most frequently used starting material for the production of aluminum. However, the bauxite cannot be used in its natural state as it contains several impurities. The latter first have to be removed from the ore.

26

To accomplish this the bauxite ore is initially roasted. The net effect of the roasting is the splitting off of one water molecule from every two molecules of bauxite thus:

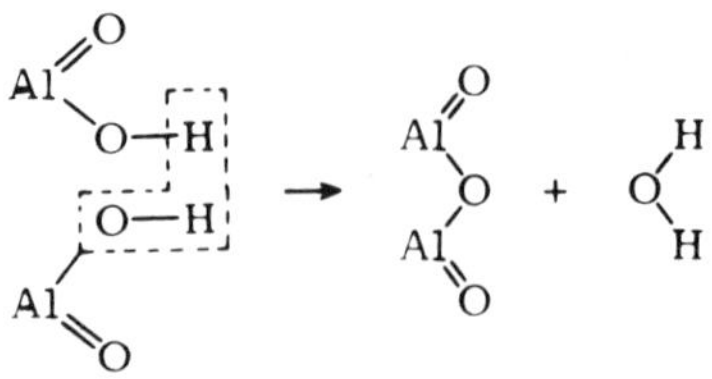

Write down this equation, but use ordinary chemical formulas instead of the above structural formulas.

Now continue your reading at [47] .

---

[28] In electrolysis, metals are always deposited at the *cathode.* We may recall here, for instance, the manufacture of sodium from molten NaOH.

Now complete the sentence below:

In electrolysis, the aluminum (like all other metals) is deposited on the cathode, the .... electrode. → [38]

---

[29] Either you are uncertain as to the valences of Al (trivalent) and Mn (tetravalent here) or you have miscalculated.

- Would you like to make one more attempt? → [8]
- Would you prefer to have some guidance on setting up the desired equation? → [39]

---

[30] The factors in the equation you have written down are not correct.

You should now proceed as follows:

Write down on the left hand side the starting materials: $Al(OH)_3$ and H_2SO_4, and on the right hand side the end products: $Al_2(SO_4)_3$ and H_2O.

Then make the following comparisons:

How many Al atoms are there on the right? And on the left? How many SO_4 groups are there on the right? And on the left? Continue in this way.

See if you are now able to get the equation in [44] right.

---

31 Aluminum is trivalent whereas oxygen is divalent. To determine the ratio in which these two substances combine together to form aluminum oxide, consider the following:

It is *not* possible for one atom of aluminum to combine with one atom of oxygen, for then one valence bond on the aluminum would be left over.

Two aluminum atoms provide a total of 2 x 3 = 6 valence bonds. These 6 bonds can be joined to 3 oxygen atoms, since 3 oxygen atoms will also have 3 x 2 = 6 valence bonds. The chemical formula for aluminum oxide must therefore be ....

Continue your reading at 26 .

32 Your answer is quite right.

In an aluminothermic process the aluminum acts as a *reducing agent.* Here we repeat the example of a typical aluminothermic reaction: aluminum reacting with iron oxide.

$$Fe_2O_3 + Al \rightarrow Fe + Al_2O_3$$

The factors are missing from this equation. Write out now the balanced equation, and then turn to 21 .

33 Let's work out the equation for the combustion of aluminum stepwise.

We start with the substances involved:

Aluminum reacts with oxygen to form aluminum oxide thus:

$$Al + O_2 \rightarrow Al_2O_3$$

Note that on the right hand side of this equation there is an uneven number of *oxygen atoms. But, as there must be an O_2 on the left hand side, there will always* be an *even number* of oxygen atoms on that side.

To reach an *even* number of oxygen atoms on the right, it is necessary to take 2 Al_2O_3:

$$Al + O_2 \rightarrow 2\ Al_2O_3$$

You should not now find it difficult to determine the correct factors for the left hand side of the equation.

Write out the complete version of the equation and then check it at 14 .

---

34 Your answer is wrong.

Use keywords to write out the various procedures involved in the purification of bauxite. Give the relevant equation for each procedure.

Then check your answer at 71 .

---

35 Your equation is not correct.

Continue your reading at 79 .

---

36 You have either miscalculated here or your formula was wrong.

Would you like to make one more attempt?

- Yes → 67
- No, I should like some help → 56

---

37 How did you arrive at the conclusion that $(NH_4)MgPO_4$ is not a double salt?

The negative charges on the phosphate ion are balanced here by the positive charges of two different metal ions, namely $NH_4^{\oplus}$ and $Mg^{2\oplus}$.

Or were you unwilling to regard "ammonium" as a metal? It is a fact, however, that the $NH_4^{\oplus}$ ion behaves in salts as though it were a metal cation.

Go back now to 69 and focus on the precise meaning of the definition given there.

---

38 Yes, like all metals, aluminum is deposited on the *cathode,* the *negative* electrode. The aluminum is run off from a tapping hole. (Refer here to your diagram!)

Oxygen is formed at the *anode,* the *positive* electrode. The anode consists of a carbon electrode. The oxygen which forms attacks the carbon and combines with it to form carbon monoxide. At the high operating temperature, this gas ignites and burns to yield carbon dioxide. Thus, during the electrolysis of aluminum oxide, flames of burning carbon monoxide appear. Because of this, the *anodes* are gradually *consumed* as the electrolysis proceeds.

In this section we have mentioned two chemical processes which take place at the anode. Write down now the relevant chemical equations for these two processes, and then continue your reading at 27 .

---

39 Let's work out this equation stepwise.

We start by writing down only the substances involved: the starting materials and the end products. Manganese (IV) oxide (manganese dioxide) reacts with aluminum to form manganese and aluminum oxide:

$$MnO_2 + Al \rightarrow Mn + Al_2O_3$$

Now, there are two O atoms on the left as against three O atoms on the right. To attain the same number of oxygen atoms (6) on both sides, we have to take 3 MnO_2 on the left and 2 Al_2O_3 on the right:

$$3\ MnO_2 + Al \rightarrow Mn + 2\ Al_2O_3$$

It remains now to find the number of Al and Mn atoms required to balance the equation. What factors have to be placed in front of the Al and the Mn?

	In front of Al	In front of Mn		
•	2	3	→	10
•	4	6	→	49
•	4	3	→	25

---

40 Let's work out the equation for the combustion of aluminum stepwise.

We start with the substances involved:

Aluminum reacts with oxygen to form aluminum oxide thus:

$$Al + O_2 \rightarrow Al_2O_3$$

Note that on the right hand side of this equation there is an uneven number of oxygen atoms. But, as there must be an O_2 on the left hand side, there will *always* be an *even number* of oxygen atoms on that side.

To reach an *even* number of oxygen atoms on the right, it is necessary to take 2 Al_2O_3:

$$Al + O_2 \rightarrow 2\ Al_2O_3$$

You should not now find it difficult to determine the correct factors for the left hand side of the equation.

Write out the balanced version of this equation and then check it at [14] .

---

41 Na_2HPO_4 is *not* a double salt, but rather a hydrogen salt.

Phosphoric acid is certainly a polybasic acid, yet the definition of a double salt involves something quite different:

A double salt is a compound in which the negative charges on the acid radical of a polybasic acid are balanced by the positive charges of several different metal ions.

Are the two salts below double salts?

$$(NH_4)MgPO_4 \text{ and } Ca(HCO_3)_2$$

- Yes, both are double salts → [73]
- No, neither is a double salt → [37]
- Only $(NH_4)MgPO_4$ is a double salt → [58]
- Only $Ca(HCO_3)_2$ is a double salt → [61]

---

42 Although it is true that a great deal of heat is generated in an aluminothermic reaction, you were asked rather about the *chemical role* of the aluminum in such a reaction.

Read through [14] once again.

---

43 $Al(OH)_3 + 3\ CH_3COOH \rightarrow Al(CH_3COO)_3 + 3\ H_2O$

aluminum hydroxide	acetic acid	aluminum acetate	water

26

If you are able to answer all of the questions below confidently and correctly, you can be sure that you have worked through all the material contained in this

Program carefully and that you have retained the essential parts. (It would be best to write out your answers in full!)

1. Is Al a light or a heavy metal?
2. Is Al attacked by water and air?
   If so, why? If not, why not?
3. Is Al attacked by acids and alkalis?
   If so, why? If not, why not?
4. If you answered questions 2 or 3 above with a "yes", can you write down the relevant equations to support your answer?
5. What is meant by an aluminothermic process and for what purpose is it used? Give two examples.
6. How is aluminum manufactured? Give a diagram of the apparatus used and explain the processes taking place at the electrodes.
7. How is the raw material used in the manufacture of aluminum purified? Give all four steps!
8. What is meant by a double salt?
9. Write out the chemical formula for each of the following compounds: aluminum oxide, bauxite, aluminum sulfate, potassium alum, aluminum acetate.
10. How is aluminum sulfate prepared?

Now check your answers at [80] .

---

[44] Your answer is quite right. The chemical formula for aluminum sulfate is

$Al_2(SO_4)_3$

Aluminum sulfate is prepared by reacting aluminum hydroxide with sulfuric acid. Write down the relevant equation for this process, and then answer the question below.

What are the factors which appear in this equation?

	In front of $Al(OH)_3$	In front of H_2SO_4	In front of $Al_2(SO_4)_3$	In front of H_2O		
•	2	4	1	3	→	[9]
•	2	3	1	6	→	[59]
•	2	3	0	6	→	[16]
•	1	3	1	3	→	[30]
•	I need some guidance on this				→	[70]

---

45 *oxygen, heat, thermite*

Metal oxides, such as Fe_2O_3 and MnO_2, can be reduced with aluminum. In addition to the Al_2O_3 which forms, the free .... is also formed. The process is described as an .... process.

Jot down the missing words, and then turn to 54 .

---

46 The protective layer on aluminum consists of Al_2O_3!

When this layer is dissolved away by caustic soda the product formed is $Na[Al(OH)_4]$!

What are the names of the two substances mentioned here only by formula?

Jot down their names and then continue your reading at 55 .

---

47 Now check your equation:

$$2\ AlO_2H \rightarrow Al_2O_3 + H_2O$$

When *bauxite* is roasted *aluminum oxide* and *water* are formed.

The aluminum oxide is purified by treating it with hot caustic soda. The aluminum oxide goes into *solution* by forming a new aluminum compound known as *sodium aluminate.*

Sodium aluminate has the chemical formula $Na[Al(OH)_4]$.

Do you think you could manage to write out the equation for the dissolution of aluminum oxide in caustic soda?

If so, go ahead and then check your equation at 72 .

If you would prefer some further clarification, however, on how to set up this equation, turn instead to 62 .

---

48 You still seem to be uncertain about the bauxite purification procedures.

26

Repeat now from 27 .

Note particularly the sequence of the various steps involved.

49 You should not be making such mistakes at this stage!

In the equation:

$$3\ MnO_2 + Al \rightarrow Mn + 2\ Al_2O_3$$

there are 3 Mn on the left hand side, and so there must be 3 Mn on the right hand side too. On the right hand side there are 4 Al and so there must be 4 Al on the left hand side.

Now write out the balanced equation and check your answer at 25 .

50 *A catalyst is a substance which speeds up a chemical reaction, but which at the end of the reaction is present in the same form as at the beginning.*

Here is an example: When oxygen is prepared by heating potassium chlorate, the reaction can be catalysed by manganese dioxide.

In an *aluminothermic reaction,* however, the aluminum is converted into aluminum oxide. Because the aluminum clearly changes in such a reaction, it cannot be a catalyst!

Copy down now the above definition in italics, and then work through 14 once again.

51 After it has been roasted, *bauxite* has to be dissolved in order that the insoluble impurities it contains can be filtered off.

How is bauxite taken up into solution? The two steps involved are:

Step 1: Roasting

Step 2: Dissolution

Do you feel able to write down equations for these two steps? See if you can!

If you can manage this without difficulty, check your equations at 74 . If you find this too difficult, repeat now from 27 .

52 $Al_2(SO_4)_3$ is *not* a double salt.

Sulfuric acid is certainly a polybasic acid, yet the definition of a double salt involves something quite different:

A double salt is a compound in which the negative charges on the acid radical of a polybasic acid are balanced by the positive charges of several different metal ions.

Are the two salts below double salts?

$(NH_4)MgPO_4$ and $Ca(HCO_3)_2$

- Yes, both are double salts → 73
- No, neither is a double salt → 37
- Only $(NH_4)MgPO_4$ is a double salt → 58
- Only $Ca(HCO_3)_2$ is a double salt → 61

---

53 Copy down the following two sentences:

Clay is chemically *aluminum oxide.*

Bauxite is an hydroxide of aluminum which lacks one molecule of water.

Go back now to 75 .

---

54 metal, an *aluminothermic* process.

Aluminum is produced from the *electrolysis* of molten Al_2O_3.

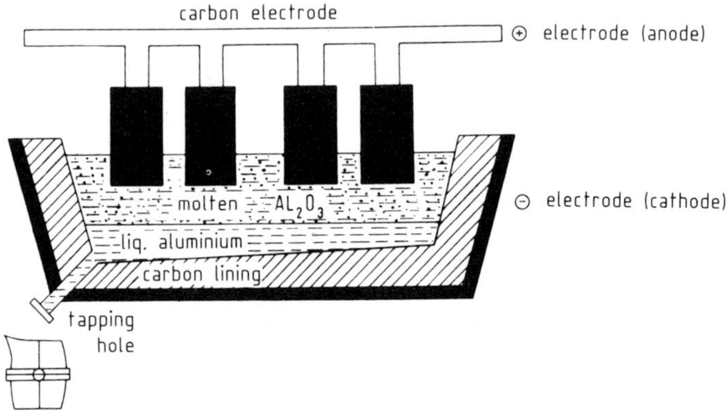

The carbon lining of the melting bath forms the cathode (indicated here as the negative electrode). Graphite electrodes, which dip into the melt, function as the anode (indicated as the positive electrode).

Now copy down this diagram *with all the descriptive labels,* and then answer the question below:

On which electrode will the aluminum be deposited?

- On the cathode (the negative electrode) → 38
- On the anode (the positive electrode) → 28
- I don't know → 64

---

55 Your answer is quite correct. The protective layer consists of *aluminum oxide.* When this layer is dissolved away by caustic soda, the product formed is *sodium aluminate.* Sodium aluminate is formed from 1 mole of aluminum hydroxide and 1 mole of sodium hydroxide.

Write out the relevant equation for this reaction, and then check your answer at 67 .

---

56 When aluminum salts dissociate in water they form triply positive aluminum ions, $Al^{3\oplus}$.

The sulfate ion bears a double negative charge, $SO_4^{2\ominus}$.

The formula for aluminum sulfate contains as many aluminum and sulfate ions as are necessary to equalize the number of positive and negative charges in the molecule:

2 aluminum ions bearing 2 x 3 = 6 positive charges

3 sulfate ions bearing 3 x 2 = 6 negative charges

Now see if you can write down the formula for aluminum sulfate. → 44

---

57 Your answer is only partially correct.

Bauxite has the formula AlO(OH).

Now copy down the following sentence:

Clay is chemically *aluminum oxide.*

What will be the chemical formula for clay? → 76

---

58 Your answer is quite right; $Na_2(NH_4)PO_4$ and $NaFe(SO_4)_2$ are double salts.

Now complete the definition of a double salt below:

A double salt is a compound in which the .... charges on the acid radical of a polybasic acid are balanced by the positive charges of .... different metal ions. → 76

---

59 Your answer is quite right (though the number 1 is never written in front of the $Al_2(SO_4)_3$!):

$$2\ Al(OH)_3 + 3\ H_2SO_4 \rightarrow Al_2(SO_4)_3 + 6\ H_2O$$

The equation you have written down here represents a neutralization reaction.

Aluminum sulfate is very soluble in water. It crystallizes out with 18 molecules of water of crystallization.

Write down now the formula for crystalline aluminum sulfate containing water of crystallization. → 69

---

60 Step 4: $2\ Al(OH)_3 \xrightarrow{\text{roasting}} Al_2O_3 + 3\ H_2O$

What was the third step in the purification of bauxite?

- The dissolution of the Al_2O_3 → 48
- The crystallizing out of the $Al(OH)_3$ → 34
- The roasting of the bauxite → 20

---

61 How did you arrive at the conclusion that $(NH_4)MgPO_4$ is not a double salt? It is indeed a double salt!

The negative charges on the phosphate radical are balanced by the positive charges of two metal ions. Or were you unwilling to regard "ammonium" as a metal? It is a fact, however, that the $NH_4^{\oplus}$ ion behaves in salts as though it were a metal cation.

$Ca(HCO_3)_2$ is however not a double salt, but rather a hydrogen salt.

Go back now to 69, and focus on the precise meaning of the definition given there.

---

62 Aluminum oxide dissolves in caustic soda to form sodium aluminate.

Let's work out the equation for this reaction stepwise. We start by writing down the starting materials and end products. (You should also copy this down!)

$$Al_2O_3 + NaOH \rightarrow Na[Al(OH)_4]$$

On the left we have two Al, and so there must be two Al on the right too:

$$Al_2O_3 + NaOH \rightarrow 2\ Na[Al(OH)_4]$$

Now, we have two Na on the right, and so we shall have to take 2 NaOH on the left:

$$Al_2O_3 + 2\ NaOH \rightarrow 2\ Na[Al(OH)_4]$$

If you count up the O and H atoms at this point, you will discover that there are fewer O and H atoms on the left than on the right. On the left we have 2 H, but on the right 8 H.

On the left twice as many H atoms as O atoms are missing, which points to the fact that some water molecules are missing! See if you can work out the missing number of water molecules and thereby complete the equation below:

$$Al_2O_3 + 2\ NaOH + H_2O \rightarrow 2\ Na[Al(OH)_4]$$

Check again to ensure that the number of atoms of each type is the same on both the left and right hand sides. Then turn to 72 to see if your version of the equation is correct.

---

63 The protective layer on aluminum consists of Al_2O_3!

When this layer is dissolved away by caustic soda the product formed is $Na[Al(OH)_4]$.

What are the names of the two substances mentioned here only by formula?

Jot down their names and then continue your reading at 55 .

---

64 In electrolytic reactions, metals are always deposited *on the cathode.* Remember here, for instance, the manufacture of sodium from molten sodium hydroxide!

Now copy down the following sentence:

During electrolysis, aluminum (like all metals) is deposited on the cathode, the .... electrode. → 38

65 Copy down the sentences below:

Clay is chemically *aluminum oxide.*

Bauxite is aluminum hydroxide from which one molecule of water is missing.

Go back now to 75 .

66 Now check your formula for potassium iron alum:

$KFe(SO_4)_2 \cdot 12\ H_2O$

Here we repeat the formulas of the alums we have discussed so far:

$KAl(SO_4)_2 \cdot 12\ H_2O$

$NaAl(SO_4)_2 \cdot 12\ H_2O$

$(NH_4)Al(SO_4)_2 \cdot 12\ H_2O$

$KFe(SO_4)_2 \cdot 12\ H_2O$

You should now have no difficulty in inserting the missing words in the sentences below:

In the formula for any alum there is a singly positively charged metal ion and a .... positively charged metal ion together with two doubly negatively charged .... and twelve molecules of .... → 75

67 $Al(OH)_3 + NaOH \rightarrow Na[Al(OH)_4]$

26

Another important compound of aluminum is aluminum sulfate. Write out now the chemical formula for aluminum sulfate, and use this formula to calculate its molecular mass. The relevant atomic masses are: Al = 27, S = 32, O = 16.

What molecular mass do you obtain for aluminum sulfate?

- 123 → 3
- 219 → 19
- 246 → 36
- 342 → 44
- I need some help on this → 56

---

68 Aluminum acetate is a salt of *acetic acid.*

The salts of acetic acid are known as acetates.

Now see if you can still write down the chemical and structural formulas for acetic acid. Bear in mind here that acetic acid is an organic acid which contains carbon, hydrogen and oxygen. In organic compounds, carbon always has a valence of four, oxygen is divalent and hydrogen is monovalent.

If you feel confident that your formulas are correct, go on to 77 .

If you have any doubts or feel the question is too difficult for you, go instead to 11 .

---

69 $Al_2(SO_4)_3 \cdot 18\ H_2O$

Among its other uses, aluminum sulfate is used in the paper industry. Its role in this context is as a "size" for paper. Paper which has not been sized cannot be written on with ink as the ink would simply run off.

Another important compound of aluminum is known as alum and this has the chemical formula:

$$KAl(SO_4)_2$$

In alum the four negative charges of the two sulfate ions are balanced by the three positive charges on the aluminum ion and the single positive charge on the potassium ion. $KAl(SO_4)_2$ is a double salt.

A double salt is a compound in which the negative charges on the acid radical of a polybasic acid are balanced by the positive charges of several different metal ions. Ammonium ions ($NH_4^{\oplus}$) are regarded here as equivalent to singly positively charged metal ions ($Na^{\oplus}$, $K^{\oplus}$).

You should memorize this definition of a double salt. Now see if you can answer the following question:

Which of the salts listed below are double salts?

$$Na_2HPO_4, Na_2(NH_4)PO_4, NaFe(SO_4)_2, Al_2(SO_4)_3$$

- All four are double salts → 12
- The first three salts are double salts → 41
- The latter three salts are double salts → 52
- $Na_2(NH_4)PO_4$ and $NaFe(SO_4)_2$ are double salts → 58
- Only $NaFe(SO_4)_2$ is a double salt → 18

---

70 Let's work out this equation step by step.

When aluminum hydroxide reacts with sulfuric acid the products formed are aluminum sulfate and water thus:

$$Al(OH)_3 + H_2SO_4 \rightarrow Al_2(SO_4)_3 + H_2O$$

On the right hand side of our equation there are 2 Al atoms. Accordingly, we shall have to start out with 2 molecules of aluminum hydroxide on the left hand side:

$$2\ Al(OH)_3 + H_2SO_4 \rightarrow Al_2(SO_4)_3 + H_2O$$

This equation, of course, is still not balanced.

What are the factors which are missing from the equation?

	In front of H_2SO_4	In front of $Al_2(SO_4)_3$	In front of H_2O		
•	2	1	4	→	35
•	3	1	6	→	59
•	3	0	6	→	16
•	I shall need some further explanation			→	79

71

The purification of bauxite involves:

1. Roasting the bauxite: $2\ AlO(OH) \rightarrow Al_2O_3 + H_2O$

2. Dissolution of the Al_2O_3 : $Al_2O_3 + 2\ NaOH + 3\ H_2O \rightarrow 2\ Na[Al(OH)_4]$

3. Crystallizing out the $Al(OH)_3$

4. Roasting the $Al(OH)_3$: $2\ Al(OH)_3 \rightarrow Al_2O_3 + 3\ H_2O$

The equation which represents step 2 is interesting for another reason. It reveals why metallic aluminum is attacked by alkalis in spite of a protective oxide layer.

The thin protective layer which forms on aluminum is dissolved away by the alkali.

Now answer the two questions below:

a) What does the protective layer consist of?

b) What product is formed when the protective layer is dissolved by caustic soda?

- a) Aluminum hydroxide b) Sodium aluminate → 63
- a) Aluminum oxide b) Aluminum hydroxide → 78
- a) Aluminum oxide b) Sodium aluminate → 55
- I don't know → 46

72

$$Al_2O_3 + 2\ NaOH + 3\ H_2O \rightarrow 2\ Na[Al(OH)_4]$$

In this way the bauxite can be brought into solution after it has been converted into Al_2O_3 by roasting.

The *impurities* present in the roasted bauxite remain *undissolved*. They may be filtered off. From the filtrate containing the sodium aluminate, it is now possible to crystallize out aluminum hydroxide, $Al(OH)_3$.

Write out your answer to the question below:

Why is it necessary to bring the roasted bauxite into solution? → 51

73 $Ca(HCO_3)_2$ is a hydrogen salt, but not a double salt!

A double salt consists of several different metal ions and the acid radical of a polybasic acid.

Go back again to 69 and focus your attention on the precise meaning of the definition given there.

74 The purification of bauxite involves:

Step 1: Roasting: $2\ AlO(OH) \rightarrow Al_2O_3 + H_2O$

Step 2: Dissolution: $Al_2O_3 + 2\ NaOH + 3\ H_2O \rightarrow 2\ Na[Al(OH)_4]$

Step 3: Crystallizing out the $Al(OH)_3$

By roasting the $Al(OH)_3$ formed, it is then possible to prepare the very pure Al_2O_3 needed for the manufacture of aluminum by electrolysis.

You should now write out the equation for Step 4 (the roasting of aluminum hydroxide to yield aluminum oxide).

There should be no difficulty here if you remember Step 1 where a similar reaction was carried out. → 60

75 The missing words are:

triply

sulfate ions

water

Another important compound of aluminum is *clay*. Chemically, pure clay consists of *aluminum oxide*. Like bauxite, clay is found in nature. In fact, clay and bauxite are the two most important naturally occurring compounds of aluminum.

What are the formulas for the two most important naturally occurring compounds of aluminum?

	Clay	*Bauxite*		
•	$AlO(OH)$	$Al(OH)_3$	→	53
•	$Al(OH)_3$	Al_2O_3	→	65
•	Al_2O_3	$AlO(OH)$	→	81
•	$Al(OH)_3$	$AlO(OH)$	→	57

76 A double salt is a compound in which the *negative* charges on the acid radical of a polybasic acid are balanced by the positive charges of *several* different metal ions.

$KAl(S)_4)_2$, potassium aluminum sulfate (or alum for short) is a double salt.

There are many other ***alums***, e.g. sodium alum, ammonium alum, etc. They all have a similar composition and all of them crystallize out with 12 molecules of water. The alums mentioned here have the formulas indicated:

Potassium alum: $KAl(SO_4)_2 \cdot 12\ H_2O$; Sodium alum: $NaAl(SO_4)_2 \cdot 12\ H_2O$; Ammonium alum: $(NH_4)Al(SO_4)_2 \cdot 12\ H_2O$

The aluminum present in an alum compound can be replaced by some other trivalent metal, e.g. by trivalent iron. In this case we would then obtain potassium iron alum.

Write out now the chemical formula for the latter compound. → 66

77 Now compare the formula you have drawn out with that below:

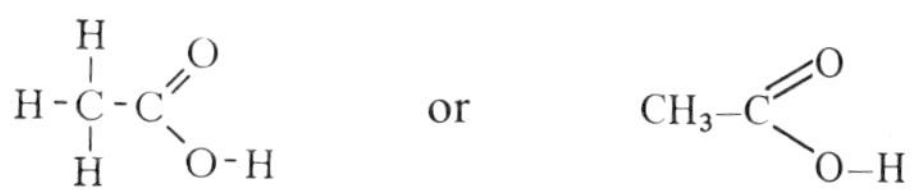

The acid radical of acetic acid is the singly negatively charged acetate ion:

$CH_3COO^{\ominus}$

Here is the formula for aluminum acetate:

$(CH_3{-}COO)_3Al$

which can also be written (since we are used to writing the metal in front of the acid radical) as $Al(CH_3COO)_3$.

Aluminum acetate may be prepared by reacting aluminum hydroxide with acetic acid.

Write out the relevant equation for this reaction – a neutralization reaction – and then also write in the names of the starting materials and end products.

→ 43

---

78 The protective layer on aluminum consists of Al_2O_3.

When this layer is dissolved away by caustic soda, the product formed is $Na[Al(OH)_4]$.

What are the names of the two substances mentioned here only by formula?

Jot down their names and then continue your reading at 55 .

---

79 Here once again is the unbalanced equation for the dissolution of aluminum hydroxide in sulfuric acid:

$$2\ Al(OH)_3 + H_2SO_4 \rightarrow Al_2(SO_4)_3 + H_2O$$

Now, the aluminum atoms are balanced on each side. However, there are 3 SO_4 groups on the right hand side but only one SO_4 group on the left hand side. To balance these up we must write in 3 molecules of sulfuric acid on the left:

$$2\ Al(OH)_3 + 3\ H_2SO_4 \rightarrow Al_2(SO_4)_3 + H_2O$$

But the number of water molecules is still not balanced in this equation.

On the left we have 6 OH groups. Does this hint help you?

Now write out the balanced version of the equation, and then check your answer at 59 .

---

80 1. Al is a *light* metal.

2. No, Al is *not* attacked by air and water because it becomes coated with a *protective layer of aluminum oxide.*

3. Yes, Al *is* attacked by alkalis and strong acids because these *dissolve away the protective layer* of Al_2O_3.

4. Here are two examples:

$$Al_2O_3 + 6\ HCl \rightarrow 2\ AlCl_3 + 3\ H_2O$$

$$Al_2O_3 + 2\ NaOH + 3\ H_2O \rightarrow 2\ Na[Al(OH)_4]$$

5. In an aluminothermic process *metal oxides are reduced* by *aluminum*. In the process *a great deal of heat* is generated.

   Here is an example:

   $Fe_2O_3 + 2\ Al \rightarrow 2\ Fe + Al_2O_3$

   This reaction is used, among other things, to weld rails together.

6. Al is manufactured by the electrolysis of molten Al_2O_3.

   The Al is deposited on the *cathode*. The *anodes*, which consist of *carbon*, are *attacked* by the *oxygen* liberated on them and so are gradually consumed. CO is formed and this burns to yield CO_2 at the surface.

   $2\ C + O_2 \rightarrow 2\ CO$

   $2\ CO + O_2 \rightarrow 2\ CO_2$

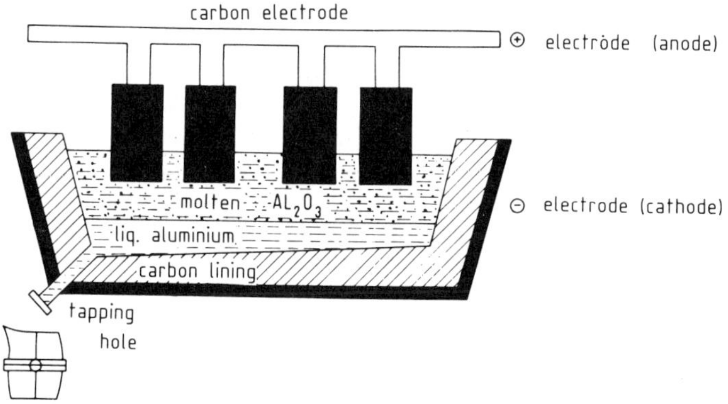

7. Step 1: The raw material bauxite is roasted.

   $2\ AlO(OH) \rightarrow Al_2O_3 + H_2O$

   Step 2: The Al_2O_3 is dissolved in caustic soda.

   $Al_2O_3 + 2\ NaOH + 3\ H_2O \rightarrow 2\ Na[Al(OH)_4]$

   Step 3: The $Al(OH)_3$ is crystallized out.

   Step 4: The $Al(OH)_3$ is roasted and *pure* Al_2O_3 is formed.

   $2\ Al(OH)_3 \rightarrow Al_2O_3 + 3\ H_2O$

8. A double salt is a compound in which the negative charges on the acid radical of a polybasic acid are balanced by the positive charges of several different metal ions.

9. Al_2O_3	$AlO(OH)$	$Al_2(SO_4)_3$	$KAl(SO_4)_2 \cdot 12\ H_2O$	$Al(CH_3COO)_3$
Aluminum oxide	Bauxite	Aluminum sulfate	Potassium alum	Aluminum acetate

10. Aluminum sulfate is prepared by dissolving Al_2O_3 or $Al(OH)_3$ in sulfuric acid:

$$Al_2O_3 + 3\ H_2SO_4 \rightarrow Al_2(SO_4)_3 + 3\ H_2O$$

$$2\ Al(OH)_3 + 3\ H_2SO_4 \rightarrow Al_2(SO_4)_3 + 6\ H_2O$$

End of Program 26.

---

81 Here once again we repeat the two correct formulas:

Bauxite: $AlO(OH)$ Clay: Al_2O_3

The last of the aluminum compounds that we shall be discussing here is an organic aluminum compound known as *aluminum acetate*.

Aluminum acetate is a salt of aluminum. What is the name of the corresponding acid?

Complete the sentence below:

Aluminum acetate is a salt of .... → 68

Program 27

Magnesium

1 Magnesium has the symbol Mg. Magnesium is a light metal. It is not attacked by either air or water as it becomes coated with a firmly-adhering, insoluble layer of oxide. This is able to protect the metal from further attack by air or water. However, dilute acids are capable of attacking magnesium.

Which of the following metals does magnesium resemble most?

- Sodium → 31
- Aluminum → 22
- Zinc → 10

2 It is just as well that you did not guess here. You should now memorize the sentence in italics below.

In electrolysis, metals are always deposited on the cathode (the negative electrode).

End of Program 27.

3 Your answer is quite right.

When aluminum reacts with caustic soda the products formed are *sodium aluminate* and *hydrogen.* Initially, the protective layer of oxide is dissolved away.

Magnesium, however, does not react in this way with caustic soda because its protective layer of oxide is *not* attacked.

Magnesium is not found in the free state in nature. Examples of important naturally occurring magnesium compounds include:

$MgCO_3$

$MgCl_2$

$MgSO_4$

Jot down the names of these magnesium compounds, and then continue your reading at 28 .

4 In addition to magnesium, the following metals are also prepared by molten electrolysis:

Sodium
Potassium
Aluminum

For which of these metals is a *salt* used in the molten electrolysis?

- For sodium → 12
- For potassium → 20
- For magnesium → 36
- For aluminum → 29

---

5 In the case of dolomite a *mixed* carbonate is involved. Write down the names of the two carbonates contained in this mineral. Then continue your reading at 15 .

---

6 Magnesium hydroxide: $Mg(OH)_2$

Now write out the equation for the reaction of magnesium with water.

Continue your reading at 37 .

---

7 Even at room temperature sodium reacts very violently with water. In fact, sodium is attacked by the moisture contained in ordinary air.

Because of its great reactivity towards water sodium has to be stored under naphtha.

Now complete the sentences below:

Magnesium reacts with water only if heat is supplied. The products which then form are magnesium .... and ....

Continue your reading at 18 .

27

---

8 Now check your answers:

Magnesium chloride, magnesium sulfate, magnesium nitrate.

In aqueous solution, these salts dissociate into doubly positively charged magnesium ions and negatively charged acid radicals.

Write out the equations for the dissociation of magnesium chloride, magnesium sulfate and magnesium nitrate. → 14

---

9 In fact they are!

Both Al and Mg are attacked by dilute acids!

Continue reading now at 27 .

---

10 Zinc and magnesium certainly resemble one another in their behavior towards water and dilute acids (both metals do not react with water, yet both are attacked by dilute acids). However, magnesium is a *light metal* whereas zinc is a *heavy* metal.

Read through 1 once again.

---

11 *Molten electrolysis* of *magnesium chloride* yields *chlorine* at the *anode* and metallic *magnesium* at the *cathode.*

Magnesium is of great importance to industry. Normally, however, the pure metal itself is not used but rather a variety of magnesium alloys. One commonly used alloying metal is aluminum. Magnesium-aluminum alloys are used in the construction of door and window frames and handles, machine parts, and automobile and aircraft bodies.

Let's now summarize what we have covered so far.

Copy down and complete the sentences below:

Aluminum and magnesium are .... metals having the symbols .... and .... Both of these metals are .... attacked by air or water because they are coated with a protective layer of aluminum/magnesium ....

Now continue your reading at 23 .

---

12 Use is made of sodium hydroxide to produce sodium by means of molten electrolysis. But sodium hydroxide is a base! → 4

13 *Magnesium does not react* with caustic soda whereas *aluminum is attacked* and forms sodium aluminate and hydrogen.

The reason for this is that the protective layer of magnesium oxide is not attacked by alkalis. The protective layer of aluminum oxide is, however, dissolved away by alkalis.

Write out the completed version of the sentence below:

When aluminum reacts with caustic soda, the products formed are .... and ....

Now continue your reading at 3 .

14 Now check your equations:

$$MgCl_2 \rightarrow Mg^{2\oplus} + 2\ Cl^{\ominus}$$

$$MgSO_4 \rightarrow Mg^{2\oplus} + SO_4^{2\ominus}$$

$$Mg(NO_3)_2 \rightarrow Mg^{2\oplus} + 2\ NO_3^{\ominus}$$

Unlike aluminum, magnesium is *not attacked by alkalis*. The reason for this is that the protective layer of magnesium oxide does not dissolve in alkalis, whereas the protective layer of aluminum oxide does dissolve in alkalis and is thereby removed from the surface of the metal. Please take careful note of this important difference between these two light metals.

Magnesium is not found in the free state in nature.

Some important examples of naturally occurring magnesium compounds are:

Magnesium carbonate
Magnesium sulfate
Magnesium chloride

Copy down the names of these three magnesium compounds and write alongside each its chemical formula.

Now continue your reading at 33 .

15 Your answer is quite right.

Dolomite has the chemical composition $MgCO_3 \cdot CaCO_3$, which implies that for each $MgCO_3$ there is (roughly) one $CaCO_3$.

At this point we start our discussion on the production and uses of magnesium metal.

Industrially, magnesium metal is produced from the *molten electrolysis of magnesium chloride.*

You may recall that certain other metals are also produced by means of molten electrolysis. What metals are referred to here?

Write down the names of these metals, and then check your answers at [4] .

16 Dolomite is a mixed carbonate consisting of magnesium and calcium carbonates, namely $MgCO_3 \cdot CaCO_3$.

Which magnesium compound is used in the production of the free metal?

Write down its name and then proceed to [24] .

17 Your answer is quite right.

In electrolysis, metals are always deposited on the cathode (the negative electrode).

End of Program 27.

18 Magnesium is attacked by water only if heat is applied. The products then formed are magnesium *hydroxide* and *hydrogen.*

By contrast, sodium reacts violently with water even at low temperatures. Because of its great reactivity towards water sodium has to be stored under naphtha.

In its reactivity towards acids magnesium behaves very much like aluminum.

Are magnesium and aluminum attacked by dilute acids?

- Yes → [27]

- No → 9
- I don't know → 38

19 Now check your equations and make any corrections which are necessary:

$$Mg + 2\ HCl \rightarrow MgCl_2 + H_2$$

$$Mg + H_2SO_4 \rightarrow MgSO_4 + H_2$$

$$Mg + 2\ HNO_3 \rightarrow Mg(NO_3)_2 + H_2$$

Write down the names of the three magnesium salts formed in the above equations, and then continue reading at 8 .

20 Use is made of potassium hydroxide to produce potassium means of molten electrolysis. But potassium hydroxide is a base! → 4

21 Now check your equation:

$$2\ Mg + O_2 \rightarrow 2\ MgO$$

The observations that magnesium burns with a very intense flame and that a great deal of heat is generated by the reaction, indicate that it combines very readily with oxygen. It may be said that magnesium displays a great *affinity* for oxygen.

Magnesium is very resistant to attack by water. Only after a long heating period is it possible to observe a chemical reaction in damp magnesium powder. The products then formed are magnesium hydroxide and hydrogen.

Jot down the formula for magnesium hydroxide, and then continue your reading at 6 .

22 Your answer is quite right.

Magnesium is very similar to aluminum.

Both of these metals are light metals. The presence of an oxide layer on each protects both from further attack by either air or water. Moreover, both are attacked by dilute acids.

In its compounds magnesium has a valence of *two.*

Write down now the formula for magnesium oxide, and then continue reading at [30] .

23 Now check your answers:

Aluminum and magnesium are *light* metals having the symbols *Al* and *Mg.* *Neither* of these metals is attacked by air or water because they are coated with a protective layer of aluminum/magnesium *oxide.*

Are aluminum and magnesium attacked by caustic soda?

- Neither is attacked by caustic soda → [35]
- Both are attacked by caustic soda → [13]
- Aluminum is attacked by caustic soda whereas magnesium is not → [3]

24 To produce magnesium metal, one starts out from *magnesium chloride.* As in the case of aluminum, the magnesium is produced by means of *molten electrolysis,* though the difference is that with aluminum one starts from the oxide.

On which electrode will the aluminum/magnesium be deposited?

- On the anode (the positive electrode) → [32]
- On the cathode (the negative electrode) → [17]
- I am not sure → [2]

25 In the case of dolomite a *mixed* carbonate is involved. Write down the names of the two carbonates contained in this mineral. Then continue your reading at [15] .

26 Now check your equation:

$$2\ Na + 2\ H_2O \rightarrow 2\ NaOH + H_2$$

How does magnesium compare to sodium in terms of its reactivity towards water?

- Magnesium is much less reactive → 18
- Magnesium is more reactive → 7
- Both metals are equally reactive towards water → 34

---

27 Your answer is quite correct.

Mg and Al are both attacked by dilute acids.

When dilute acids act on magnesium, the corresponding magnesium salt and hydrogen are formed.

Write out now the three equations for the reaction of magnesium with:

1. Dilute hydrochloric acid
2. Dilute sulfuric acid
3. Dilute nitric acid.

Then continue your reading at 19 .

---

28 Now check your answers:

$MgCO_3$ magnesium carbonate

$MgCl_2$ magnesium chloride

$MgSO_4$ magnesium sulfate

Another important naturally occurring compound of magnesium is known as dolomite. Write down now the components of dolomite.

Continue your reading at 16 .

---

29 Use is made of aluminum oxide to produce aluminum by means of molten electrolysis. But oxides are not salts! → 4

---

30 Magnesium oxide: MgO

27

In air, magnesium burns with an exceptionally bright white light. The flash light bulbs used by photographers contain thin magnesium ribbon. After the ribbon has been burnt, a white deposit of magnesium oxide is to be seen on the inside of the bulb.

Write out now the equation for the combustion of magnesium, and then check your answer at [21] .

---

[31] Although magnesium and sodium are both light metals, *they differ markedly in their behavior towards water.* In air, magnesium becomes coated with a layer of oxide which is not attacked by water. Accordingly, magnesium does not react with water. By contrast, sodium reacts with water very violently. In fact, it is necessary to protect sodium from even the moisture in air. This is done by storing the sodium under naphtha.

Read through [1] once again.

---

[32] Your answer is wrong.

In electrolysis, metals are always deposited on the cathode (the negative electrode).

Please memorize this sentence.

End of Program 27.

---

[33] Now check your answers:

Magnesium carbonate	$MgCO_3$
Magnesium sulfate	$MgSO_4$
Magnesium chloride	$MgCl_2$

Magnesium carbonate is usually found in nature together with calcium carbonate in the form of a mixed carbonate known as *dolomite.* Whole mountain ranges, such as the Dolomites in Italy, are known to consist of this particular mineral.

What is the chemical composition of dolomite?

- $MgCO_3$ → [5]
- $CaCO_3$ → [25]
- $CaCO_3 \cdot MgCO_3$ → [15]

34 Even at room temperature sodium reacts very violently with water. In fact, the moisture contained in ordinary air attacks sodium metal.

Because of its great reactivity towards water sodium has to be stored under naphtha.

Now copy down and complete the sentences below:

Magnesium is attacked by water only if heat is applied.

The products then formed are magnesium .... and .... → 18

---

35 *Magnesium is not attacked* by caustic soda. By contrast, *aluminum is attacked* and forms sodium aluminate and hydrogen in the process.

The reason for this is that the protective layer of magnesium oxide does not react with alkalis, whereas the protective layer of aluminum oxide is dissolved away by alkalis.

Now complete the sentence below:

When aluminum reacts with caustic soda, the products formed are .... and ....

Continue your reading at 3 .

---

36 Your answer is quite right.

In the production of metals by means of molten electrolysis the following compounds are used:

For sodium:	sodium hydroxide
For potassium:	potassium hydroxide
For magnesium:	magnesium chloride
For aluminum:	aluminum oxide

Thus, a *salt* is used in the production of magnesium, namely magnesium chloride.

Now write out your answer to the questions below:

What product is formed at the cathode?

What product is formed at the anode? → 11

37 Now check your equation and make any necessary corrections:

$$Mg + 2\ H_2O \xrightarrow{100\ ^\circ C} Mg(OH)_2 + H_2$$

We have already discussed a very similar reaction in the case of sodium. Write out now the equation for the reaction of sodium with water, and then continue your reading at 26.

38 *Like magnesium, aluminum is attacked by dilute acids.*

Continue your reading at 27 .

Program 28

The Periodic Table of the Elements

1 106 elements have so far been discovered; the last three are little known, however, and will not receive further treatment here.

The elements are ordered according to nuclear charge, also known as the atomic number, in the *periodic table.*

The usefulness of this table will become clear in the course of this program.

Now find Table 3 at the end of the book and mark it with a book mark. This shows you an abbreviated version of the periodic table, in which

> the horizontal lines are called *rows* or *periods;*
> the vertical columns are called *groups.*

How many periods and how many groups are there in the periodic table?

Complete the sentence:

The periodic table of the elements consists of .... periods and .... groups. → 13 .

---

2 Yes, silver, cadmium, yttrium, and zirconium are the elements of the first, second, third and fourth sub-groups in the fifth period.

The transition elements are placed between the main group elements in the fourth and fifth period. Furthermore, in the sixth period there is an additional series of 14 elements, the *lanthanides,* within the transition elements. This is shown in the complete periodic table in Table 4 (end of the book). The position of the lanthanides is indicated in the table, but for the sake of clarity, the elements themselves are listed below the table.

The lanthanides have the atomic numbers ....
Complete the sentence. → 14

---

3 Chromium, *Cr,* and copper, *Cu,* have only *one* electron in the outer shell. All the other transition elements in the table have *two* such electrons.

Up to now we generalized:
The number of electrons in the outer shell is responsible for the chemical properties of an atom.

Since the transition elements have similar, but not identical properties, a more precise formulation would be:

The chemical properties of the elements of the transition series are determined not only by the number of electrons in the outer shell, but also by the number of electrons in the partly filled inner shells.

The knowledge that all transition elements have either one or two outer shell electrons allows an important classification of these elements. In order to decide whether they are metals or non-metals, answer the problem:

Main group elements which have one or two outer shell electrons are

- metals → 26
- non-metals → 19

4

The lanthanides have
2 electrons in the 1st shell
8 electrons (up till Gd) in the 2nd shell
18 electrons in the 3rd shell.

Therefore, the following table gives the compositions of the 4th, 5th, and 6th shells only.

Atomic number	Element		4th shell N-shell	5th shell O-shell	6th shell P-shell
58	cerium	Ce	20	8	2
59	praseodymium	Pr	21	8	2
60	neodymium	Nd	22	8	2
61	promethium	Pm	23	8	2
62	samarium	Sm	24	8	2
63	europium	Eu	25	8	2
64	gadolinium	Gd	25	9	2
65	terbium	Tb	27	8	2
66	dysprosium	Dy	28	8	2
67	holmium	Ho	29	8	2
68	erbium	Er	30	8	2
69	thulium	Tm	31	8	2
70	ytterbium	Yb	32	8	2
71	lutetium	Lu	32	9	2

In this series, the fourth shell is filled up to a total of 32 electrons, starting initially with 18 electrons at lanthanum (element Nr. 57).

Continue at [14] .

5 Your knowledge of the way the atoms are built up is still insufficient, and your answer is wrong. You should have a closer look at Table 3, end of the book, before continuing with the present program.

6 The following elements are in the fourth period:

K, Ca, Sc, Ti, V, Cr, Mn, Fe, Co, Ni, Cu, Zn, Ga, Ge, As, Se, Br, Kr.

The ten transition elements, Sc to Zn, can be divided into eight groups due to differing chemical properties, just as is the case for the main group elements. Doing this, groups 1 to 7 receive one transition element each, and the remaining three go into group 8.

In order to distinguish between main and sub-group elements in the periodic table, the main groups are given the letter a, and the sub-groups the letter b (see Table 4, end of the book).

The fourth period starts with the elements K and Ca, which belong to the first and second main groups, respectively. Scandium belongs to the third sub-group, titan to the fourth sub-group, and so on:

1a	2a	3b	4b
K	Ca	Sc	Ti

Study this with the aid of Table 4 (Program 7) and then answer the following question:

Which elements of the fourth period belong to the

4th sub-group? ....
5th main-group? ....
6th sub-group? ....
7th sub-group? ....

Give the chemical symbols of the elements and designate the groups 4b, 5a, etc. → [18] .

7 Your answer is wrong. Repeat from 17.

---

8 Correct, the maximum number of outer electrons is *eight;* therefore there are eight groups in the periodic table.

There are particular family names for the elements within each group:

First group:	the alkali metals
second group:	the alkaline earth metals
third group:	the boron group
fourth group:	the carbon group
fifth group:	the nitrogen group
sixth group:	the oxygen group
seventh group:	the halogens
eighth group:	the inert or noble gases

Complete the following statements:
The alkali metals have .... outer electrons.
The alkaline earth metals have .... outer electrons.
The halogen have .... outer electrons.
The noble gases have .... outer electrons. → 20

9

1b	Cu
2b	Zn
3a	Ga
4a	Ge

The arrangement of groups in the periodic table may be summarized as follows:

The transition elements ("b-groups") are inserted between the second and third main groups in such a way that the second main group (2a) is followed by the third sub-group (3b). Then follow the fourth to the eighth b-groups, and only then 1b and 2b, as shown below:

1a, 2a, 3b, 4b, 5b, 6b, 7b, 8b, 1b, 2b, 3a, 4a, 5a, 6a, 7a, 8a

The b-groups containing the transition elements are framed for clarity.

Examine the long form of the periodic table in Table 4 (end of the book). Write down the elements in the 4th, 5th and 6th periods which belong to the first main group and the first sub-group, respectively → 33

10 The periodic table was discovered in 1869 by D. J. Mendeleeff and L. Meyer, who were working independently.

This concludes Program 28.

11 Your number is too low. Examine Table 3 again to see how many groups there are in the periodic table. For the elements considered so far it holds that

first group elements have *one* outer electron,
second group elements have *two* outer electrons,
third group elements have *three* outer electrons,
etc.

Thus, elements of the eighth group will have the highest number of outer electrons.

Repeat 17 .

12 Wrong. Repeat from 17 .

13 The periodic table of the elements consists of

seven periods and
eight groups.

Look up the elements carbon, sulfur, and potassium in the table:

carbon, C, in the second period,
sulfur, S, in the third period,
potassium, K, in the fourth period.

In Table 3 you will see that the atomic number is given beside each element.

Write down the atomic numbers of C, S, and K.

Remember the way the atoms are built up, and then answer the following question with the aid of Table 1 at the end of the book.

Which electron shells contain the *outer* electrons in C, S, and K, respectively?

C	S	K			
• first	second	third	shell	→	28
• second	second	third	shell	→	5
• second	third	fourth	shell	→	22
• fourth	sixth	first	shell	→	32

---

14 The lanthanides are the elements with atomic numbers 58 to 71. The name "lanthanide" comes from element Nr. 57, lanthanum. Like lanthanum, the lanthanides all have two electrons in the outer shell (the P-shell). There can be either 8 or 9 electrons in the next inner shell – the O-shell. The main differences between the lanthanides are in the N-shell, the population of which increases from 18 to 32 electrons through this series. A table showing this is given under 4 .
Since the elements 58 to 71 are electronically almost identical with lanthanum except for the N-shell, the chemical properties strongly resemble those of lanthanum.
In the same way that the lanthanides follow lanthanum in the sixth period, the actinides follow actinium in the seventh period.

Examine Table 4 (end of the book) and write down the atomic numbers of the actinides. → 86

---

15 In the series of transition elements of atomic number 21 to 30 each new electron is added to the M-shell. This is shown in the following table:

		Number of electrons				
	Atomic number	K-shell	L-shell	M-shell	N-shell	
Ca	20	2	8	8	2	Element of main group 2
Sc	21	2	8	9	2	
Ti	22	2	8	10	2	
V	23	2	8	11	2	
Cr	24	2	8	13	1	sub-group elements
Mn	25	2	8	13	2	
Fe	26	2	8	14	2	
Co	27	2	8	15	2	
Ni	28	2	8	16	2	
Cu	29	2	8	18	1	
Zn	30	2	8	18	2	
Ga	31	2	8	18	3	main group element

You will see that there are two jumps in the M-shell in this table, i.e. in two places the M-shell receives an additional electron at the expense of the N-shell. Consequently, these two elements have only *one* outer-shell electron.

Find these two elements. → [3] .

---

[16] In your answer you gave the period-five elements of groups 1a to 4a, which are *main groups*. However, the *sub-group* elements were required. These elements stand between strontium (Sr) and indium (In). You will see that their atomic numbers are 39 to 48.

The sub-group or transition elements are shown hatched in Table 4 for clarity.

Try [33] again.

---

[17] The electron shell containing the outer electrons determines the period in which the element should be placed.

The *number* of outer-shell electrons determines the *group* in which the element should be placed.

One outer electron: first group,
two outer electrons: second group,
three outer electrons: third group,
four outer electrons: fourth group.

From this rule determine the maximum number of outer electrons possible for any element.

- Maximum number 4 → [11]
- Maximum number 8 → [8]
- Maximum number 16 → [21]
- Maximum number 18 → [34]
- Maximum number 32 → [29]

18

Compare:

sub-group 4 = 4b Ti
main group 5 = 5a As
sub-group 6 = 6b Cr
sub-group 8 = 8b Fe, Co, Ni

In the periodic table in the long form, sub-groups 1b and 2b follow sub-group 8b.

Now note down the elements of the fourth period which belong to

sub-group 1 ....
sub-group 2 ....
main group 3 ....
main group 4 ....

Use Table 4 (end of the book) for help, and indicate the group number as well as the symbols of the elements. → 9

19

Elements possessing one or two outer-shell electrons belong to the first and second groups in the short periodic system (first and second main groups in the long form).

Examine the short periodic table once more (Table 3) and try to determine whether the elements of the first and second groups are

metals or non-metals? → 26

20

Compare:

Alkali metals have *one* outer electron;
alkaline earth metals have *two* outer electrons;
halogens have *seven* outer electrons;
noble gases have *eight* outer electrons.

What are groups 4,5, and 6 called? Use Table 3 (end of the book) for help.

The third group is the boron group
the fourth group is the .... group
the fifth group is the .... group
the sixth group is the .... group. → 30

21 Wrong.

You should alrcady know that:

one	outer electron	= group 1
two	outer electrons	= group 2
three	outer electrons	= group 3

Conversely:

Group 1 = 1 outer electron
group 2 = 2 outer electrons
group 3 = 3 outer electrons

With the aid of Table 3, answer 17 once more.

22 Correct. The outer electrons of

carbon are in the *second* shell; of
sulfur in the *third* shell; and of
potassium in the *fourth* shell.

Check the periodic system once more:

carbon is in the *second* period;
sulfur is in the *third* period;
potassium is in the *fourth* period.

Thus, the *period* to which an element belongs is determined by the shell containing the outer electrons.

Since there are *seven shells,* there are *seven periods.*

In other words, the number of the period equals the number of the shell.

The electron shells are designated by letters and note the following

First shell	= K-shell
second shell	= L-shell
third shell	= M-shell
fourth shell	= N-shell
fifth shell	= O-shell
sixth shell	= P-shell
seventh shell	= Q-shell

→ 37

---

23 Correct; the outer electron shell determines the period of the element.

Now, what determines the *group* in which a main group element is placed?

- The number of outer electrons → 39
- The outer electron shell → 12
- The number of protons → 7

---

24 It is correct that HF is an acid, but CH_4 is not normally called a hydride since it doesn't dissociate.

Try 63 again.

---

25 The elements whose outer electrons are in the P-shell belong to the *sixth* period.

The period of each element is determined by the electron shell containing the outer ....

Complete and continue at 17 .

---

26 Yes, all elements with one or two electrons in the outer shell are *metals*.

This knowledge, together with the fact that the number of outer-shell electrons determines the chemical properties makes it clear why all transition elements are metals.

To start our discussion of the transition metals, we commence with the fourth period of the periodic table.

The ten transition elements in the period are inserted between calcium and gallium in the order of increasing atomic number. Examine Table 4 at the end of the book and write out the entire fourth period. → 6

---

27 The electronegativity of the elements increases from the left to the right through the periods (i.e. increasing atomic number), and from the bottom to the top of each group (i.e. decreasing atomic number). The same is *not* true

of the metallic character of the elements or the basicity of their hydroxides.
Repeat from → 39

28 Your knowledge of the building-up of the atoms is still insufficient. Your answer is wrong. Have a closer look at Table 3 (end of the book) before continuing with the present program.

29 The number 32 has another significance: it is the maximum number of electrons in the N-shell (2 · 4 · 4). However, the question was concerned with the maximum number of *outer electrons*. Examine Table 3 (end of the book) once more in order to determine the number of groups in the periodic table. For the elements discussed so far, it holds that

elements of group 1 have one outer electron;
elements of group 2 have two outer electrons;
elements of group 3 have three outer electrons;
etc.

Hence, elements of group 8 have the highest number of outer electrons.

Repeat 17 .

30 Third group: boron group
fourth group: carbon group
fifth group: nitrogen group
sixth group: oxygen group

Remember that the noble gases have eight outer-shell electrons. The following is a general rule:

An element never has more than eight outer electrons.

How is this possible?

The increasing number of electrons in the shells seems to contradict the rule; as we already know, there is room for

2 electrons in the K-shell,
8 electrons in the L-shell,
18 electrons in the M-shell, and
32 electrons in the N-shell.

In order to understand this, we will discuss some elements individually.

What is the name of the element with atomic number 9, and how many electrons does it have in the K- and L-shells?

Element 9: .... (name)

K-shell: .... electrons
L-shell: .... electrons → [40]

---

[31] Wrong. Repeat from [39] .

---

[32] Your knowledge of the building-up principle of the atoms is insufficient, and your answer is wrong. You should have a closer look at Table 3 (end of the book) before going on.

---

[33] Compare:

1a	1b
K	Cu
Rb	Ag
Cs	Au

If you had any mistakes, study the long periodic system (Table 4, end of the book) once more before continuing.

The comparison of groups 1a and 1b is intended to illustrate a significant difference between main group and transition elements. Although all the transition elements are metals, their chemical properties are quite different from those of the main group metals. Thus, the alkali metals (group 1a) are very reactive, whereas two of the coinage metals, namely the noble metals silver and gold (group 1b) are very unreactive. More details of the chemistry and use of transition metals will be given in Program 29.

Let us practice the periodic table of the elements again:

Four rows of elements from the fifth period are given below. Select the row which contains elements from sub-groups 1, 2, 3, and 4.

- Rb, Sr, In, Sn → [16]
- Ag, Cd, Y, Zr → [2]
- Rb, Sr, Y, Zr → [56]
- Ag, Cd, In, Sn → [96]

---

34 The M-shell (third shell) can accept a total of $2 \cdot 3 \cdot 3 = 18$ electrons. The question was a different one, namely the maximum number of *outer* electrons in any element. In Table 3 you will see how many groups there are in the periodic table. So far, we have seen that

group 1 elements have *one* outer electron
group 2 elements have *two* outer electrons
group 3 elements have *three* outer electrons.

Seeing that there are only eight groups, you should be able to answer [17] .

35 There is room for another 10 electrons in the M-shell. The total

occupancy is:	$2 \cdot 3 \cdot 3 =$ 18 electrons.
With element 18 there are:	8 electrons.
There is still room for:	10 electrons.

The M-shell is being filled up to a total of 18 electrons as we go from element 21 to element 30. These elements are not shown in the short periodic table (Table 3, Program 7) as they are transition elements. Therefore, ^{31}Ga follows ^{20}Ca in the short table.

From the way the electron shells are filled up we can distinguish two major families of elements:

a) Those where each new element contains one more *outer-shell* electron than its predecessor; these are

the *main group elements.*

b) Those in which an *inner,* incompleteley filled shell contains one more electron than in the preceding element; these are the

sub-group elements.

Complete the following statements:

Any *main group element* has one electron more than the element of preceding atomic number in the .... electron shell.

Any *sub-group element* has one electron more than the element of preceding atomic number in .... electron shell. → [48] .

36 It is true that NaH is a hydride, but NH_3 is not an acid. Therefore, repeat 63 .

37 Compare:

First shell	= K-shell
second shell	= L-shell
third shell	= M-shell
fourth shell	= N-shell
fifth shell	= O-shell
sixth shell	= P-shell
seventh shell	= Q-shell

There are seven and only seven electron shells, numbered K to Q as shown.

Any element whose outer electron is in the K-shell belongs to the first period.

When the outer electron is in the L-shell, the element belongs to the second period.

When the outer electron is in the M-shell, the element belongs to the third period, and so on.

Complete:
When the outer electron is in the P-shell, the element belongs to the .... period. → 25 .

38 Your answer is wrong. Repeat from 35 .

39 Correct, the number of electrons in the outer shell determines the group in the periodic table to which a main group element belongs.

Before continuing with the sub-group elements, let us look at the periodicities of chemical and physical properties of main group elements which can be derived from the periodic table.

First of all, we can distinguish metals from non-metals: *Metals* have metallic lustre and conduct electric current. Under normal conditions they are solid (an exception is the sub-group element mercury, which is a liquid).

Non-metals can be solid, liquid or gaseous, and they are poor conductors of electricity.

Many elements are on the borderline between metals and non-metals and are thus moderate conductors.

The following rule applies to the short periodic table (Table 3):

> The metallic character of the elements decreases from the left to the right within each period. Conversely, the non-metallic character increases.

In the second period the elements lithium and beryllium are metals; boron, carbon, oxygen, fluorine, and neon are non-metals.

Now write down the metals in the third period. → 49

40

Element 9 = *fluorine, F*
K-shell: = *two* electrons
L-shell: *seven* electrons.

Now write down the name and electron distribution of element 18:

element 18 = ... (name)
K-shell: = ... electrons
L-shell: = ... electrons
M-shell: = ... electrons → 50

41

You haven't yet fully mastered the use of the periodic table. Repeat from 95 .

42

Your answer is wrong.

Repeat this program carefully from 39 .

43

Your answer is wrong. Repeat from 22 .

44

H_2O: water
H_2S: hydrogen sulfide

The acidic character increases markedly:

H_2O is neutral.
H_2S is a weak acid.

The corresponding hydrides of selenium and tellurium are stronger acids than H_2S.

The increasing acidity of hydrides within periods and groups can be illustrated by arrows as follows:

Periods →
Increasing acidity of hydrides ↓ groups

Consider the three acids

H_2S, HCl and HBr. Which is the strongest?

- Hydrogen sulfide, H_2S → 83
- Hydrogen chloride, HCl → 76
- Hydrogen bromide, HBr → 63

---

45 Correct, element 19, potassium, has the following electron distribution:

K-shell	L-shell	M-shell	N-shell	
2	8	8	1	electrons

Note:
Even though the M-shell for argon contains only 8 electrons, the filling of the N-shell has already commenced at potassium.

In general:

> As soon as an electron shell reaches a population of 8 electrons, the filling of the next shell commences.

Thus, the incompletely filled M-shell in potassium becomes an inner shell.

There is room for how many more electrons in the M-shell?

Write down the number. → 35

---

46 No, antimony, Sb, is not the most typical metal in group 5, although it does have metallic lustre and conducts electricity. The metallic character increases downward in the group.

So which element stands at the boottom of group 5? → 68

47 Wrong. Repeat from 35 .

48 Compare:

main group elements: *outer* shell
sub-group elements: *inner* shell

In the short version of the periodic table only the main group elements are shown. All other elements belong to the sub-groups.

Look for the atomic numbers of the elements in the fourth, fifth, and sixth periods in the short periodic table, and you will see which elements are missing. These are the sub-group elements. Which are they?

In the fourth period, the elements 21 to 30 are missing.
In the fifth period, thc elements .... are missing.
In the sixth period, the elements .... are missing.

Complete the last two sentences. → 57

49 The metals in the third period are:

sodium, magnesium, and aluminum.

The third period is a good example of the gradual change from metals to non-metals.

The metallic character of the elements increases to the left from silicon; the non-metallic character to the right:

metals ← Si → non-metals.

Silicon itself is on the borderline. It looks metallic, conducts electricity, but behaves chemically as a non-metal.

Na	Mg	Al	Si	P	S	Cl	Ar
←		solids at room temperature			→	← gases	→
←	metallic lustre		→				
← good conductors	→	← conductors →		←	insulators (non-conductors)		→

The compound to the far right in the third period, argon, is a noble gas, a complete non-metal. Chlorine, too, is gaseous and non-metallic. Although sulfur and phosphoros are solid crystalline materials at room temperature, they are non-metals.

A further rule concerning the metallic character:

> The metallic character of the elements increases from the top to the bottom within each group.

Look up group 4 in Table 3 (end of the book) and write down the metals in this group. → [59]

---

[50] Element 18: *argon, Ar.*
K-shell: *two* electrons
L-shell: *eight* electrons
M-shell: *eight* electrons

How many electrons are there in each shell of element 19, potassium?

	K-	L-	M-	N-shell			
•	2	8	9	0	electrons	→	[70]
•	2	8	8	1	electrons	→	[45]
•	2	8	7	2	electrons	→	[60]
•	2	8	6	3	electrons	→	[65]

---

[51] HBr, hydrogen bromide, is an acid.
CaH_2, calcium hydride, is, as the name suggests, a hydride.

Let us now continue the description of the periodic table and see how the transition elements (sub-group elements) fit in.

As we have seen, ten elements are "missing" between Ca and Ga in the fourth period.

The electron distribution in calcium and gallium is shown below:

	Atomic number	Number of electrons			
		K-shell	L-shell	M-shell	N-shell
Ca	20	2	8	8	2
Ga	31	2	8	18	3

As we go through the transition elements 21 to 30, which electron shell accepts each additional electron? → 15 .

52 You have not yet grasped the full significance of the periodic system. Repeat from 95 .

53 $Al(OH)_3$ is amphoteric (the emphasis is on the e in amphoteric).

$Si(OH)_4$ is not a base at all, but rather an acid. Therefore, it is better written as H_4SiO_4. When one molecule of water is lost from this compound, we obtain silicic acid, H_2SiO_3.

The tendency of the elements of the third period to form oxy-acids increases in the following sequence:

P → H_3PO_4 (phosphoric acid; medium strength)
S → H_2SO_4 (sulfuric acid; strong)
Cl → $HClO_4$ (perchloric acid; strong).

We have now learned about three properties which change in a systematic manner within the periodic table of the elements:

1. the metallic character of the elements,
2. the electronegativity of the elements,
3. the basicity of the hydroxides of the elements.

How do these properties change on going from the left to the right within the periods (i.e. increasing atomic number), and from the bottom to the top within the groups (i.e. decreasing atomic number)?

1.	2.	3.		
● increases	increases	increases	→	27
● decreases	decreases	decreases	→	74
● decreases	increases	decreases	→	64
● increases	decreases	increases	→	42

---

54 The first four elements in group 7 combine with hydrogen fo form the acids:

HF,	hydrogen fluoride
HCl,	hydrogen chloride
HBr,	hydrogen bromide
HI,	hydrogen iodide

These acids have different strengths. Hydrogen chloride is already a strong acid, but HI is stronger still.

The following rule is general:

The acid strength of the hydrogen compounds increases from the top to the bottom within a group.

You will notice the validity of this rule applied to the first two elements in group 6.

Write down the formulas and names of the hydrogen compounds formed by the first two elements in group 6. → 44

---

55 Your answer is wrong. Repeat from 22 .

---

56 You selected the row containing the first four elements in period 5. However, the first two of these are main group elements.

The transition elements are shown in the hatched area of the long periodic table (Table 4).

Now answer the question at 33 once more.

---

57 The following elements are missing in the short periodic table:
elements Nos. 21 – 30 in period 4
elements Nos. 39 – 48 in period 5
elements Nos. 57 – 80 in period 6

or, in other words,

10 elements in period 4,
10 elements in period 5, and
24 elements in period 6.

The seventh period is also incomplete as given in the short periodic table. However, in this case, the incompleteness is due in part to the fact that the missing main group elements (Nos. 104 – 109) are mostly unknown. Twenty elements are known with certainty in the seventh period. Two of these are main group elements, and the rest belong to the sub-groups. (Elements Nos. 104 – 106 are not shown in the tables.)

To repeat:

The elements can be classified into two major groups:

a) the elements of the main groups, and
b) the elements of the sub-groups.

What is the essential difference between these two classes of elements?

- Different atomic masses. → 47
- Number of neutrons. → 73
- The position of the last valence electron within the electron shells → 67
- There is no essential difference; the classification is arbitrary → 38

58 Most of the elements which form cations have one, two, or three electrons in the outer electron shell. The elements are transformed into ions by the release of these electrons. The resulting ions will then have eight electrons in the outer shell (octet shell, or noble gas configuration).

In the periodic table of the elements we have:

elements with *one* outer electron in group 1,
elements with *two* outer electrons in group 2,
elements with *three* outer electrons in group 3.

Look up the elements of the first three groups in the shortened periodic system in Table 3 (end of the book).

Are these elements mainly metals or non-metals?

- Metals → 81
- Non-metals → 31

---

59 The group 4 metals are *lead* (Pb) and *tin* (Sn). Germanium (Ge) is also considered a metal although in some of its chemical properties it behaves more like a non-metal.

Starting with the non-metal carbon in group 4, we see that the metallic character increases through silicon and germanium to the typical metals tin and lead.

A similar change from non-metals to metals is also clearly observable in group 5:

Nitrogen,	N	gaseous		non-conductors
Phosphorus,	P	solids	increasing metallic lustre ↓	
Arsenic,	As			
Antimony,	Sb			conductors of electricity
Bismuth,	Bi			

From these data, which element of group 5 do you expect to be the most metallic?

- Bi → 68
- As → 75
- Sb → 46

---

60 The filling of the N-shell cannot begin until the M-shell has eight electrons. Therefore, your answer is wrong.

Repeat the question at 50 .

---

61 Your answer is wrong. Read the detailed explanation at 87 .

---

62 Compare:

$Al(OH)_3 + 3\ HCl \rightarrow AlCl_3 + 3\ H_2O$

Aluminum hydroxide reacts as a base, forming a salt of the cation $Al^{3\oplus}$.

In contrast, aluminum hydroxide behaves as an acid in the reaction with the strong base, sodium hydroxide:

$NaOH + Al(OH)_3 \rightarrow Na[Al(OH)_4]$

Aluminum hydroxide dissolves in sodium hydroxide solution forming a salt of the complex anion $[Al(OH)_4]^{\ominus}$ in which the central atom is aluminum.

We repeat the sequence of the hydroxides just considered:

NaOH:	strong base
$Mg(OH)_2$:	weak base
$Al(OH)_3$:	....

How is the acid-base character of $Al(OH)_3$ described? Write down the word which implies reaction with both acids and bases. → 53

63 Correct; of the elements S, Cl, and Br, bromine is the furthest to the right and toward the bottom in the periodic table. For the same reason, hydrogen bromide is a stronger acid than either HCl or H_2S.

To the left of oxygen we come to nitrogen and then carbon in the periodic table. These two elements form the following compounds with hydrogen:

NH_3 ammonia
CH_4 methane

Ammonia does not dissociate with the formation of protons. Likewise, methane does not dissociate. Methane is the simplest of the organic compounds.

As you will appreciate, the properties of the hydrogen compounds change dramatically within the second period. The same is true of the other periods.

Below are given four rows of two hydrogen compounds each. Find the row which commences with an acid and ends with a hydride.

- NH_3, NaH → 36
- HF, CH_4 → 24
- HBr, CaH_2 → 51
- HI, SiH_4 → 85

---

64 Your answer is correct.

There are many more characteristics which change in a periodic manner within the periodic table. When going toward higher atomic numbers in the periods, and smaller atomic numbers in the groups,

the metallic character of the elements decreases,
the non-metallic character correspondingly increases,
the electronegative character of the elements increases,
the electropositive character correspondingly decreases,
the basicity of the hydroxides decreases, and
the acidity of the oxy-acids increases.

Using these laws, you can answer the question:

Is calcium hydroxide, $Ca(OH)_2$, a stronger base than barium hydroxide, $Ba(OH)_2$?

- Yes → 89
- No → 80

---

65 The filling of the N-shell cannot start until the M-shell has eight electrons. Therefore, your answer is wrong.

Try the question at 50 once more.

---

66 Wrong. Read the detailed explanation at 87 .

---

67 Correct. The position of the last electron determines the classification of the elements:

a) in a main group, the last outer-shell electron;
b) in a sub-group, the last electron in an inner shell.

The periodic table of the elements consists of periods and groups.

Which characteristics determine the period of any one element of the main groups?

- Number of outer shell electrons → 55
- Number of protons → 43
- Number of the outer electron shell → 23

68

Yes, bismuth (Bi) is the last element in group 5. It has the most pronounced metallic character.

The increase in metallic character is recapitulated below:

periods ←

groups ↓ increasing metallic character

We learned about the ions:

cations, positively charged;
anions, negatively charged.

Now which elements form cations more readily?

- Metals → 81
- Non- metals → 88
- I don't know → 58

69

Of all the elements, cesium (Cs) forms a cation most readily.

In other words, cesium readily releases an electron, and it is the most *electropositive* of the elements.

periods ←

groups ↓ increasing electropositivity

Conversely, elements which easily accept electrons are said to be *electronegative*. The electronegativity of the elements increases toward the top within the groups, and toward the right in the periods:

periods →

groups ↑ increasing electronegativity

So which is the most electronegative element?

In answering this question you should ignore group 8 and period 7.

Write down the symbol of the element. → 79

---

70 Your answer is wrong.

It is not possible to have nine electrons in the M-shell as long as the N-shell is empty. As soon as the M-shell contains eight electrons, the filling of the N-shell will commence with the next element.

Try the question at 50 again.

---

71 N_2O_5, nitrogen pentoxide

The highest valences of the elements in their compounds with oxygen are:

in group 1: 1
in group 2: 2
in group 3: 3
in group 4: 4
in group 5: 5
in group 6: 6
in group 7: 7

These *highest* valences *can* apply to the oxygen compounds, but there are many such compounds having lower valences of the elements.

Thus, not only N_2O_5 exists, but also NO (nitric oxide) and NO_2 (nitrogen dioxide). Besides SO_3, also SO_2 (sulfur dioxide) exists, and chlorine forms not only Cl_2O_7, but also Cl_2O (chlorine monoxide) and ClO_2 (chlorine dioxide).

Write down these five new oxides together with the valences of the elements other than oxygen. → 95

---

72 CsOH, cesium hydroxide, is the strongest base of all the hydroxides.

The basicity of the hydroxides of the elements increases downward within each group. However, in group 1 LiOH is still a strong base.

The changing basicity of the hydroxides within a *period* is well illustrated by the hydroxides of the third period. The first four are

$NaOH$
$Mg(OH)_2$
$Al(OH)_3$
$Si(OH)_4$

Now, among these hydroxides select:

a) the strongest base
b) the weakest base

and check your answer at [82] .

---

[73] Wrong. Repeat from [35] .

---

[74] The metallic character of the elements decreases toward the right in the periods (increasing atomic number) and toward the top in the groups (decreasing atomic number). The same is true of the basicity of the hydroxides, but *not* of the electronegativity of the elements. Repeat from → [39]

---

[75] No. Although arsenic has metallic lustre and conducts electricity, it is not the most typical metal in group 5. The metallic character increases downward in the groups.

So which element will it be in group 5? → [68] .

---

[76] Your answer is wrong. Read the detailed explanation at [94] .

---

[77] Compare:

LiH	lithium hydride
BeH_2	beryllium hydride
BH_3	boron hydride (borane)
CH_4	methane
NH_3	ammonia
OH_2 or H_2O	water
FH or HF	hydrogen fluoride

Borane, BH_3, cannot be isolated. It exists as the dimeric molecule B_2H_6, diborane.

The acidity of the hydrogen compounds increases with the increasing atomic number within this period.

Which of the compounds shown above dissociate with the formation of protons? → 91

---

78 a) Hydrogen belongs to the first group but is not a metal.
b) The noble gases form the null-group.

You have now experienced some of the rules and periodicities which apply to the elements, and seen that the principle of the building-up of the atoms is the key to the ordering in the periodic table.

Therefore, you will appreciate the magnitude of the discovery in 1869 by D. J. Mendeleeff and L. Meyer of the principle of the periodic table. These scientists, working independently, had no knowledge of the electronic building-up of the atoms. They ordered the elements in the periodic table on the basis of atomic weights and known chemical properties.

Write down and complete the following sentence:

The periodic table of the elements was discovered in 1869 by .... and .... → 10 .

---

79 Fluorine (F) is the most electronegative element. It readily accepts an electron, thus forming the fluoride anion ($F^{\ominus}$).

The most electropositive element (cesium) is in the lower left hand corner of the periodic table; the most electronegative one (fluorine) in the upper right hand corner. The non-metals are also collected in the upper right hand part of the table, whereas the metals are concentrated in the lower left hand corner and extending upward and to the right.

The borderline between metals and non-metals is found along the elements boron, silicon, arsenic, tellurium, and the radioactive and short-lived astatine (At). These elements are counted among the non-metals.

Check this borderline in Table 3 and write down the bordering elements on the metallic side. → 90

---

80 Correct; barium hydroxide is a stronger base than calcium hydroxide.

Compare the four acids

H_2CO_3, H_2SiO_3, HNO_3, and H_3PO_4.

Which is the weakest of these four acids?

- Carbonic acid, H_2CO_3 → 92
- Silicic acid, H_2SiO_3 → 97
- Nitric acid, HNO_3 → 66
- Phosphoric acid, H_3PO_4 → 61

81

Metals form cations.

The tendency to form cations increases as shown below:

periods ←

groups ↓ tendency to form cations

Which of the elements in the periodic table has the highest tendency to form a positively charged ion? (In answering this question, you should ignore period 7.)

Write down the symbol of the element. → 69

82

Of the four hydroxides mentioned,

NaOH is the strongest and
$Si(OH)_4$ the weakest base.

In more detail,

NaOH is a strong base. It forms salts with acids, e.g. with HCl:

$$NaOH + HCl \rightarrow NaCl + H_2O$$

$Mg(OH)_2$ is a weak base. It also has low solubility. However, it does form salts with acids, e.g. with hydrochloric acid:

$$Mg(OH)_2 + 2\ HCl \rightarrow MgCl_2 + H_2O.$$

$Al(OH)_3$ is *amphoteric* (emphasis on the e). That is, it reacts with acids as well as with bases and it is itself both a base and an acid. The way it will react depends on the reaction partner:

with acids it reacts as a base;
with bases it reacts as an acid.

Write out the equation for the reaction of aluminum hydroxide with hydrochloric acid. → 62

---

83 Your answer is wrong. Read the detailed explanation at 94 .

---

84 The elements in the third period form the oxides:

Na_2O	sodium oxide
MgO	magnesium oxide
Al_2O_3	aluminum oxide
SiO_2	silicon dioxide
P_2O_5	phosphorus pentoxide
SO_3	sulfur trioxide
Cl_2O_7	chlorine heptoxide
	Ar, a noble gas, does not form an oxide.

The valence toward oxygen always increases with the increasing atomic number of the elements within a period. The elements attain their highest valences in their compounds with oxygen.

Which will be the highest valence oxide that the element nitrogen can form? → 71

---

85 As far as the acid is concerned, you are right. HI is an acid. Silicon stands below carbon in group 4 of the periodic table. Like CH_4, SiH_4 should not be described as a hydride: it does not dissociate at all.

Go back to 63 and try again.

---

86 The actinides are the elements of atomic numbers 90–103. Like all the elements in the seventh period, they are radioactive.

We will terminate with a consideration of the transition elements in the context of the periodic table. As with the main group elements, there are certain systematic relationships between the groups of transition elements, even though they are not always as clearly defined.

The elements of the main groups attain their maximum valences in the products of reaction with oxygen; this valence equals the group number. This simple relationship applies also to the transition elements.

From this, which are the maximum valences of chromium (Cr) and manganese (Mn)?

You will be able to answer this question when you find the two elements in the periodic table (end of the book, Table 4). → 93

---

87 The *metal hydroxides* are *bases* and their basicities increase according to the position of the metals downward and to the left in the periodic table. In contrast, the *hydroxides* of the *non-metals* (or their dehydrated forms) are *acids*. The strengths of these acids increase in opposite directions within the periodic table:

periods ←
groups ↓ increasing basicity of metal hydroxides

periods →
groups ↑ increasing acidity of oxy-acids

The question was:

Which of the four acids

H_2CO_3, H_2SiO_3, H_3PO_4, and HNO_3

is the weakest one?

The non-metals from which these acids are derived are C, Si, P, and N. Look up these four elements in the shortened periodic table (Table 3, end of the book). The one which is lowest and farthest to the left in this table will form the weakest oxy-acid.

Which element is it?

- C → 92
- Si → 97
- P → 66
- N → 61

88 The non-metals form anions. Therefore, your answer is wrong. Continue reading at 58 .

89 Your answer is wrong. The basicity increases within a group with the increasing atomic number of the elements. Calcium, Nr. 20, stands above barium, Nr. 56. Hence, calcium hydroxide ($Ca(OH)_2$) is a weaker base than $Ba(OH)_2$.

Repeat from 72

90 Compare:

Beryllium, Be	(2. period)
aluminum, Al	(3. period)
germanium, Ge	(4. period)
antimony, Sb	(5. period)
polonium, Po	(6. period)

Each of these elements has a non-metal as neighbor with the next highest atomic number.

The basicity of the hydroxides changes in the same manner as the electropositive character of the elements:

periods
←
increasing basicity
groups ↓

From this, which metal hydroxide will be the strongest base of all? To answer this question, consider only periods 1 to 6. → 72

91 Hydrogen fluoride dissociates upon dissolution in water into a proton and a fluoride ion:

$$HF \rightleftarrows H^{\oplus} + F^{\ominus}$$

Water, too, dissociates:

$$H_2O \rightleftarrows H^{\oplus} + OH^{\ominus}$$

Thus, both compounds form protons upon dissociation.

In LiH, however, hydrogen is in an anionic form, $H^{\ominus}$, the hydride ion.

As you see, hydrogen can form two types of compounds with the elements of the second period:

in the form of $H^{\ominus}$ with elements to the left, and
in the form of $H^{\oplus}$ with elements to the right in the periodic table.

The elements in the other periods can also form either hydrides or acids.

Which are the acids formed by the first four elements of group 7 in combination with hydrogen?

Write them down. → [54] .

---

[92] Your answer is wrong. Read the detailed explanation at [87] .

---

[93] The maximum valence is *six* for chromium (Cr) and *seven* for manganese (Mn).

You will learn more about this in Program 29.

Since group 8 contains *noble gases* only, and these do not undergo normal chemical reactions, this group is also known as the *null-group*.

The element hydrogen is also exceptional. Although it is placed in group 1, it is not a metal like the other members of the group, but rather a gas.

Which are the two peculiarities which we have learned about here concerning

a) the position of hydrogen in the periodic table;
b) the position of the noble gases in the periodic table?

Answer this question by writing down the appropriate key-words. → [78]

---

[94] The acidity of the hydrogen compounds increases toward the right in the periods and downward in the groups. In order to answer the question correctly, look up the elements S, Cl, and Br in the periodic table (Table 3).

Which of the three elements is closest to the lower right-hand corner of the periodic table?

- Br → 63
- Cl → 52
- S → 41

---

95 Compare:

NO, nitric oxide (N is divalent)
NO_2, nitrogen dioxide (N is tetravalent)
SO_2, sulfur dioxide (S is tetravalent)
Cl_2O, chlorine monoxide (Cl is monovalent)
ClO_2, chlorine dioxide (Cl is tetravalent)

Let us now consider the valences of the elements toward hydrogen.

The valence toward hydrogen increases within each period from group 1 to group 4. From there on, the valence decreases again from group to group:

Group number:	1	2	3	4	5	6	7
Valence toward H:	one	two	three	four	three	two	one.

Write down the formulas for the compounds with hydrogen formed by the elements of the second period and, if you know them, also their names. → 77 .

---

96 You chose the row containing four consecutive elements of period 5. However, you forgot that the last two of these are main group elements.

The transition elements are shown in the hatched area of the long periodic table in Table 4 (see end of the book).

Repeat the question at 33 .

---

97 Yes, silicic acid is the weakest of the four acids shown. Of the elements C, N, Si, and P, silicon is furthest to the left and deepest in the periodic table.

The following elements form strong oxy-acids:

N: nitric acid, HNO_3
S: sulfuric acid, H_2SO_4
Cl: perchloric acid, $HClO_4$.

Fluorine (F) does not form an oxy-acid.

To recapitulate:

1. Almost all the elements form oxy-acids.
2. The maximum valences are reached in the compounds with oxygen.
3. The maximum valence equals the number of the group.

Write down the formulas for the oxides of the elements of period 3 and, if known, also their names. Note: the noble gases do not form oxides. → 84

Program 29

The Transition Elements

1 The building-up principle of the atoms and the position of the transition elements within the periodic table were discussed in Program 28.

In the present program we will start with a description of the *relationships* which can be *derived from the periodic table* of the elements. This will be followed by a summary of the *occurrence, production,* and *use* of the transition elements.

Write down the italicized words and continue then at [13] .

---

2 Tin is one of the components of bronze. However, nickel is not used in making brass. → [12]

---

3 No, *aqua regia* contains hydrochloric acid, but the other component is not phosphoric acid.

Repeat [25] .

---

4 Ag, Cu, Ti, Ni, Pt, Hg, W, and U have uses as pure metals.

Ag and *Cu* are used in electronics and telecommunications due to their good electric conductivity.

Ti and *Ni* are stable toward alkali hydroxides and dilute non-oxidizing acids. For these reasons, they are used in apparatus in the chemical industry.
Since titanium is a light metal, it is also becoming an important component in the building of ships and airplanes.

Pt is a noble metal and mechanically resistant. This makes it a useful metal for laboratory vessels. For example, platinum crucibles can be heated above 1000 °C.

To repeat: which metals are used for/in

a) laboratory equipment
b) electronics and telecommunications
c) chemical apparatus
d) airplane- and shipbuilding.

Write down the symbols of the appropriate metals. → [11]

29

5 In the compounds given below, the metals are tetravalent:

PtO_2 platinum(IV) oxide, or platinum dioxide
RuO_2 ruthenium(IV) oxide
$IrCl_4$ iridium tetrachloride
$PdCl_4$ palladium tetrachloride.

Platinum, ruthenium, iridium, and palladium are tetravalent.

Do you also know the two metals which can be octavalent? If so, write them down. → 15

---

6 $2\ ZnS + 3\ O_2 \rightarrow 2\ ZnO + 2\ SO_2$

SO_2 is used in the manufacture of sulfuric acid.

Iron pyrites, FeS_2, is also used in large quantities in the manufacture of sulfuric acid.

The products of roasting of iron pyrites are shown below. However, the equation is not balanced. Write down the equation completed with the number of moles of each substance.

$$FeS_2 + O_2 \rightarrow Fe_2O_3 + SO_2$$

→ 41

---

7 Silver and gold are noble metals, but copper is not. → 34 .

---

8 The answer is: 83,5% silver.

We will now consider a special application of the transition elements.

The transition elements are used – pure or in alloys – as *catalysts* in technical processes.

Platinum is used as a catalyst for the oxidation of ammonia to NO. In this process, ammonia is burnt over a platinum net at 600 °C:

$$NH_3 + O_2 \rightarrow NO + H_2O$$

Balance this equation. → 37

9

$$MoO_3 + 3\,H_2 \rightarrow Mo + 3\,H_2O$$
$$WO_3 + 3\,H_2 \rightarrow W + 3\,H_2O$$

The third method of reduction of an oxide takes advantage of the ability of certain metals to bind oxygen. Examples are aluminum, sodium, and calcium.

Thus, chromium and manganese are produced by reduction with aluminum.

Write down the equation for the reaction of MnO_2 with Al. → 61

10

HgO	WO_3
Mercury oxide	tungsten oxide

You are now familiar with the valences of the various metals in the metal oxides and know that they correspond to the group numbers. Just as in the case of the main group elements, the transition elements too can occur with other valences in their compounds.

The tendency to change valence in chemical reactions is more pronounced for the transition elements.

Roman numerals in brackets after the metal are used to designate the valences.

Example are:

Cr_2O_3	chromium(III) oxide
MnO_2	manganese(IV) oxide
$CrCl_3$	chromium(III) chloride
$Cr_2(SO_4)_3 \cdot 12\,H_2O$	chromium(III) sulfate with 12 mol of water of crystallization
$MnSO_4$	manganese(II) sulfate.

Write down the corresponding names of HgO, WO_3 and CrO_3. → 48

11

a) Laboratory equipment:	Pt
b) electronics and telecommunications:	Ag, Cu
c) chemical apparatus:	Ni, Ti
d) airplane- and shipbuilding:	Ti

Hg is the only metal which is a liquid at room temperature. Its melting point is −38.8 °C. It is used in the laboratory in thermometers and manometers.

Mercury vapor is poisonous. Spilt mercury must be collected meticulously.

29

W Because of its high melting point (3380 °C) tungsten is used as filaments in electric bulbs.

U is the nuclear fuel used in fission reactors. It undergoes radioactive decay under certain conditions.

To recapitulate, complete the following sentences:

a) metals used for filaments in electric bulbs: ....
b) metals used in thermometers: ....
c) metals used in nuclear reactors: ....
d) metals used in electronics: ....
e) metals used in chemical apparatus: .... → 80

12

Cu, Ag, Au	1. sub-group
Zn	2. sub-group
Ni, Pt	8. sub-group

Copper is used in virtually all alloys except those containing iron.

Copper is the main constituent in the *bronzes* and *brass.*

Bronzes are made from tin or aluminum, brass from zinc together with copper.

Which of these alloys contains a main group element? → 33 .

13

1. Relationships derived from the periodic table.
2. Occurrence.
3. Production.
4. Use.

We will treat these four subjects separately.

First, the *relationships* which can be derived from the periodic table.

It was shown in the previous program that the valences of the main group elements with respect to oxygen increase from group 1 to group 7. This is also true for sub-group elements. Thus, the transition metals in sub-group 1 are monovalent in their oxides; those in sub-group 2 are divalent, etc.

Derive the formulas for the oxides of the following metals:

Cu, Zn, Ti, Cr, Mn.

(You will find the positions of these metals in the periodic table in Table 4 at the end of the book.) → 26

14

Gold dissolves in *aqua regia.*
Silver dissolves in nitric acid.
Platinum dissolves in *aqua regia.*

Do you know from which acids *aqua regia* is prepared?

Agua regia is a mixture of

- concentrated hydrochloric acid and concentrated sulfuric acid → 30
- concentrated nitric acid and concentrated sulfuric acid → 21
- concentrated nitric acid and concentrated phosphoric acid → 3
- concentrated nitric acid and concentrated hydrochloric acid → 36

15

Osmium and ruthenium are octavalent.

The metals iron, cobalt, and nickel have lower valences:

Fe: di- and trivalent
Co: divalent; in complexes trivalent
Ni: divalent.

Write down the empirical formulas for iron(III) sulfate, nickel(II) nitrate, and cobalt(II) chloride. → 34

16

The metals are octavalent in the compounds:

OsO_4, osmium(VIII) oxide (or osmium tetroxide)
RuO_4, ruthenium(VIII) oxide.
OsF_8, osmium octafluoride.

Thus, *osmium* and *ruthenium* are *octavalent.*

29

Below you have a row of compounds from which you should select those containing *tetravalent* metals: $Fe_2(SO_4)_3$, PtO_2, $PdCl_2$, RuF_5, RuO_2, $RhCl_3$, $FeCl_3$, $IrCl_4$, $PdCl_4$, $OsCl_2$.

Check your answer at [5] .

---

[17] Vanadium is in the fifth sub-group.

Therefore, its maximum valence is five. Vanadium changes its valence quite easily, however; di-, tri-, tetra-, and pentavalent compounds are known. This property may well be responsible for the catalytic power of vanadium.

The metals nickel, palladium, and platinum are capable of absorbing large quantities of hydrogen. This hydrogen can be transferred to unsaturated organic compounds; i.e. the metals function as hydrogenation catalysts. Such reactions are important in organic chemistry.

Complete the following statements:

The metals .... are hydrogenation catalysts.
.... are oxidation catalysts. → [65]

---

[18] Maybe you guessed the answer. At any rate, you should examine the periodic table again and locate all the elements which can be alloyed with iron.

You have already written these elements down.

Now answer the question at [45] once more.

---

[19] Aluminum is a component of bronze. However, another metal also exists which forms bronze in combination with copper.

In the case of brass you did not know the metal with which copper is alloyed. → [12]

---

[20] *Sodium amalgam*

The noble metals silver and gold are quite soft, but already the addition of small amounts of copper causes an appreciable increase in hardness. Silver and gold coins usually contain ca. 10% copper. The percentage of copper can be higher in other articles. The *standard* – which is defined numerically below – tells how much silver or gold a particular article contains:

Pure silver and gold have standard	1000.
Gold containing 10% copper has standard	900.
Gold containing 15% copper has standard	850.

The standard is engraved on the articles.

When a silver article has standard 835, what percentage of pure silver does it contain? → 8

21 No, *aqua regia* contains concentrated hydrochloric acid but not sulfuric acid.

Repeat 14 .

22 You have confused the alloying metals used in brass and bronze. → 33

23 The answer ist:

Cu is *di*valent
Au is *tri*valent

How is the situation in sub-group 8?

In fact, some of the elements in sub-group 8 are octavalent, but the maximum valence is usually less. The elements in the eigth sub-group are at least divalent.

Some examples are:
$FeCl_3$, RuO_2, OsO_4, $RhCl_3$, $PdCl_4$, $NiSO_4$, RuO_4, $Co(NO_3)_2$, OsF_8.

From among these compounds select those in which the metal is octavalent.

Compare your list with 16 .

24 Correct. Note that the elements which influence the properties of steel are those which are close to iron in the periodic table.

These elements influence the hardness, elasticity, tenacity, and extensibility.

In order to make steel alloys it is not necessary to admix the pure metals as additives in the melt. Instead, iron ore is mixed with ores containing the additives before being reduced to the metal.

29

For example, molybdenum steel is prepared by reduction of a mixture of iron oxide and molybdenum oxide with coal.

In addition to these, many *iron free alloys* are known. Copper, silver, gold, zinc, nickel, and platinum may be used in such alloys.

In which sub-groups do you find the elements copper, silver, gold, zinc, nickel, and platinum?

Write them down. → 12

---

25 Yes, silver, gold, and platinum are the best known noble metals.

Noble metals are stable toward oxygen and do not corrode.

The transition from "ordinary" to "noble" metals is a gradual one. Therefore, there will also be differences among the "noble" metals themselves.

Generally, within each group in the periodic table *the elements with the higher atomic numbers have the most pronounced noble metal character.*

A distinction between the noble metals can be made on the basis of reaction with nitric acid, which attacks only silver, forming silver nitrate. The reaction equation is given in Program 30.

Gold and platinum do not dissolve in nitric acid, but in a mixture of nitric and hydrochloric acids. This acid mixture is known as *aqua regia* because it dissolves the "royal" metal, gold.

Complete the following sentences:

Gold dissolves in ....
Silver dissolves in ....
Platinum dissolves in .... → 14

---

26 Compare with the series

Cu_2O, ZnO, TiO_2, CrO_3, Mn_2O_7

and correct any mistakes you made.

Now write down the formulas for mercury oxide (mercury = Hg) and tungsten oxide (tungsten = W). → 10

---

27 Compare:

Color	Pigment	Color	Pigment
white	zinc oxide (ZnO)	red	iron oxide (Fe_2O_3)
	titanium dioxide (TiO_2)	green	chromium oxide (Cr_2O_3)
yellow	cadmium sulfide (cadmium yellow) (CdS)	black	iron oxide (iron hydroxide)

The synthetic pigments surpass the naturally occurring ones in brightness and constancy of tone.

This is the end of Program 29.

28 Correct. Silver and gold occur in native form in nature; copper and iron are found as oxides and sulfides.

Dependent on the form in which the metals occur in nature, the following methods can be used for preparing the pure elements:

1. Neat: collect and smelt
2. Sulfides: convert to oxides by roasting
3. Oxides: reduce to metal

First, we will consider the process of roasting by which the sulfide ores are converted to oxides. Roasting implies reaction of the sulfide with oxygen (air) in the heat. The metal sulfide is converted into sulfur dioxide and metal oxide.

"Zinc blende" (zinc sulfide, ZnS) is transformed into zinc oxide (ZnO) by roasting.

Write down the equation for this reaction. → 6

29 Platinum and palladium are noble metals; nickel is not. → 34

30 No, *aqua regia* contains concentrated hydrochloric acid but not sulfuric acid. What is the other component? Repeat 25 .

31 Your answer is wrong. Only one of the metals mentioned, namely titanium, is a member of sub-group 4. Three metals are found in another group.

If you look up the periodic table (end of the book) and locate the elements given, you can easily answer the question. → 47

---

32 The problem was to find the element which is *not* a member of a *sub-group*. Tungsten, however, belongs to sub-group 6. Look for the elements in the periodic table and repeat 66 .

---

33 Bronzes contain copper as the main constituent together with *tin* or *aluminum*.

Bronzes are used in statues and bells.

Brass is used in certain machinery.

Apart from copper, which metals are used in

brass?	*bronze?*		
• Tin	Zinc	→	22
• Nickel	Tin	→	2
• Zinc	Tin	→	44
• Tin	Aluminum	→	19

---

34 Compare:

$Fe_2(SO_4)_3$
$Ni(NO_3)_2$
$CoCl_2$

The noble metals belong to the sub-groups and are found in group 1 as well as group 8. Surely you know some noble metals.

Below you have five rows of metals containing three each. Pick out the row which contains noble metals only.

- Copper, silver, and gold → 7
- Nickel, palladium, and platinum → 29
- Chromium, nickel, and platinum → 52

- Silver, platinum, and gold → 25
- Mercury, gold, and copper → 46

---

35 Maybe you guessed the answer. Look up the periodic table again and find the elements which can form alloys with iron. You have already written these elements down.

Then answer the question at 45 again.

---

36 Your answer is correct.

Due to their great inertness, the noble metals are found *neat* in nature.

The other transition elements are found mostly as *oxides* or *sulfides*.

Write down the three habitual forms in which the transition elements occur in nature:

1. ....
2. ....
3. .... → 49

37

$$4\ NH_3 + 5\ O_2 \xrightarrow{(Pt)} 4\ NO + 6\ H_2O$$

In this reaction platinum acts as an oxidation catalyst. Vanadium pentoxide (V_2O_5) is another oxidation catalyst. SO_2 is oxidized to SO_3 over a V_2O_5 contact catalyst in a technical synthesis of sulfuric acid:

$$2\ SO_2 + O_2 \rightarrow 2\ SO_3$$

Vanadium is in which sub-group? → 17

---

38 None of the metals mentioned are in the third sub-group. Hence, your answer is wrong.

Look up the periodic table and locate the metals in question. Then you will easily solve the problem. → 47

29

39 The question was: which of the elements does *not* belong to a *sub-group?* Nickel is, however, a member of sub-group 8. Locate each element in the periodic table before answering the question at 66 .

40 Cadmium yellow has the formula CdS.

Other pigments are chromium oxide (Cr_2O_3) and iron oxide.

Chromium oxide is green; *iron oxide* (Fe_2O_3) is red, but can exist also in hydrated forms as hydroxides which can be yellow or black.

Make a table of the pigments and their colors mentioned so far. Give the chemical or trivial names of the compounds and, when possible, also the formulas:

Color	Pigment
white	....
yellow	....
red	....
green	....
black	....

→ 27

41 Compare:

$$4\ FeS_2 + 11\ O_2 \rightarrow 2\ Fe_2O_3 + 8\ SO_2.$$

The Fe_2O_3 formed here is then reduced to free iron. Most of the iron production is based on the reduction of oxidic iron ore.

There are many ways of *isolating individual transition elements.* We will consider the four general methods:

1. reduction of an oxide with coal;
2. reduction of an oxide with hydrogen;
3. reduction of an oxide with another metal;
4. electrolysis of a salt solution,

and give examples applied to selected transition elements.

An example of method 1 is the reduction of Fe_2O_3 with coal in blast-furnaces. Iron and carbon monoxide are the reaction products.

Complete the equation:

$Fe_2O_3 + C \rightarrow$

→ 55

---

42 Yes, by treatment with sulfuric acid.

$$ZnO + H_2SO_4 \rightarrow ZnSO_4 + H_2O$$

$ZnSO_4$ is electrolyzed as an aqueous solution.

Having described the methods of isolation, let us now present the *uses of the transition metals.*

The uses can be summarized as follows:

1. use as *pure metal*
2. use as *coating* (plating)
3. use as *alloy*
4. use as *catalyst*
5. use as *pigment.*

Write down these five major applications.

Only a few metals are widely used in the pure state. These are: silver, copper, titanium, nickel, platinum, mercury, tungsten, and uranium.

Write down the chemical symbols of the elements which have applications as native metals. → 4

---

43 Although zinc sulfide (ZnS) is white, it isn't used as a pigment in paints.

Zinc oxide (ZnO), in contrast, is used as a white pigment in paints (Zinc white). → 51

---

44 Correct.

The alloys of *mercury* have a particular name: *amalgams.*

Mercury forms amalgams with many metals. Depending on the mercury content, these amalgams can be liquid or solid. The well-known *silver amalgam* is used for fillings in dentistry. The freshly prepared amalgam is plastic but hardens rapidly.

Write down the name of the amalgam of mercury and sodium. → 20

---

45 Correct. Apart from silicon, all the elements alloyed with iron are transition metals. The alloyed steels take their names from the elements used; e.g. chromium steel is an alloy of iron and chromium. Similarly for other metals (Mn, Ni, Mo, W, V, Co):

Manganese steel, nickel steel, etc.

These alloyed steels have different hardness, elasticity, tenacity, or extensibility, and are used for different purposes.

Locate these alloying components (Cr, Mn, Ni, Mo, W, V, Co) in the periodic table.

Where are these transition elements found in the periodic table?

- Near iron → 24
- Not near iron → 56
- To the left in the periodic table → 18
- To the right in the periodic table → 35

---

46 Only gold is a noble metal. Copper and mercury are not. → 34

---

47 Yes, the hydrogen dissolved in the catalyst is activated and more reactive. It reacts with the compound to be hydrogenated.

The importance of catalysts is increasing in modern technical chemistry. In addition to the metals already considered, Cu, Ag, Ti, Cr, Mo, W, Mn, and Co are also used as catalysts in the form of certain compounds.

Locate the groups in the periodic table to which these metals belong.

In which group are most of these elements found?

- Sub-group 1 → 73
- Sub-group 2 → 78
- Sub-group 3 → 38
- Sub-group 4 → 31

- Sub-group 5 → 58
- Sub-group 6 → 83
- Sub-group 7 → 68
- Sub-group 8 → 63

48

HgO	mercury(II) oxide
WO_3	tungsten(VI) oxide
CrO_3	chromium(VI) oxide

Chromium is in sub-group 6. Therefore, it can be hexavalent, but trivalent chromium is also known.

Manganese is in sub-group 7. Therefore, it can be heptavalent, but di- and tetravalent manganese are also known.

When a metal can have different valences in its compounds, it usually applies that the *lower valences* are found in *cations,* e.g.

$Cr^{3\oplus}$
$Mn^{2\oplus}$

The *higher valences* are found in *complex anions,* e.g.

$CrO_4^{2\ominus}$
$MnO_4^{\ominus}$

Write down salts of the following ions

$Cr^{3\oplus}$
$Mn^{2\oplus}$
$CrO_4^{2\ominus}$
$MnO_4^{\ominus}$

using sulfate ($SO_4^{2\ominus}$) as the counterion in the first two cases, and $Na^{\oplus}$ in the last two. → 59

49

1. native
2. as oxides
3. as sulfides

Native:
silver, gold, and platinum

As oxides:
copper, titanium, chromium, uranium, manganese, and iron

As sulfides:
copper, silver, zinc, mercury, and iron.

It would be too much to describe all the elements in this way. The above listing is intended only to present the more important metals.

Below are given five rows of four metals each. Select the row which contains two metals that are found neat in nature, and two metals which occur both as sulfides and as oxides.

- Cu, Ti, Ag, Pt → 57
- Cu, Ag, Fe, Au → 28
- Au, Pt, Mn, Fe → 64
- U, Zn, Ag, Ti → 69
- Pt, Cu, Cr, Ag → 74

---

50 It is possible but combersome to reduce zinc oxide whith coal to the free metal and then dissolve the metal in acid. Zinc oxide itself dissolves in dilute sulfuric acid. Complete the equation:

$ZnO + H_2SO_4 \rightarrow$ → 42

---

51 Yes, zinc oxide is used as a white pigment (zinc white).

Zinc sulfide is also white but in not used as a pigment.

Cadmium sulfide is an important yellow pigment, known as cadmium yellow. It can be precipitated from the solution of a cadmium salt by the addition of hydrogen sulfide.

Derive the formula for cadmium yellow from this information. → 40

---

52 Although chromium and nickel are used as coatings on other metals, they are not noble metals. Platinum is the only noble metal in the row you picked. → 34

---

53 The treatment of the substance to be hydrogenated with hydrogen alone does not result in any reaction. The hydrogenation starts only upon addition of a finely divided metal catalyst. It is likely that only the molecules in the neighborhood of the catalyst are hydrogenated. Thus, there is a difference in reactivity between pure hydrogen and hydrogen dissolved in a metal.

The metal catalyst increases the reactivity of hydrogen. → 47

---

54 The problem was to find the element which is *not* a member of a *sub-group*. However, cobalt is a member of sub-group 8. Look for the elements in the periodic table once more and then repeat the question at 66 .

---

55 $Fe_2O_3 + 3\ C \rightarrow 2\ Fe + 3\ CO$

This is certainly one of the most important reduction reactions. It was described in detail in Program 23.

In a similar manner zinc is obtained by the reduction of zinc oxide with coal.

Write out the equation for the latter reaction. → 75

---

56 Look up the periodic table and locate each of the elements which can form alloys with iron. You should have them written down already. Then repeat the question at 45 .

---

57 Silver (Ag) and plantinum (Pt) occur neat. Copper (Cu) is found both as an oxide and as a sulfide, but titanium (Ti) occurs only as the oxide.

Go back to 49 and select the row containing two noble metals and two other metals which occur both as oxides and as sulfides.

---

58 None of the metals mentioned are found in sub-group 5, so your answer is wrong.

Use the periodic table to determine the groups to which the elements belong and then answer 47 .

59

Compare with the following and correct any mistakes:

$Cr_2(SO_4)_3$	chromium(III) sulfate
$MnSO_4$	manganese(II) sulfate
Na_2CrO_4	sodium chromate
$NaMnO_4$	sodium permanganate.

Write down the names of these salts also.

Sodium chromate is a salt of *chromic acid.*
Sodium permanganate is a salt of *permanganic acid.*

You will see the analogy to main group elements by comparing with the following compounds of sulfur (group 6) and chlorine (group 7):

Na_2SO_4	sodium sulfate
$NaClO_4$	sodium perchlorate.

Sodium sulfate is a salt of sulfuric acid.
Sodium perchlorate is a salt of perchloric acid.

Now derive the formulas for chromic acid and permanganic acid. → 85

60

Your answer is only partially correct. The main advantage of plating is the protection of the coated metal, mostly iron, against attack by water and oxygen. → 82

61

$$3\ MnO_2 + 4\ Al \rightarrow 3\ Mn + 2\ Al_2O_3$$

Manganese and aluminum oxide are formed.

This reaction is the basis of the thermite or aluminothermic process which was discussed in "Chemistry made easy – a Programmed Course for Self-Instruction I".

We have now seen the following reactions:

1. reduction with coal	→	Fe, Zn
2. reduction with hydrogen	→	W, Mo
3. reduction with aluminum, sodium, or calcium	→	Mn, Cr

A fourth method is the electrolysis of salt solutions. For example, copper and silver are purified electrolytically. The impure metals are dissolved in acid and the resulting salt solutions electrolyzed. Zinc, too, is often prepared by electrolysis of a solution of its salt.

How would you prepare a solution of a zinc salt starting from zinc oxide?

- By treatment with dilute sulfuric acid → 42
- By reduction with coal and then dissolving the metal in sulfuric acid → 72
- I don't know → 50

---

62 Compare:

$TiCl_4 + O_2 \rightarrow TiO_2 + 2\ Cl_2$

We have got to know two white pigments based on zinc and titanium, respectively.

What is the white pigment based on zinc?

- Zinc oxide → 51
- Zinc sulfide → 43

---

63 Of the metals mentioned, only cobalt is in sub-group 8. Three metals belong to another group, so your answer is wrong.

Use the periodic table to determine the groups to which the elements belong and then answer the question at 47 .

---

64 Gold (Au) and platinum (Pt) are found neat in nature. Iron (Fe) occurs both as on oxide and a sulfide, but manganese (Mn) is found as the oxide only.

Try the question at 49 again.

---

65 Hydrogenation catalysts: Ni, Pd, Pt
Oxidation catalysts: Pt, V_2O_5

In order to be active, the hydrogenation catalysts must be finely divided. Such finely divided metals have specific names; in the case of

nickel: Raney nickel
palladium: palladium black
platinum: platinum black.

29

The solubility of hydrogen in palladium is as follows: solid palladium dissolves a 600-fold volume of H_2; finely divided palladium dissolves a 1200-fold volume of H_2.

In practise a hydrogenation is carried out by adding the catalyst to a solution of the compound to be hydrogenated, and then treating the resulting mixture with hydrogen gas, usually under pressure. Part of the hydrogen gas dissolves in the catalyst from where it is transferred to the compound in question during the reaction.

What can you conclude from this?

- The hydrogen dissolved in the catalyst is more reactive than hydrogen gas → 47
- There is no difference in reactivity between gaseous hydrogen and hydrogen dissolved in a catalyst → 53

---

66 Ni, Cr, Zn, and Cd are used as coatings.

Alloys are formed by adding various metals or non-metals to the melt of another metal.

The properties of the pure metal are changed due to the additives in the alloy.

The iron which is used in large quantities in various constructions is not pure. In the form of steel it contains at least 0.5% carbon. The brittle raw iron can contain as much as 4% carbon.

The properties of steel are changed significantly by the addition of *chromium, silicon, manganese, nickel, molybdenum, tungsten, vanadium, or cobalt.* Memorize these *steel additives* and write down their chemical symbols.

One of these elements is not a member of a sub-group. Which one?

- Cr → 84
- Si → 45
- Mn → 77
- W → 32
- Ni → 39
- Co → 54

---

67 The anhydride of chromic acid is CrO_3.

The anhydride of permanganic acid is Mn_2O_7.

To summarize:

S	sulfur	main-group 6.	Na_2SO_4	sodium sulfate
Cr	chromium	sub-group 6.	Na_2CrO_4	sodium chromate
Cl	chlorine	main-group 7.	$NaClO_4$	sodium perchlorate
Mn	manganese	sub-group 7.	$NaMnO_4$	sodium permanganate.

As you see, important compounds of the transition metals have compositions similar to those of main group elements of the same group number.

Locate the elements tugnsten (W) and vanadium (V) in the periodic table (Table 4 at the end of the book). Try and determine the formulas for the oxygen acids of these elements, and then give the formulas for

sodium tungstate and sodium vanadate. → 76

68 Only one of the metals, namely manganese, is in sub-group 7. You will find three metals in another group. Use the periodic table to locate each element, and you will be able to answer 47 .

69 Silver (Ag) occurs in native form, but uranium (U) and titanium (Ti) only as oxides, whereas zinc (Zn) is found mostly as the sulfide.

Go back to 49 and select the row containing two noble metals and two metals which can occur as oxides as well as sulfides.

70 *Compare:*

Hg:	liquid	*thermometers, manometers*
W:	*high melting point*	filaments
Pt:	noble metal mechanically resistant	*crucibles; laboratory equipment*
Ti:	*light metal*	airplane construction
Cu:	*good conductivity*	electronics; telecommunications
Ti and Ni:	stable toward dilute acids and bases	*chemical apparatus*
U:	radioactive decay	atomic reactors

29

Let us consider now the use of the various metals in coating or plating.

Coatings of *zinc, cadmium, chromium* and *nickel* protect the coated metal – mostly iron – against corrosion.

To plate with zinc (i.e. to galvanize), the object is simply dipped into molten zinc. Cadmium coatings are made in the same way. Gutters are often made from galvanized iron sheets. Cadmium-plated metal can be coated once more with chromium by electrolytic deposition, giving a particularly durable skin. Likewise, nickel coatings are applied electrolytically. Nickel plated articles have a characteristic lustre.

Why are iron articles plated with zinc, cadmium, chromium, or nickel?

- To protect the iron against oxygen and water → 82
- In order to obtain a better lustre → 60

---

71 $2\ Zn + O_2 \rightarrow 2\ ZnO$

ZnO is zinc oxide.

Another white pigment of increasing importance is *titanium dioxide.* Paints containing titanium dioxide give a particularly bright white appearance. TiO_2 can be produced in several ways, the most modern of which consists of burning titanium tetrachloride.

Write out the equation for the combustion of titanium tetrachloride. Note: Chlorine is one of the products. → 62

---

72 It is possible but cumbersome to reduce zinc oxide to the free metal with carbon, and then dissolve the metal in acid. Zinc oxide itself dissolves in dilute sulfuric acid. Complete the equation:

$ZnO + H_2SO_4 \rightarrow$ → 42

---

73 Only the metals copper (Cu) and silver (Ag) are found in sub-group 1. There is another group containing three of the metals given. Locate each of the elements with the aid of the periodic table, and then answer the question at 47 .

---

74 Platinum (Pt) and silver (Ag) are found neat in nature. Copper (Cu) occurs both as the oxide and the sulfide, but chromium (Cr) is found as the oxide only.

Go back to 49 and select the row containing two noble metals and two metals which occur as oxides as well as sulfides.

---

75 $ZnO + C \rightarrow Zn + CO$

This reaction is carried out in a shaft furnace at a temperature higher than the melting point of zinc (908 °C). Therefore, the zinc distills off as it is formed and is passed through condensers before being collected as the metal.

Although iron and zinc are both prepared by reduction with carbon, the industrial plants required are quite different.

The reduction of an oxide with coke is the most economical of the reduction methods mentioned.

Which were the four methods of producing metals from their oxides? Write down and complete:

1. Reduction of an oxide with coke
2. Reduction of an oxide with ....
3. Reduction of an oxide with ....
4. Electrolysis of .... → 81

---

76 Sodium tungstate: Na_2WO_4
Sodium vanadate: Na_3VO_4

Tungsten and vanadium are in sub-groups 6 and 5, respectively.

Analogous compounds of the corresponding main group elements are

Na_2SO_4, sodium sulfate, and
Na_3PO_4, sodium phosphate.

Some of the elements of sub-group 1 can have valences higher than the group number. Examples are:

copper and gold.

Typical compounds of these elements are

$CuSO_4$ and $AuCl_3$.

Which valences do copper and gold have in these compounds?

Cu is .... valent
Au is .... valent → 23

---

77 The problem was to find the one element which is *not* a member of a *sub-group*. However, manganese is a member of sub-group 7. Locate each element in the periodic table and repeat the question at 66 .

---

78 None of the metals mentioned are found in sub-group 2, so your answer is wrong.

Use the periodic table to locate each element, and you will easily answer the question at 47 .

---

79 The reaction

anhydride + water → acid

is one way of preparing acids.

In order to derive the formula of an anhydride knowing that of the acid, simply subtract an equivalent of water:

$$H_2SO_4 - H_2O \rightarrow SO_3$$

Thus, SO_3 is the anhydride of sulfuric acid.

When the acid contains only one hydrogen atom, a molecule of water is eliminated from two molecules of the acid

$$2\ HClO_4 = H_2Cl_2O_8$$
$$H_2Cl_2O_8 - H_2O \rightarrow Cl_2O_7$$

Thus, the anhydride of perchloric acid is Cl_2O_7.

Derive the anhydrides of

chromic acid, H_2CrO_4: ....
permanganic acid, $HMnO_4$: .... → 67

80

a) filaments:	W
b) thermometers:	Hg
c) nuclear reactors:	U
d) electronics:	Ag, Cu
e) chemical apparatus:	Ti, Ni

The properties and applications of the elements are summarized below in an incomplete manner.

Complete the table:

Ag:	good conductivity;	electronics and telecommunications
Hg:	liquid at room temperature;	....
W:	....	filaments
Pt:	noble, metal, mechanically resistant;	....
Ti:	....	airplane construction
Cu:	....	electronics and telecommunications
Ti and Ni:	stable toward dilute acids and bases;	....
.............:	radioactive decay;	atomic reactors

→ 70

81 Compare:

1. Reduction of an oxide with coke or carbon.
2. Reduction of an oxide with hydrogen.
3. Reduction of an oxide with a metal.
4. Electrolysis of a salt solution.

In the second method – reduction with hydrogen – hydrogen gas is led through heated tubes containing the oxide.

The reduction with hydrogen can in principle be applied to virtually all oxides. However, it is used only when no other method is economically advantageous.

Tungsten and molybdenum are examples.

Formulate the equations for the reduction of MoO_3 and WO_3 with hydrogen. → 9

82 Correct, the coating protects the iron against corrosion. The mechanical properties of iron remain unchanged.

Point out those of the following metals which are used for plating:

W, Ni, Cu, Cr, Zn, Ti, Cd, Co

There are four in all. Write down their symbols. → 66

---

83 Yes, three of the elements given are in sub-group 6. These are Cr, Mo, and W.

Considering that we have already described most of the elements of group 8 and selected only the most important catalysts, you may appreciate that:

almost all the transition elements in periods 4, 5, and 6 have applications as catalysts.

To terminate, we will describe the use of some transition metal compounds as *pigments*.

Pigments are used in the manufacture of paints, as well as colored synthetics, concrete, and plaster.

As a *white* pigment zinc oxide is commonly used in paints under the name "zinc white". It is produced by the combustion of zinc vapor. Write down the equation for the production of zinc oxide. → 71

---

84 The problem was to find the element which is *not* a member of a *sub-group*. However, chromium is in sub-group 6. Use the periodic table to locate each element, and repeat the question at 66.

85

Chromic acid	H_2CrO_4
permanganic acid	$HMnO_4$

Now write down the formulas for the anhydrides of these acids. → 67

If you have difficulties deriving these formulas, go to 79.

Program 30

The Electrochemical Series

1 In this program we shall be concerned with the principles of oxidation and reduction – processes which can be understood on the basis of the electrochemical series.

Elements and ions are arranged in the electrochemical series according to their oxidation potential.

Oxidation and reduction processes are invariably coupled with each other.

An example is:

$$Fe_2O_3 + 3\ C \rightarrow 2\ Fe + 3\ Co$$

What is being oxidized during this reaction?

- Fe_2O_3? → 12
- C? → 17

---

2 You were a bit too quick with your answer. Iodine was mentioned only in examples where there was no reaction.

Study 21 more attentively.

---

3 This reaction *is* possible:

$$Cu + HgCl_2 \rightarrow Hg + CuCl_2.$$

When a sheet of copper is dipped into a solution of $HgCl_2$, it will become covered by a film of copper amalgam. At the same time some of the copper goes into solution.

Where are Cu and $Hg^{2\oplus}$ in the electrochemical series?

You will see that Cu stands three places above $Hg^{2\oplus}$; i.e. Cu can reduce $Hg^{2\oplus}$. This is the same as saying that $Hg^{2\oplus}$ can oxidize Cu.

Read 10 once more and locate each metal and each cation in the electrochemical series before answering the question.

30

4 Of the four metals given, aluminum stands just below magnesium. Both of these metals are quite good reducing agents, i.e. they readily release an electron.

Repeat 24 .

5 Chlorine *oxidizes* bromide ions to bromine:

$$2\ Br^{\ominus} + Cl_2 \rightarrow Br_2 + 2\ Cl^{\ominus}.$$

Release of an electron implies oxidation.

The reaction between sodium iodide and bromine is quite similar:

$$2\ NaI + Br_2 \rightarrow I_2 + 2\ NaBr.$$

In order to recognize the electron transfer more easily, we can rewrite the above equation in ionic form:

$$2\ Na^{\oplus} + 2\ I^{\ominus} + Br_2 \rightarrow I_2 + 2\ Na^{\oplus} + 2\ Br^{\ominus}.$$

The two sodium ions ($Na^{\oplus}$) remain unchanged. The two negatively charged iodide ions each release an electron to become molecular iodine:

$$2\ I^{\ominus} \rightarrow I_2 + 2\ e^{\ominus}.$$

The molecular bromine (Br_2) accepts the two electrons, thereby forming two bromide ions:

$$Br_2 + 2\ e^{\ominus} \rightarrow 2\ Br^{\ominus}.$$

How is Br_2 acting in this reaction?

- As an oxidizing agent. → 21
- As a reducing agent. → 30

6 The reaction:

Silver nitrate + sodium chloride = sodium nitrate + silver chloride is not a redox reaction: no valences change.

Please look for the redox reaction at [17] .

7 It isn't as simple as that.

Use Table 5 at the end of the book in order to find Al, $Ag^{\oplus}$, $Zn^{2\oplus}$, and Cu in the electrochemical series.

Then solve the problem at [61] .

8 Compare:

H_2	$\leftrightarrows$	$2\,H^{\oplus}$	+	$2\,e^{\ominus}$	$\pm$ 0.00 volt
$2\,I^{\oplus}$	$\leftrightarrows$	I_2	+	$2\,e^{\ominus}$	+ 0.53 volt
$2\,Br^{\ominus}$	$\leftrightarrows$	Br_2	+	$2\,e^{\ominus}$	+ 1.06 volt
$2\,Cl^{\ominus}$	$\leftrightarrows$	Cl_2	+	$2\,e^{\ominus}$	+ 1.36 volt
$2\,F^{\ominus}$	$\leftrightarrows$	F_2	+	$2\,e^{\ominus}$	+ 2.65 volt

Apart from hydrogen and fluorine these non-metals have already been described in this program.

The observation that chlorine is a stronger oxidant than bromine, and bromine is stronger than iodine, reflects the positions of these elements in the electrochemical series.

Which of the four halogens is the strongest oxidant? Write it down. → [56]

9 No; in the electrochemical series you will see:

Zn	$\leftrightarrows$	$Zn^{2\oplus}$	+	$2\,e^{\ominus}$	− 0.76 volt
Cu	$\leftrightarrows$	$Cu^{2\oplus}$	+	$2\,e^{\ominus}$	+ 0.34 volt

Zinc can reduce $Cu^{2\oplus}$, but Cu cannot reduce $Zn^{2\oplus}$.

Consider Al and $Ag^{\oplus}$ in the same manner, and repeat the question at [61] .

10 Hydrogen has a normal redox potential of 0 volt.

The elements above hydrogen have a negative normal redox potential; those below it a positive one.

The electrochemical series allows the prediction wheter or not certain redox reactions will be possible.

30 For the elements below hydrogen we have:

Any element or anion on the left side of the double arrow will reduce a cation or element on the right, provided that the latter are at least one place lower in the electrochemical series.

You already know the following examples:

1. $2\ NaBr + Cl_2 \rightarrow 2\ NaCl + Br_2$

   where the bromide ion ($Br^{\ominus}$) is two places above Cl_2. Here, the bromide ion can reduce chlorine to chloride or, in other words, chlorine oxidizes bromide ion to bromine.

2. $2\ NaI + Br_2 \rightarrow 2\ NaBr + I_2$

   where $I^{\ominus}$ is three places above Br_2.

Consider now the three equations:

1. $Cu + 2\ AgNO_3 \rightarrow 2\ Ag + Cu(NO_3)_2$
2. $Cu + HgCl_2 \rightarrow Hg + CuCl_2$
3. $Hg + 2\ AgNO_3 \rightarrow 2\ Ag + Hg(NO_3)_2$.

Look up the metals and cations for each equation in the electrochemical series (Table 5 at the end of the book) in order to determine whether the reactions are possible. One of the reactions will *not* take place. Which one?

- Reaction 1 will not take place → 58
- Reaction 2 will not take place → 3
- Reaction 3 will not take place → 61

---

11 Although bromine will oxidize iodide ion, it cannot oxidize the other halides, so it is not the strongest oxidant.

Repeat 21 .

12

No, the question was, which element is being oxidized in the reaction

$$Fe_2O_3 + C \rightarrow 2\ Fe + 3\ CO$$

Fe_2O_3 is not being further oxidized, but the reaction $C \rightarrow CO$ is an ....? → 17

13

Correct; only reaction 1 is possible:

$$Al + 3\ AgNO_3 \rightarrow 3\ Ag + Al(NO_3)_3.$$

Write down the essential part of this reaction in ionic form in order to visualize the electron transfer from aluminum to silver. → 22

14

Correct; the reducing agent *releases electrons.*

We shall now consider the different oxidizing powers of the halogens.

First the reaction

$$2\ NaBr + Cl_2 \rightarrow 2\ NaCl + Br_2.$$

When chlorine is introduced into a solution of sodium bromide, the solution turns brown due to the formation of molecular bromine.

Bromide ions are transformed into bromine,

$$2\ Br^{\ominus} \rightarrow Br_2 + 2\ e^{\ominus}$$

while chlorine becomes chloride,

$$Cl_2 + 2\ e^{\ominus} \rightarrow 2\ Cl^{\ominus}$$

So the overall process is

$$2\ Br^{\ominus} + Cl_2 \rightarrow Br_2 + 2\ Cl^{\ominus}$$

Complete the statement:

Bromide ions are .... (oxidized or reduced) to bromine by chlorine. → 5

---

15 Your answer is wrong. Therefore, let us consider the normal potentials in more detail.

Mg	→	$Mg_2^{\oplus} + 2\,e^{\ominus}$	− 2.34 volt
$2\,Cl^{\ominus}$	→	$Cl_2 + 2\,e^{\ominus}$	+ 1.36 volt

Accordingly, magnesium stands above chlorine in the electrochemical series. Magnesium can lose electrons which are taken up by chlorine. Magnesium reduces Cl_2 to $2\,Cl^{\ominus}$.

Consider now the normal potentials of the systems engaged in reaction 2:

Fe	→	$Fe^{2\oplus} + 2\,e^{\ominus}$	− 0.44 volt
H_2	→	$2\,H^{\oplus} + 2\,e^{\ominus}$	± 0.00 volt

Knowing this, will the following reaction be possible or not?

$$Fe + H_2SO_4 \rightarrow FeSO_4 + H_2$$

- Yes → 39
- No → 25

---

16 Iron does not have the lowest reduction potential of the four metals given. Strong reducing agents lose electrons easily; weak ones not so readily.

Study 24 again.

---

17 Yes, C is oxidized to CO in the reaction

$$Fe_2O_3 + 3\,C \rightarrow 2\,Fe + 3\,CO.$$

At the same time Fe_2O_3 is reduced to Fe.

Reduction and oxidation are invariably coupled.

This fact is reflected in the term *redox* reaction formed by the contraction of *red*uction and *ox*idation.

The reaction $Fe_2O_3 + 3\,C \rightarrow 2\,Fe + 3\,CO$ is an example of a *redox* reaction.

Which of the three reactions below is a *redox* reaction?

1. 2 $NaOH + H_2SO_4 \rightarrow Na_2SO_4 + 2\ H_2O$
2. $AgNO_3 + NaCl \rightarrow NaNO_3 + AgCl$
3. $CuO + H_2 \rightarrow Cu + H_2O$.

The redox reaction is shown in

- Equation 1 → 32
- Equation 2 → 6
- Equation 3 → 26

18 *Lithium* is the strongest reducing agent;
fluorine is the strongest oxidizing agent.

The strongest reductant is a metal. The strongest oxidant is a non-metal.

Which non-metals do you find in the electrochemical series? Write down the equilibria and the normal potentials for these elements. → 8

19 No, the reaction between Cu and H_2SO_4 is not possible. Copper is below hydrogen in the electrochemical series and cannot, therefore, reduce $H^{\oplus}$.

Consider now reaction 1, the reaction between zinc and sulfuric acid. Is it possible? → 57

20 No, the reducing agent causes reduction. It releases electrons and is itself oxidized during the process. The oxidizing agent takes up electrons and is thus reduced. → 14

21 Correct; Br_2 acts as an *oxidizing agent.*

Bromine takes up electrons which are being released from iodide ions. The iodide ions are oxidized to iodine by bromine.

Chlorine will also oxidize iodide ions to iodine. All the possible combinations of halide ions with the free halogens (except fluorine) are given below, and in each case it is indicated whether or not a reaction will occur.

$NaCl + Br_2$ no reaction
$NaCl + I_2$ no reaction

$NaBr + Cl_2$ reaction
$NaBr + I_2$ no reaction

$NaI + Cl_2$ reaction
$NaI + Br_2$ reaction

Which of these halogens is the strongest oxidant?

- Iodine → 2
- Bromine → 11
- Chlorine → 31

---

22 $Al + 3\,Ag^{\oplus} \rightarrow Al^{3\oplus} + 3\,Ag$

Aluminum is a much stronger reducing agent than silver. Electrons are transferred from aluminum to $Ag^{\oplus}$, and metallic silver is formed.

Another set of reactions:

1. $Mg + Cl_2 \rightarrow MgCl_2$
2. $Fe + H_2SO_4 \rightarrow FeSO_4 + H_2$

Which of these two reactions will take place?

- Neither → 15
- Reaction 1 → 33
- Reaction 2 → 43
- Both → 39

---

23 Wrong. The reducing agent is being oxidized in the reaction. But MnO_2 is reduced. → 36

---

24 Correct. The equation is

$$Zn + 2\,Ag^{\oplus} \rightarrow Zn^{2\oplus} + 2\,Ag$$

Zinc is a stronger reducing agent than either copper or silver. A longer series of metals with decreasing reducing ability is shown below:

Na
Mg
Al here, sodium is the strongest,
Zn silver the weakest reducing agent.
Fe
Pb
Cu
Ag

Let us consider just four of these elements, Pb, Al, Fe, and Mg. Which of these has the lowest tendency to release electrons?

- Fe → 16
- Al → 4
- Pb → 35
- Mg → 48

25

As you are not yet very confident, repeat in detail from 46 .

26

Yes, the equation

$$CuO + H_2 \rightarrow Cu + H_2O$$

represents a redox reaction in which CuO is reduced and H_2 is oxidized.

Since hydrogen causes reduction, it is referred to as the *reducing agent.* The reduction that it effects is

$$CuO \rightarrow Cu$$

At the same time, H_2 is oxidized to H_2O:

$$H_2 \rightarrow H_2O$$

Any redox reaction starts out with a reducing and an oxidizing agent.

Which is the oxidizing agent in the reaction

$CuO + H_2 \rightarrow Cu + H_2O$? → 36

---

27 No, reaction 1 between Zn and H_2SO_4 is possible.

Zinc stands above hydrogen in the electrochemical series and can reduce 2 $H^{\oplus}$ to H_2.

How is it with reaction 2; Cu + H_2SO_4? Is it possible? → 57

---

28 The following metals are below hydrogen in the electrochemical series:

copper, silver, mercury, platinum, and gold.

None of them are attacked by sulfuric acid since they are unable to reduce $H^{\oplus}$.

As you know, some of these are noble metals. In order to dissolve any of the metals above, it is necessary to use strongly oxidizing acids or acid mixtures. In this case the oxidant will not be $H^{\oplus}$ but other ions or molecules.

The reaction between silver and nitric acid is an example:

$$Ag + 2\ HNO_3 \rightarrow AgNO_3 + H_2O + NO_2$$

Not only silver nitrate, but also H_2O and nitrogen dioxide are formed.

Redox reactions are treated in more detail in Program 31.

This is the end of Program 30.

---

29 Correct; Na is oxidized in the reaction

$$2\ Na + Cl_2 \rightarrow \underbrace{2\ Na^{\oplus} + 2\ Cl^{\ominus}}_{2\ NaCl}$$

An electron is transferred from each sodium atom to chlorine, forming two negative chloride ions and – consequently – two positively charged sodium ions.

Chlorine is the oxidizing agent in this reaction. It takes up the electrons which are released by sodium.

Which of the following statements is correct?

- The reducing agent takes up electrons. → 20
- The reducing agent releases electrons. → 14

30 No, Br_2 accepts two electrons in this reaction. Thus, it is reduced. The oxidizing agent is always itself reduced.

Hence, in the reaction

$$2\ NaI + Br_2 \rightarrow I_2 + 2\ NaBr,$$

bromine is the (oxidant or reductant?) ....

Complete this statement. → 21

31 Of the three halogens Cl_2, Br_2, and I_2, chlorine is the strongest oxidizing agent.

Chlorine oxidizes bromide and iodide ions.

Bromine oxidizes iodide ion only.

(Hence, iodine oxidizes *neither* chloride *nor* bromide ions.)

In other words,

bromine is a stronger oxidant than iodine, and chlorine is stronger than bromine. Chlorine is the strongest of the three.

The decreasing oxidizing potentials of the halogens,

F_2
Cl_2
Br_2
I_2

parallel the order of these elements in the periodic table (Program 28).

Redox reactions can also take place between metals and metal ions.

If, as an example, a zinc rod is placed in a copper sulfate solution, zinc ions will go into solution, while copper ions will become discharged and separate as metallic copper:

$$Zn + Cu^{2\oplus} \rightarrow Zn^{2\oplus} + Cu$$

We can write this equation in two steps for clarity:

$Zn \rightarrow Zn^{2\oplus} + 2\ e^{\ominus}$	zinc releases electrons
$Cu^{2\oplus} + 2\ e^{\ominus} \rightarrow Cu$	copper ions accept electrons

What happens to the $Cu^{2\oplus}$ ions in this reaction?

30

- $Cu^{2\oplus}$ is reduced → 45
- $Cu^{2\oplus}$ is oxidized → 38

---

32 The reaction:

Sodium hydroxide + sulfuric acid = sodium sulfate + water

is a neutralization and not a redox reaction.

Go back to 17 .

---

33 Your answer is only partially correct; reaction 1, the reduction of chlorine by magnesium is possible, but so is reaction 2. Compare:

$$Fe \rightarrow Fe^{2\ominus} + 2\ e^{\ominus}\ -0.44\ \text{volt}$$
$$H_2 \rightarrow 2\ H^{\oplus} + 2\ e^{\ominus}\ \pm 0.00\ \text{volt}$$

As iron is above hydrogen in the electrochemical series, it can reduce $H^{\oplus}$.

Write down the equations for reactions 1 and 2 in ionic form. → 57

---

34 No, $Na^{\oplus}$ isn't oxidized. Study the definition of oxidation and reduction again and then repeat the question at 50 .

---

35 Correct; lead (Pb) has the lowest reduction potential and the lowest tendency to lose electrons, forming $Pb^{2\oplus}$.

The equations for the reduction of lead and iron are:

$$Pb \rightarrow Pb^{2\oplus} + 2\ e^{\ominus}$$
$$Fe \rightarrow Fe^{2\oplus} + 2\ e^{\ominus}$$

Write down the analogous equations for the elements Ca and Zn. → 44

---

36 *CuO* is the oxidizing agent. It oxidizes hydrogen:

$$H_2 \rightarrow H_2O$$

being itself reduced to copper:

$$CuO \rightarrow Cu.$$

The oxidizing agent oxidizes and is itself reduced. The reducing agent reduces and is itself oxidized.

Another example of a redox reaction:

$$3\ MnO_2 + 4\ Al \rightarrow 3\ Mn + 2\ Al_2O_3.$$

Point out the reducing agent!

- MnO_2 → 23
- Al → 51
- Mn → 47
- Al_2O_3 → 42

---

37 Indeed, there will be a reaction taking place.

As you should already know, zinc reduces $Cu^{2\oplus}$:

$$Zn + Cu^{2\oplus} \rightarrow Zn^{2\oplus} + Cu$$

You have also just learned that copper metal reduces silver ions.

$$Cu + 2\ Ag^{\oplus} \rightarrow Cu^{2\oplus} + 2\ Ag$$

Since copper already reduces $Ag^{\oplus}$, Zn will certainly do the same. → 24

---

38 You do not understand the basic interrelationship between oxidation and reduction, and the electron transfer processes taking place here.

Repeat from 51

---

39 Correct. Reactions 1 and 2 are both possible.

Complete the two equations in ionic form:

1. $Mg + Cl_2 \rightarrow$
2. $Fe + 2\ H^{\oplus} + SO_4^{2\ominus} \rightarrow$ → 57

30

40 No, atoms, molecules, or ions which release electrons are being oxidized. Chlorine is not such a molecule: it accepts electrons.

Repeat the question at 50 .

41 No, $e^{\ominus}$ symbolizes an electron.

In the partial reaction

$$Cl_2 + 2\,e^{\ominus} \rightarrow 2\,Cl^{\ominus}$$

two electrons are taken up by chlorine, thereby forming two chloride ions.

Taking up electrons = reduction.

When this is clear, go on to 50

42 Your answer is wrong. Al_2O_3 is not the reducing agent. The reducing agent is one of the starting materials. Which one? → 36

43 Your answer is only partially correct.

Both reactions are possible. You will find the normal potentials in Table 5 (at the end of the book)

$$Mg \rightarrow Mg^{2\oplus} + 2\,e^{\ominus} \quad -2.34 \text{ volt}$$
$$2\,Cl^{\ominus} \rightarrow Cl_2 + 2\,e^{\ominus} \quad +1.36 \text{ volt}$$

Since magnesium stands above chlorine in the electrochemical series, electrons can be transferred from Mg to Cl_2.

Magnesium reduces Cl_2 to $2\,Cl^{\ominus}$.

Complete the two ionic equations:

1. $Mg + Cl_2 \rightarrow$
2. $Fe + 2\,H^{\oplus} + SO_4^{2\ominus} \rightarrow$ → 57

44 Compare and correct any mistakes:

$$Ca \rightarrow Ca^{2\oplus} + 2\,e^{\ominus}$$
$$Zn \rightarrow Zn^{2\oplus} + 2\,e^{\ominus}$$

These two processes are *reversible,* and we will designate them, therefore, with double arrows:

$$Ca \rightleftarrows Ca^{2\oplus} + 2\ e^{\ominus}$$
$$Zn \rightleftarrows Zn^{2\oplus} + 2\ e^{\ominus}$$

Write these equations down.

When the reactions proceed from the left to the right, the metals are acting as reducing agents:

They give up electrons and are oxidized to cations.

If the reaction takes place in the opposite direction, from the right to the left, the cations are acting as oxidizing agents:

They (take up or release) electrons and are (oxidized or reduced) to metals.

Complete this statement by appropriate deletion. → 55

---

45 Correct; copper ions are reduced by zinc.

$$Zn + Cu^{2\oplus} \rightarrow Zn^{2\oplus} + Cu$$

Zinc itself is oxidized to zinc ions in the process.

The electrons released from zinc are taken up by the copper ions. Zinc releases its electrons quite readily.

Comparing zinc and copper, which is the strongest reducing agent?

- Zinc → 59
- Copper → 54

---

46 Yes, the more negative the normal redox potential, the stronger is the reducing ability of the element.

The *electrochemical series* is the arrangement of the elements according to increasing normal redox potential.

The electrochemical series comprising the most important elements is shown in Table 5 at the end of the book.

In this series, the elements or ions *to the left* of the double arrows are the reduced forms. Those to the right of the double arrows are the oxidized forms. The reduced forms are reducing agents. The oxidized forms are oxidizing agents.

The strongest *reducing agent* (reductant) is found to the upper left.

The strongest *oxidizing agent* (oxidant) is found to the lower right.

Write down the

strongest reducing agent: ....
strongest oxidizing agent: .... → 18

---

47 You are wrong. The question was about the reducing agent. This is one of the starting materials, so it must be on the left-hand side of the equation. → 36

---

48 Magnesium stands quite high in this series and is thus a strong reducing agent; it easily loses electrons. However, the problem was to find the element which gives up electrons with the greatest difficulty. Go back to 24 .

---

49 Only reaction 1 is possible. If you did not hesitate in giving this answer, you understand the subject well. Here is the reaction again:

$$Zn + H_2SO_4 \rightarrow ZnSO_4 + H_2$$

or

$$Zn + 2\,H^{\oplus} \rightarrow Zn^{2\oplus} + H_2$$

Zinc metal is oxidized to $Zn^{2\oplus}$ by 2 $H^{\oplus}$. At the same time, the two $H^{\oplus}$ are reduced by Zn to H_2. Zinc is the reducing agent, $H^{\oplus}$ the oxidizing agent.

In contrast, reaction 2, the treatment of copper with sulfuric acid does not proceed at all.

Why?

Copper is below hydrogen in the electrochemical series. Therefore, it cannot reduce $H^{\oplus}$.

Write down all the metals that are below hydrogen in the electrochemical series. → 28

---

50 The chlorine molecule accepts two electrons, thereby being transformed into two chloride ions. This is a reduction.

Where do the electrons come from?
They must come from other atoms, molecules, or ions.

In the reaction

sodium + chlorine → sodium chloride

the sodium atom releases an electron and becomes a positive ion:

$$Na \rightarrow Na^{\oplus} + e^{\ominus}$$

This electron can be taken up by chlorine, thus completing the reaction:

$$2\ Na + Cl_2 \rightarrow \underbrace{2\ Na^{\oplus} + 2\ Cl^{\ominus}}_{2\ NaCl}$$

What is being *oxidized* in this reaction?

- Na → 29
- Cl_2 → 40
- $Na^{\oplus}$ → 34
- $Cl^{\ominus}$ → 60

51 Correct; Al is the reductant. It reduces MnO_2 to Mn and is itself oxidized to Al_2O_3.

In the reactions discussed so far an oxide was always reduced by a free element, and the element was oxidized to a new oxide. It would be wrong to think that oxygen is needed in a redox reaction!

The *essential* feature of a *redox* reaction is not a transfer of oxygen, but rather a *change of valence*.

Consider the reaction

$$3\ MnO_2 + 4\ Al \rightarrow 3\ Mn + 2\ Al_2O_3.$$

Which valence changes take place here?

30

Write down the valences of

Mn in MnO_2: ....   Mn (elemental): ....
Al in Al_2O_3: ....   Al (elemental): ....

and the *valence changes* during reaction:

Mn: .... → ....
Al.: .... → .... → 62

---

52 No, reaction 2 is not possible. Copper is below hydrogen in the electrochemical series, and cannot, therefore, reduce $H^{\oplus}$.

How is it with reaction 1? Will the reaction between zinc and sulfuric acid take place? → 57

---

53 It is not as simple as that.

Locate Al, $Ag^{\oplus}$, $Zn^{2\oplus}$, and Cu in the electrochemical series (Table 5 at the end of the book) and you should be able to answer the question at 61 .

---

54 No, zinc, not copper, is the stronger reductant.

When a piece of zinc is dipped into a copper salt solution, it immediately reduces copper ions:

$$Zn + Cu^{2\oplus} \rightarrow Zn^{2\oplus} + Cu$$

but copper does not reduce zinc ions.

So, zinc is the strongest .... agent.

Complete the above sentence. → 59

---

55 They *take up* electrons and are *reduced* to metals.

An example is

$$Cu^{2\oplus} + 2\ e^{\ominus} \rightarrow Cu$$

Copper ions are reduced to elemental copper.

Strong reducing agents are those which give up electrons with particular ease.

The readiness to release electrons can be measured experimentally. The measurement gives a tension value in volts. These tensions have been measured for all the elements and are known as the *normal redox potentials.*

The normal redox potentials of zinc, copper, and silver – elements with which we are already familiar – are given below:

$$\begin{array}{llll} Zn & \leftrightarrows Zn^{2\oplus} & + 2\,e^{\ominus} & -0.76 \text{ volt} \\ Cu & \leftrightarrows Cu^{2\oplus} & + 2\,e^{\ominus} & +0.34 \text{ volt} \\ Ag & \leftrightarrows Ag^{\oplus} & + \;\; e^{\ominus} & +0.80 \text{ volt} \end{array}$$

As you see, the redox potential can be either positive or negative.

What is the sign of the normal redox potential of the strongest reducing metals?

- Negative (−) → 46
- Positive (+) → 63

56

Fluorine is the strongest oxidant of the halogens (F_2, Cl_2, Br_2, I_2).

Fluorine is in fact the strongest oxidizing agent of all.

What is the normal redox potential of hydrogen?

Look it up in Table 5 (end of the book) and write down. → 10

57

1. $Mg + Cl_2 \rightarrow Mg^{2\oplus} + 2\,Cl^{\ominus}$
2. $Fe + 2\,H^{\oplus} + SO_4^{2\ominus} \rightarrow Fe^{2\oplus} + SO_4^{2\ominus} + H_2$.

In reaction 1 magnesium is oxidized to $Mg^{2\oplus}$ by chlorine. Chlorine is reduced to $2\,Cl^{\ominus}$ by magnesium.

In reaction 2 iron is oxidized to $Fe^{2\oplus}$ by $2\,H^{\oplus}$. The $2\,H^{\oplus}$ are reduced to H_2 by iron.

Which of the following two reactions will occur?

1. $Zn + H_2SO_4 \rightarrow$
2. $Cu + H_2SO_4 \rightarrow$

30

- None → 27
- Only reaction 1 → 49
- Only reaction 2 → 19
- Both → 52

---

58 You should already know the reaction

$$Cu + 2\ AgNO_3 \rightarrow 2\ Ag + Cu(NO_3)_2.$$

When a piece of copper is immersed in a silver nitrate solution, it becomes coated with silver. Some of the copper dissolves as $Cu^{2\oplus}$ during the reaction.

Look up Cu and $Ag^{\oplus}$ in the electrochemical series. As you see, Cu stands two places above $Ag^{\oplus}$ and can therefore reduce $Ag^{\oplus}$. In other words, $Ag^{\oplus}$ can oxidize Cu.

Repeat 10 and look up each of the metals and cations in the electrochemical series before answering again.

---

59 Correct; zinc is the strongest *reducing agent.*

Another example is the silver-plating of copper by immersing it in a solution of silver nitrate:

$$Cu + 2\ Ag^{\oplus} \rightarrow Cu^{2\oplus} + 2\ Ag$$

In this case copper reduces silver ions. Metallic silver separates out and copper ions go into solution.

$$2\ Ag^{\oplus} + 2\ e^{\ominus} \rightarrow 2\ Ag$$
$$Cu \rightarrow Cu^{2\oplus} + 2\ e^{\ominus}$$

To summarize:

In the reaction

$$Zn + Cu^{2\oplus} \rightarrow Zn^{2\oplus} + Cu,$$

zinc is reducing copper ions.

In the reaction

$$Cu + 2\ Ag^{\oplus} \rightarrow Cu^{2\oplus} + Ag,$$

copper is reducing silver ions.

Problem:

What happens when a piece of zinc is put into a silver nitrate solution?

- Zinc dissolves and silver separates out. → [24]
- Nothing happens → [37]

---

60 No, $Cl^{\ominus}$ isn't oxidized. It couldn't be, since it is on the right-hand side of the equation, i.e. it is a reaction product.

Look up the definition of oxidation and reduction before answering the question at [50] again.

---

61 Correct; the reaction

$$Hg + 2\ AgNO_3 \rightarrow 2\ Ag + Hg(NO_3)_2$$

will *not* take place.

The reason is that $Ag^{\oplus}$ stands above Hg in the electrochemical series. Reaction will only take place when the reducing agent is higher in the series than the oxidizing agent.

Let us continue now with the elements *above* hydrogen in the electrochemical series.

All elements above H_2 will reduce $H^{\oplus}$ and any of the elements and ions below $H^{\oplus}$ (i.e. metal cations and halogens).

Consider two experiments:

1. A piece of aluminum is placed in a silver nitrate solution.
2. A piece of copper is placed in a zinc sulfate solution.

In which of these experiments – if any – will a reaction take place? (use Table 5, end of the book).

1. $Al + 3\ AgNO_3 \rightarrow 3\ Ag + Al(NO_3)_3$
2. $Cu + ZnSO_4 \rightarrow Zn + CuSO_4$

- None of the reactions will occur. → 7
- Only reaction 1 occurs. → 13
- Only reaction 2 occurs. → 9
- Both reactions will occur. → 53

---

62 Compare:

Mn: $4 \rightarrow 0$	$MnO_2 \rightarrow Mn$
Al: $0 \rightarrow 3$	$Al \rightarrow Al_2O_3$

Elemental manganese is zerovalent; in MnO_2 Mn is tetravalent.

Elemental aluminum is zerovalent; in Al_2O_3 Al is trivalent.

The valence of a given element in a specific compound is invariant, but the same element may form other compounds with a different valence. The valence of the elements in the free (elemental) state is always zero.

When an element undergoes a valence change, electrons are always either taken up or released.

> Thus, we can define oxidation and reduction as follows:
> *Oxidation* is the *release* of electrons from atoms, molecules, or ions.
> *Reduction* is the *taking up* of electrons by atoms, molecules, or ions.

Write down these definitions!

Problem:

Is the following partial reaction

$Cl_2 + 2\ e^{\ominus} \rightarrow 2\ Cl^{\ominus}$, where $e^{\ominus}$ = electron

- an oxidation? → 41
- or a reduction? → 50

---

63 No, zinc is the strongest reductant of the three metals, zinc, copper, and silver. Zinc causes separation of elemental copper from a copper salt solution, and silver from a silver salt solution. In these reactions, zinc itself goes into solution as $Zn^{2\oplus}$.

The normal potentials are:

Zn	−0.76 volt
Cu	+0.34 volt
Ag	+0.80 volt

There are many other reducing metals, also some that are more strongly reducing than zinc.

The more strongly reducing a metal, the more .... (positive/negative) is the normal redox potential.

Write down the completed sentence. → 46

Program 31

Oxidation Numbers and Redox Equations

31

1 This program will deal with oxidation numbers and their use in formulating complicated redox equations.

What is the oxidation number?

There are four simple rules which allow the determination of the oxidation number of any atom in any compound. You must memorize these rules. Here are the first two:

1. *Elements* always have oxidation number zero.
2. In *monoatomic ions* the oxidation number equals the *electric charge* of the ion.

Examples:	sodium	oxidation number 0
	sodium ion, $Na^{\oplus}$	oxidation number +1
	sulfide ion, $S^{2\ominus}$	oxidation number −2

Problem:
What are the oxidation numbers of elemental iron, the chloride ion, the proton, and sulfur?

Oxidation number 0: .... and ....
Oxidation number −1: ....
Oxidation number +1: .... → 17

2 *NO_2* and *H_2O* (which are formed in addition to silver nitrate).

$$Ag + 2\,HNO_3 \rightarrow AgNO_3 + NO_2 + H_2O$$

Since H_2 is not formed, $H^{\oplus}$ isn't being reduced.

We shall examine this reaction in more detail. First the question:

What is the oxidation number of nitrogen in NO_2? Write it down. → 63

3 Your answer is wrong. Read the explanation at → 31

31

4 Correct; 5 mol of $FeSO_4$ are oxidized by 1 mol of $KMnO_4$.

Mn: $+7 \rightarrow +2$ (uptake of five electrons)
Fe: $+2 \rightarrow +3$ (release of one electron)

Hence, 5 Fe can be oxidized by 1 Mn in this reaction:

$$\begin{array}{ll} \text{Mn:} & +7 \rightarrow +2 = -5 \\ \text{5 Fe:} & +10 \rightarrow +15 = \underline{+5} \\ & \qquad\qquad\quad 0 \end{array}$$

and the molar ratio of $KMnO_4$ to $FeSO_4$ is 1:5.

The equation should therefore be written

$$KMnO_4 + 5\ FeSO_4 + H_2SO_4 \rightarrow MnSO_4 + Fe_2(SO_4)_3 + H_2O + K_2SO_4$$

where the molar ratio on the left is now correct. To balance the equation, we should then write $2\frac{1}{2}\ Fe_2(SO_4)_3$ on the right.

Instead of using fractions, we can multiply the whole equation by 2:

$$2\ KMnO_4 + 10\ FeSO_3 + H_2SO_4 \rightarrow MnSO_4 + 5\ Fe_2(SO_4)_3 + H_2O + K_2SO_4$$

However, the equation is still not balanced with respect to manganese. How many Mn should there be on the right-hand side? → 28

5 Compare:

$$Na_2Cr_2O_7 + 6\ FeSO_4 + H_2SO_4 \rightarrow Cr_2(SO_4)_3 + 3\ Fe_2(SO_4)_3 + H_2O + Na_2SO_4.$$

3. Balance the stoichiometry of the other ions or molecules.

First count the number of sodium atoms: it is the same on both sides of the equation.

Then count the number of sulfate ions on the right, subtract from this the number delivered by $FeSO_4$ from the left-hand side, and the difference will be the number of sulfate ions needed in the form of sulfuric acid on the left.

How many moles of sulfuric acid are needed? → 64

31

6 You already know the oxidation number of the sulfide ion; it is −2.

The sum of all the oxidation numbers in a compound equals zero. Compute from this the oxidation number of iron in FeS. → 26

---

7 You have 2 $KMnO_4$ on the left and 5 Cl_2 on the right-hand side of the equation. The 5 Cl_2 were formed from 10 HCl. However, you have 16 HCl in the equation.

What happens to the other six HCl? → 20

---

8 You miscalculated. Try the question at 28 again, or, if you need help, go to 45 .

9

$$\overset{0}{Cu} \rightarrow \overset{+2}{Cu^{2\oplus}} \qquad 0 \rightarrow +2 = +2$$

$$\overset{+6}{H_2SO_4} \rightarrow \overset{+4}{SO_2} \qquad +6 \rightarrow +4 = \underline{-2}$$

$$0$$

Copper is oxidized; sulfur is reduced. To reduce 1 gram-atom of copper, 1 mol of H_2SO_4 is required.

Thus, the change in oxidation numbers gives us the ratio of oxidant to reductant.

The equation – balanced in oxidation numbers only – is then

$$\overset{0}{Cu} + \overset{+6}{H_2SO_4} \rightarrow \overset{+2}{Cu^{2\oplus}} + \overset{+4}{SO_2}$$

but the stoichiometry of this equation is not yet balanced. Knowing that H_2O is also formed, and that $Cu^{2\oplus}$ must exist as a sulfate, you can complete the equation.

Write down the balanced equation. → 51

---

10 Your answer is wrong. Read the explanation at 29 .

31

11

$$\overset{0}{Zn} + 2\ H\overset{+1}{Cl} \rightarrow \overset{+2}{Zn}Cl_2 + \overset{0}{H_2}$$

The changes in oxidation number are:

$$\overset{0}{Zn} \rightarrow \overset{+2}{Zn^{2\oplus}} \qquad 0 \rightarrow +2 = +2$$

$$2\ \overset{+1}{H^{\oplus}} \rightarrow \overset{0}{H_2} \qquad +2 \rightarrow 0 = \underline{-2}$$

$$0$$

Let us now consider more complicated redox equations. The reaction between silver and nitric acid is an example:

$$Ag + 2\ HNO_3 \rightarrow AgNO_3 + NO_2 + H_2O$$

The normal course of reaction of a metal with an acid is the formation of H_2 and a salt of the metal. In the present case, the reaction proceeds in a different manner.

What is the main difference between this reaction and the normal mode of reaction between a metal and an acid? The difference lies in some of the reaction products, namely
.... and ....

Complete this sentence. → 2

---

12 Wrong. Only free elements have oxidation number zero. Any atom in a compound has an oxidation number different from zero.

Repeat the rules which apply to oxidation numbers as from 1 .

---

13 You miscalculated. Repeat the question at 28 , or, if you need help, go to 45 .

---

14 Your answer is wrong. Read the explanation at 29 .

---

15 Your answer is wrong. Read the explanation at 31 .

16 1 mol of $KMnO_4$ oxidizes 5 mol of HCl to chlorine.

The changes in oxidation numbers are:

$$\overset{+7}{KMnO_4} \rightarrow \overset{+2}{MnCl_2} \qquad +7 \rightarrow +2 = -5$$

$$5\,\overset{-1}{HCl} \rightarrow 2\tfrac{1}{2}\,\overset{0}{Cl_2} \qquad -5 \rightarrow 0 = \underline{+5}$$

$$0$$

→ 23

17 Compare:

iron and sulfur have oxidation number	0
the chloride ion has oxidation number	−1
the proton has oxidation number	+1

Iron and sulfur being elements, their oxidation number equals zero. The chloride ion being negatively charged, its oxidation number is −1. The proton ($H^{\oplus}$) being positively charged, its oxidation number is +1.

The third rule is:

3. *The sum of the oxidation numbers of the atoms in a compound equals zero.*

Examples:

$$\overset{+1}{H}\text{–}\overset{-1}{Cl} \qquad +1-1 = 0$$

$$\overset{+1}{H_2}\overset{-2}{S} \qquad 2(+1)-2 = +2-2 = 0$$

The third rule permits the calculation of individual oxidation numbers of the atoms in a molecule.

We know that the oxidation number of $Cl^{\ominus}$ equals −1. What is the oxidation number of Fe in $FeCl_3$?

The sum of the oxidation numbers of the three chloride ions in $FeCl_3$ is $3 \cdot (-1) = -3$. Since the compound has an overall oxidation number of zero (see rule 3 above), Fe must have the oxidation number +3 in $FeCl_3$.

$$\overset{+3}{Fe}\overset{-1}{Cl_3} \qquad +3 + 3 \cdot (-1) = +3 - 3 = 0$$

Now, what is the oxidation number of Fe in FeS?

If you know it → [26]
If you don't know it → [6]

---

[18] Wrong. Apparently you didn't take the number of moles of each salt into account.

Repeat [28] or, if you need help, go to [45] .

---

[19] Your oxidation number for manganese after the reaction is wrong.

Read the explanation at [39] .

---

[20] Compare:

the six other HCl molecules are needed as follows:

4 $Cl^{\ominus}$ in the form of 2 $MnCl_2$;
2 $Cl^{\ominus}$ in the form of 2 KCl.

The hydrogen from the 16 HCl and the oxygen from $KMnO_4$ combine to form 8 H_2O.

The formulation of the equation for this complicated redox reaction is carried out in three steps:

1. determination of the molar ratio of oxidant to reductant, using the changes in oxidation numbers;
2. balancing the stoichiometry of the redox partners;
3. balancing the stoichiometry of the remaining ions or compounds.

The formulation of other redox equations will be facilitated if you proceed in this order.

In many redox reactions manganese goes from oxidation number +7 to +2. These reactions always take place in acidic solution.

We can exercise the stepwise balancing of a redox reaction using an oxidation of iron(II) to iron(III), i.e. increase of the oxidation number from +2 to +3:

The incomplete equation is:

$$\overset{+7}{KMnO_4} + \overset{+2}{FeSO_4} + H_2SO_4 \rightarrow \overset{+2}{MnSO_4} + \overset{+3}{Fe_2(SO_4)_3} + H_2O + K_2SO_4$$

Write down this equation which we will now balance. Calculate, on the basis of the oxidation number changes, how many moles of $FeSO_4$ are oxidized by one mole of $KMnO_4$.

- 1 → [10]
- 3 → [14]
- 5 → [4]
- I need assistance → [29]

---

[21]

$$\overset{0}{Fe} + \overset{+1}{H_2}\overset{+6}{S}\overset{-2}{O_4} \rightarrow \overset{+2}{Fe}\overset{+6}{S}\overset{-2}{O_4} + \overset{0}{H_2}$$

In order to balance a redox equation we need only those oxidation numbers which change during the reaction, however.

Therefore, we do not need to specify the oxidation number of the sulfate ion ($SO_4^{2\ominus}$) which remains unchanged in this reaction:

$$\overset{0}{Fe} + \overset{+1}{H_2}SO_4 \rightarrow \overset{+2}{Fe}SO_4 + \overset{0}{H_2}$$

Elemental iron (oxidation number 0) releases two electrons to become the $Fe^{2\oplus}$ ion (oxidation number +2). Two protons (oxidation number $2 \cdot (+1) = +2$) take up the two electrons to become hydrogen (oxidation number 0).

$$\overset{0}{Fe} \longrightarrow \overset{+2}{Fe^{2\oplus}}; \quad 0 \longrightarrow +2 = +2 \text{ (increase of 2)}$$

$$2\,\overset{+1}{H^{\oplus}} \longrightarrow \overset{0}{H_2} \quad +2 \longrightarrow 0 = -2 \text{ (decrease of 2)}$$

Total change: $\underline{\underline{0}}$

The sum of all changes must equal zero.

Now write down the equation for the reaction between zinc and hydrochloric acid, specifying the changing oxidation numbers. → [11]

22 The correct equation is

$$Cu + 4\ HNO_3 \rightarrow Cu(NO_3)_2 + 2\ NO_2 + 2\ H_2O.$$

If you gave this answer without hesitation, you have a good understanding of the subject.

Redox reactions are not limited to reactions between metals and acids. There is a much greater number of reactions in which the ions of dissolved compounds are reduced or oxidized.

Some selected examples will be given below.

Potassium permanganate ($KMnO_4$) is an important reagent. What is the oxidation number of Mn in $KMnO_4$?

The oxidation number of Mn in $KMnO_4$ is

- + 3 → 15
- + 5 → 3
- + 7 → 57
- + 9 → 43

---

23 1 mol of $KMnO_4$ oxidizes 5 mol of HCl to chlorine.

If you had another number, read the explanation at 16 .

We now have the ratio of oxidant to reductant, but we still need to balance the stoichiometry of the redox partners.

5 HCl give $2\frac{1}{2}\ Cl_2$. Rather than use fractions in reaction equation we multiply all numbers by 2.

How many moles of $KMnO_4$ are there on the left, and how many moles of Cl_2 on the right in this equation? → 7

---

24 There are 8 H_2O on the right-hand side.

The number of oxygen atoms is also balanced. 2 $KMnO_4$ and 8 H_2O both contain eight oxygen atoms.

The correct equation is

$$2\ KMnO_4 + 10\ FeSO_4 + 8\ H_2SO_4 \rightarrow 2\ MnSO_4 + 5\ Fe_2(SO_4)_3 + 8\ H_2O + K_2SO_4$$

Sodium dichromate, $Na_2Cr_2O_7$, is another oxidizing agent comparable with potassium permanganate.

What is the oxidation number of chromium in $Na_2Cr_2O_7$?

Write down your computations and check your answer at [46] , or, if you need help, go to [32] .

---

[25] Let us balance the equation step by step.

1. Ratio of dichromate to $FeSO_4$, using changes in oxidation numbers.

Cr:	$+6 \rightarrow +3$	(uptake of three electrons)
Fe:	$+2 \rightarrow +3$	(release of one electron)

Thus, 3 Fe can be oxidized:

$$\begin{array}{ll} \text{Cr:} & +6 \rightarrow +3 = -3 \\ \text{3 Fe:} & +6 \rightarrow +9 = \underline{+3} \\ & \qquad\qquad\quad 0 \end{array}$$

The number of iron atoms oxidized is three times as large as the number of chromium atoms available. There are *two* Cr in $Na_2Cr_2O_7$. Therefore, *1* $Na_2Cr_2O_7$ can oxidize .... $FeSO_4$

Write down the number. → [56]

---

[26] *Iron* has the oxidation number +2 in FeS. The sum of the oxidation numbers equals zero.

$$\overset{+2-2}{\text{FeS}}: \quad +2 - 2 = 0$$

Now compute the oxidation numbers of

the bromine atom in HBr and

the titanium atom in $TiCl_4$.

Write down:

The oxidation numbers are:

Br in HBr ....
Ti in $TiCl_4$ .... → [36]

---

[27] You gave an erroneous oxidation number for manganese in the reaction product.

Read the explanation at [39] .

---

[28] Compare:

$$2\ KMnO_4 + 10\ FeSO_4 + H_2SO_4 \rightarrow 2\ MnSO_4 + 5\ Fe_2(SO_4)_3 + H_2O + K_2SO_4.$$

The stoichiometry of the redox partners is now balanced. We still need to balance the stoichiometry of the remaining ions and compounds.

The numbers of potassium ions on both sides of the equation are equal.

Now count the number of *sulfate ions* on the right-hand side.

How many are there?

- 5 → [18]
- 9 → [13]
- 17 → [8]
- 18 → [38]
- I need assistance → [45]

---

[29]

$KMnO_4 \rightarrow MnSO_4$: $+7 \rightarrow +2 = -5$
$FeSO_4 \rightarrow Fe_2(SO_4)_3$: $+2 \rightarrow +3 = +1$

The oxidation number of manganese changes from +7 to +2. This is the result of manganese accepting *five* electrons. The electrons come from iron which is being oxidized. Its oxidation number goes from +2 to +3. Iron gives up one electron in the oxidation process.

1 mol of $KMnO_4$ oxidizes .... mol of $FeSO_4$.

Write down the number. → 4

30 Yes, the completed equation is

$Na_2Cr_2O_7 + 6\ FeSO_4 + 7\ H_2SO_4 \rightarrow Cr_2(SO_4)_3 + 3\ Fe_2(SO_4)_3 + 7\ H_2O + Na_2SO_4$

Did you have *all* the numbers correct?

This complicated redox equation terminates our presentation of inorganic and general chemistry in this volume.

End of Program 31.

---

31 The oxidation number of Mn in $KMnO_4$ was asked for.

The oxidation number of K is +1; K = +1
The oxidation number of O is −2; $O_4 = 4 \cdot (-2) = -8$

sum: −7

In order that the sum of all the oxidation numbers remains zero in $KMnO_4$, Mn must have the oxidation number ....

Write it down. → 57

---

32 We want to derive the oxidation number of Cr in $Na_2Cr_2O_7$.

We know the oxidation numbers of sodium and oxygen in chemical compounds: +1 for sodium and −2 for oxygen.

Thus

$$\begin{aligned} Na_2 &= 2 \cdot (+1) = +\ 2 \\ O_7 &= 7 \cdot (-2) = -14 \\ & \qquad\qquad\ \ -12 \end{aligned}$$

Knowing that the sum of all the oxidation numbers in a compound must equal zero, you can compute the number for chromium. → 46

33 You gave an erroneous oxidation number for manganese in the reaction product.

Read the explanation at 39 .

34 Your answer is wrong. You did not take into account that there are *two* H-atoms in a molecule of H_2SO_4. → 62

35 You didn't take into account that there are *two* chromium atoms in each molecule of $Na_2Cr_2O_7$. Therefore, only half your amount of $Na_2Cr_2O_7$ is needed to oxidize 6 $FeSO_4$. Your amount of H_2SO_4 is also wrong.

- Correct your errors in the equation → 46
- or go to 25 for more help.

36 Compare:

Br in HBr	−1
Ti in $TiCl_4$	+4

or detailed:

$\overset{+1}{H}-\overset{-1}{Br}$: $+1 + (-1) = 1 - 1 = 0$

$\overset{+4}{Ti}\ \overset{-1}{Cl_4}$: $+4 + 4 \cdot (-1) = 4 - 4 = 0$

And now the fourth rule:

4. *The oxidation number of oxygen in its chemical compounds always equals −2.*

(There is one exception: in the peroxides oxygen has the oxidation number −1. This class of compounds will not, however, be dealt with in this book.)

Do you remember the other three rules?

Answer the following questions:

1. What is the value of the oxidation number of the elements?

2. How does one know the oxidation numbers of monoatomic ions?

3. What is the sum of all the oxidation numbers within a compound? → 47

37 Your equation is wrong.

We had seen from the oxidation numbers that 2 mol of HNO_3 were required for the oxidation of 1 gram-atom of Cu. Therefore, there must be at least 2 NO_2 on the right-hand side of the equation.

Try to work through the balancing of this redox equation once more. → 61

38 All in all there are *18* $SO_4^{2\ominus}$ ions on the right-hand side of the equation.

On the left, there are already *10* $SO_4^{2\ominus}$ ions in the form of $FeSO_4$. Thus, we need 8 H_2SO_4 on the left:

$$2\ KMnO_4 + 10\ FeSO_4 + 8\ H_2SO_4 \rightarrow 2\ MnSO_4 + 5\ Fe_2(SO_4)_3 + H_2O + K_2SO_4$$

We still have to balance the numbers of hydrogen atoms.

How many H_2O must we have on the right?

Write down the number. → 24

39 Mn exists in the form of $MnCl_2$ at the end of the reaction.

To determine the oxidation number of Mn in $MnCl_2$, we need to know that of the chloride ion. The oxidation number of the chloride ion is -1. There are two such ions in $MnCl_2$.

Using the rule that the sum of all oxidation numbers in a compound must equal zero, you should be able to answer correctly at 57 .

40 The changes in oxidation number are repeated below:

$$\overset{+7}{KMnO_4} \rightarrow \overset{+2}{MnCl_2}: \qquad +7 \rightarrow +2 = -5$$

$$\overset{-1}{HCl} \rightarrow \tfrac{1}{2}\overset{0}{Cl_2}: \qquad -1 \rightarrow \ \ 0 = +1$$

$$\overline{\qquad -4}$$

As manganese takes up *five* electrons, its oxidation number changes from +7 to +2. The electrons come from the chloride ions which each give up *one* electron when becoming molecular chlorine.

So, how many chloride ions can be oxidized by 1 $KMnO_4$? → 16

---

41 Your calculation is wrong, or you simply took the oxidation number of sulfur to be that of the sulfide ion (−2).

Repeat the rules for computing oxidation numbers as from 1 .

---

42 Your equation is wrong.

If you have four hydrogen atoms on the left (as 4 HNO_3), you must also have four on the right.

Check the numbers on both sides of the equation once more. → 61

---

43 Your answer is wrong.

Read the detailed explanation at 31 .

---

44 Compare:

$$\overset{+1}{H}\text{–}\overset{+5}{N}\overset{-2}{O_3} \quad = +1 + 5 + 3 \cdot (-2) = 1 + 5 - 6 = 0$$

Let us now look at the oxidation numbers in some simple reactions with which you are already familiar.

Iron dissolves in dilute sulfuric acid:

$$Fe + H_2SO_4 \rightarrow FeSO_4 + H_2$$

Write down this equation as well as the oxidation number of each atom. → 21

---

45 The question was concerned with the number of sulfate ions on the right-hand side of the equation.

You have

2 $MnSO_4$,	i.e.	2 $SO_4^{2\ominus}$	ions
5 $Fe_2(SO_4)_3$,	i.e.	15 $SO_4^{2\ominus}$	ions
1 K_2SO_4,	i.e.	1 $SO_4^{2\ominus}$	ion
	sum:	? $SO_4^{2\ominus}$	ions

What is the total? Continue at 38 .

---

46 Chromium has the oxidation number +6 in $Na_2Cr_2O_7$.

Also $Na_2Cr_2O_7$ will oxidize $FeSO_4$ to $Fe_2(SO_4)_3$.

The incomplete equation is:

$$\overset{+6}{Na_2Cr_2}O_7 + \overset{+2}{Fe}SO_4 + H_2SO_4 \rightarrow \overset{+3}{Cr_2}(SO_4)_3 + \overset{+3}{Fe_2}(SO_4)_3 + H_2O + Na_2SO_4.$$

Write down this equation and complete it. Proceed in exactly the same way as in the previous reaction with $KMnO_4$:

1. Determine the ratio of $Na_2Cr_2O_7$ to $FeSO_4$ using oxidation numbers.

2. Balance the stoichiometry of the redox partners.

3. Balance the remaining ions and molecules.

What are the correct numbers of molecules of

	$Na_2Cr_2O_7$	$FeSO_4$	H_2SO_4?		
•	1	6	7	→	30
•	2	6	9	→	65
•	1	3	5	→	58
•	1	3	7	→	60
•	2	6	8	→	35
•	I need more information			→	25

47 Compare:

1. The oxidation number of the elements is zero.

2. The oxidation number of monoatomic ions equals the electric charge.

3. The sum of all oxidation numbers in a compound is zero.

Do you also remember the fourth rule? Write down the oxidation number of oxygen in its compounds. → 62

---

48 The completed equation is:

$$\overset{0}{Ag} + 2\ H\overset{+5}{N}O_3 \rightarrow \overset{+1}{Ag}\overset{+5}{N}O_3 + \overset{+4}{N}O_2 + H_2O$$

Thus, two moles of HNO_3 are needed in this reaction, namely

one mole for the oxidation of silver
(the oxidation number of nitrogen goes from +5 to +4),

and

one mole as nitrate in $AgNO_3$
(in which the oxidation number of nitrogen is unchanged).

The oxidation of copper by hot concentrated sulfuric acid is a similar reaction. Copper dissolves as copper sulfate, and SO_2 is evolved.

How do the oxidation numbers of copper and sulfur change in this reaction?

Write down the following transformations complete with the changes in oxidation numbers:

$$Cu \longrightarrow Cu^{2\oplus}\text{: .... } \longrightarrow \text{ .... } = \text{ ....}$$
$$H_2SO_4 \longrightarrow SO_2\text{: .... } \longrightarrow \text{ .... } = \text{ ....}$$

.... → 9

---

49 Yes, before the reaction manganese has the oxidation number +7 ($KMnO_4$); after the reaction it is +2 ($MnCl_2$).

Manganese is reduced from an oxidation number of +7 to +2. Chlorine is oxidized from −1 to 0.

$$\overset{+7}{KMnO_4} \rightarrow \overset{+2}{MnCl_2}: \quad +7 \rightarrow +2 = -5$$

$$\overset{-1}{HCl} \rightarrow 1/2\ \overset{0}{Cl_2}: \quad -1 \rightarrow 0 = +1$$

$$\overline{\quad -4 \quad}$$

The ratio of oxidant to reductant must now be established.

How many moles of HCl can be oxidized by one mole of $KMnO_4$?

- Give the number → 23

or go to 40 for more detailed information.

---

50 Correct; sulfur has the oxidation number +6 in H_2SO_4.

The oxidation number of H is +1; $H_2 = 2 \cdot (+1) = +2$
The oxidation number of O is −2; $O_4 = 4 \cdot (-2) = -8$
Total: −6

S must have the oxidation number +6 in H_2SO_4 in order that the sum of all oxidation numbers remains zero.

$$\overset{+1}{H_2}\overset{+6}{S}\overset{-2}{O_4} = 2 \cdot (+1) + 6 + 4 \cdot (-2) = 2 + 6 - 8 = 0$$

Compute, in writing, the oxidation number of nitrogen in nitric acid, HNO_3. → 59

---

51 $Cu + 2\ H_2SO_4 \rightarrow CuSO_4 + SO_2 + 2\ H_2O$

Copper can also be oxidized by concentrated nitric acid; in this case NO_2 is evolved:

$$\overset{0}{Cu} \rightarrow \overset{+2}{Cu^{2\oplus}}: \quad 0 \rightarrow +2 = +2$$

$$H\text{–}\overset{+5}{NO_3} \rightarrow \overset{+4}{NO_2}: \quad +5 \rightarrow +4 = -1$$

$$\underline{\underline{+1}}$$

The ratio of oxidant to reductant cannot be 1:1 in this latter reaction since the sum of the changes in oxidation number is different from zero.

31

Copper releases two electrons going from $\overset{0}{Cu}$ to $\overset{+2}{Cu}{}^{2\oplus}$, but nitrogen accepts only *one* electron going from

$$H\overset{+5}{N}O_3 \text{ to } \overset{+4}{N}O_2.$$

Since nitrogen accepts only one, but copper releases two electrons, the molar ratio of nitric acid to copper must be 2:1.

Complete the following equations, specifying the changes in oxidation number, and check that the sum of the changes equals zero:

$\overset{0}{Cu} \longrightarrow \overset{+2}{Cu}{}^{2\oplus}$: .... $\longrightarrow$ .... = ...

$2\,H\overset{+5}{N}O_3 \longrightarrow 2\,\overset{+4}{N}O_2$: .... $\longrightarrow$ .... = ...

Total change: .... → 61

---

52 Your oxidation number for manganese at the end of the reaction is erroneous.

Read the explanation at 39 .

---

53 Your answer is wrong. You didn't take into account that there are *two* H-atoms and *four* O-atoms for each sulfur atom in H_2SO_4. Repeat the calculation using +1 as the oxidation number of hydrogen, and −2 for oxygen, and you will be able to answer the question at 62 .

---

54 There is still one HNO_3 missing on the left-hand side.

This extra HNO_3 is used to form $AgNO_3$; the proton goes to the formation of water.

Complete the equation. → 48 .

---

55 Your equation is erroneous.

You have 2 H_2O on the right and only 3 HNO_3 on the left, i.e. you have too few hydrogen atoms on the left. Check the numbers of the other atoms as well and carefully balance the equation. → 61

56

1 $Na_2Cr_2O_7$ oxidizes *6* $FeSO_4$.

Inserting this in the equation, we now have:

$$Na_2Cr_2O_7 + 6\ FeSO_4 + H_2SO_4 \rightarrow Cr_2(SO_4)_3 + Fe_2(SO_4)_3 + H_2O + Na_2SO_4.$$

2. Balance the stoichiometry of the redox partners.
The ratio of $Na_2Cr_2O_7$ to $FeSO_4$ on the left was already determined on the basis of oxidation numbers. We now have to find the numbers of $Cr(SO_4)_3$ and $Fe_2(SO_4)_3$ on the right-hand side of the equation.

Determine these numbers and check your equation at [5] .

57

Correct, Mn has the oxidation number +*7* in $KMnO_4$.

$$\overset{+1+7-2}{KMnO_4} = +1 +7 +4 \cdot (-2) =$$
$$+1 +7 -8 = 0$$

$KMnO_4$ is a strong oxidizing agent. It oxidizes chloride ion to chlorine:

$$2\ KMnO_4 + 16\ HCl \rightarrow 2\ KCl + 2\ MnCl_2 + 5\ Cl_2 + 8\ H_2O.$$

We will work through this equation in detail.

Write down the equation and then answer the question:

What are the oxidation numbers of manganese before and after the reaction?

Before	After		
● +7	+6	→	[19]
● +7	+5	→	[27]
● +7	+4	→	[52]
● +7	+3	→	[33]
● +7	+2	→	[49]
● I need more information		→	[39]

31

58 You forgot that there are *two* chromium atoms in $Na_2Cr_2O_7$. Therefore, twice as many $FeSO_4$ molecules can be oxidized. Your number of H_2SO_4 molecules is also wrong.

- Correct your mistakes in the equation → 46
- I need more information → 25

---

59 The oxidation number of nitrogen in nitric acid is +5.

Write down the formula for nitric acid and specify the oxidation number of each atom in the molecule. → 44

---

60 You forgot that there are *two* chromium atoms in $Na_2Cr_2O_7$. Therefore, twice as many $FeSO_4$ molecules can be oxidized. Your number of H_2SO_4 molecules is also wrong.

- Correct the errors in your equation → 46
- I need more information → 25

---

61 Compare:

$$\overset{0}{Cu} \rightarrow \overset{+2}{Cu^{2\oplus}}: \qquad 0 \rightarrow +2 = +2$$

$$2\,H\overset{+5}{N}O_3 \rightarrow 2\,\overset{+4}{N}O_2: \qquad +10 \rightarrow +8 = -2$$

$$\underline{\underline{0}}$$

The ratio of oxidant to reductant can always be determined in this manner. Here, we found that 1 gram-atom of copper is oxidized by 2 mol of nitric acid.

The still incomplete equation now has the form:

$$Cu + 2\,HNO_3 \rightarrow Cu^{2\oplus} + 2\,NO_2 + H_2O$$

Now adjust the stoichiometry of this equation. You need to consider the nitric acid needed for the formation of the salt.

What are the correct numbers of molecules of

	HNO_3	NO_2	H_2O?		
•	2	1	1	→	37
•	3	2	2	→	55
•	4	2	1	→	42
•	4	2	2	→	22

---

62 The oxidation number of oxygen in its chemical compounds *always equals −2.*

With these four rules the oxidation number of each atom in a molecule can always be determined.

Example:

What is the oxidation number of phosphorus in phosphoric acid (H_3PO_4)?

Oxidation number of H: +1; H_3: $3 \cdot (+1) = +3$
Oxidation number of O: −2; O_4: $4 \cdot (-2) = -8$

Total: −5

In order that the sum of all oxidation numbers remains zero in H_3PO_4, P must have the oxidation number +5.

$$\overset{+1}{H_3}\overset{+5}{P}\overset{-2}{O_4} = 3 \cdot (+1) + 5 + 4 \cdot (-2) = 0$$

What is the oxidation number of sulfur in H_2SO_4?

Carry out the calculation in writing.

- −2 → 41
- 0 → 12
- +1 → 53
- +6 → 50
- +7 → 34

---

63 The nitrogen atom in NO_2 has the oxidation number +4.

$$\overset{+4-2}{NO_2}: \quad +4 + 2 \cdot (-2) = 0$$

Since nitrogen has an oxidation number of +5 in nitric acid, a reduction from +5 to +4 took place during this reaction. At the same time, silver was oxidized from 0 to +1.

Here are the changes in detail:

$$\overset{+5}{HNO_3} \longrightarrow \overset{+4}{NO_2} \qquad +5 \longrightarrow +4 = -1$$

$$\overset{0}{Ag} \longrightarrow \overset{+1}{Ag^{\oplus}}: \qquad 0 \longrightarrow +1 = +1$$

$$\underline{\underline{0}}$$

The balance of the changes in oxidation number means that 1 HNO_3 oxidizes 1 Ag.

Let us start formulating this reaction by writing

$$\overset{0}{Ag} + \overset{+5}{HNO_3} \rightarrow \overset{+1}{Ag^{\oplus}} + \overset{+4}{NO_2}$$

Write down this equation which is not yet complete.
The stoichiometry isn't balanced, since we have an H and an O too many on the left-hand side.

On the right-hand side we need to neutralize $Ag^{\oplus}$ in the form of $AgNO_3$. What do we need on the left in order to balance the equation? → 54

---

64 7 H_2SO_4.

The equation now reads:

$$Na_2Cr_2O_7 + 6\ FeSO_4 + 7\ H_2SO_4 \rightarrow Cr_2(SO_4)_3 + 3\ Fe_2(SO_4)_3 + H_2O + Na_2SO_4$$

We have yet to check the number of water molecules.
Does the number of O-atoms in $Na_2Cr_2O_7$ and the number of H-atoms in H_2SO_4 tally with the one molecule of H_2O on the right?

Work out the number of H_2O molecules and check your answer at 30 .

65 You forgot there are *two* chromium atoms in $Na_2Cr_2O_7$. Therefore, only half your amount of $Na_2Cr_2O_7$ is necessary for the oxidation of 6 $FeSO_4$. Your number of H_2SO_4 molecules is also erroneous.

- Correct the errors in your equation → 46 .
- I need more information → 25 .

Appendix

Table 1. The Constitution of the Atoms.

Atomic number	Element		Number of electrons				Nucleus		Relative atomic mass
			1. shell (K-shell)	2. shell (L-shell)	3. shell (M-shell)	4. shell (N-shell)	protons	neutrons	
1	Hydrogen	H	1				1		1
2	Helium	He	2				2	2	4
3	Lithium	Li	2	1			3	4	7
4	Beryllium	Be	2	2			4	5	9
5	Boron	B	2	3			5	6	11
6	Carbon	C	2	4			6	6	12
7	Nitrogen	N	2	5			7	7	14
8	Oxygen	O	2	6			8	8	16
9	Fluorine	F	2	7			9	10	19
10	Neon	Ne	2	8			10	10	20
11	Sodium	Na	2	8	1		11	12	23
12	Magnesium	Mg	2	8	2		12	12	24
13	Aluminum	Al	2	8	3		13	14	27
14	Silicon	Si	2	8	4		14	14	28
15	Phosphorus	P	2	8	5		15	16	31
16	Sulfur	S	2	8	6		16	16	32
17	Chlorine	Cl	2	8	7		17	18	35
18	Argon	Ar	2	8	8		18	22	40
19	Potassium	K	2	8	8	1	19	20	39
20	Calcium	Ca	2	8	8	2	20	20	40

Table 2. The Chemical Elements.

Name	Symbol	Atomic No.	Atomic Mass	Name	Symbol	Atomic No.	Atomic Mass
Actinium	Ac	89	[227]	Mercury	Hg	80	200.59
Aluminum	Al	13	26.98	Molybdenum	Mo	42	95.94
Americium	Am	95	[243]	Neodymium	Nd	60	144.24
Antimony	Sb	51	121.76	Neon	Ne	10	20.18
Argon	Ar	18	39.94	Neptunium	Np	93	[237]
Arsenic	As	33	74.92	Nickel	Ni	28	58.71
Astatine	At	85	[210]	Niobium	Nb	41	92.90
Barium	Ba	56	137.36	Nitrogen	N	7	14.00
Berkelium	Bk	97	[247]	Nobelium	No	102	[254]
Beryllium	Be	4	9.01	Osmium	Os	76	190.2
Bismuth	Bi	83	208.98	Oxygen	O	8	15.99
Boron	B	5	10.82	Palladium	Pd	46	106.4
Bromine	Br	35	79.91	Phosphorus	P	15	30.97
Cadmium	Cd	48	112.41	Platinum	Pt	78	195.09
Calcium	Ca	20	40.08	Plutonium	Pu	94	[244]
Californium	Cf	98	[251]	Polonium	Po	84	[209]
Carbon	C	6	12.01	Potassium	K	19	39.10
Cerium	Ce	58	140.13	Praseodymium	Pr	59	140.90
Cesium	Cs	55	132.91	Promethium	Pm	61	[145]
Chlorine	Cl	17	35.45	Protactinium	Pa	91	[231]
Chromium	Cr	24	52.01	Radium	Ra	88	[226]
Cobalt	Co	27	58.9332	Radon	Rn	86	[222]
Copper	Cu	29	63.546	Rhenium	Re	75	186.2
Curium	Cm	96	[247]	Rhodium	Rh	45	102.90
Dysprosium	Dy	66	162.1	Rubidium	Rb	37	85.47
Einsteinium	Es	99	[254]	Ruthenium	Ru	44	101.07
Erbium	Er	68	167.27	Samarium	Sm	62	150.35
Europium	Eu	63	152.0	Scandium	Sc	21	44.95
Fermium	Fm	100	[257]	Selenium	Se	34	78.96
Fluorine	F	9	19.00	Silicon	Si	14	28.08
Francium	Fr	87	[223]	Silver	Ag	47	107.86
Gadolinium	Gd	64	157.26	Sodium	Na	11	22.98
Gallium	Ga	31	69.72	Strontium	Sr	38	87.62
Germanium	Ge	32	72.60	Sulfur	S	16	32.06
Gold	Au	79	197.0	Tantalum	Ta	73	180.94
Hafnium	Hf	72	178.50	Technetium	Tc	43	[97]
Helium	He	2	4.00	Tellurium	Te	52	127.60
Holmium	Ho	67	164.94	Terbium	Tb	65	158.92
Hydrogen	H	1	1.00	Thallium	Tl	81	204.37
Indium	In	49	114.82	Thorium	Th	90	232.03
Iodine	I	53	126.9045	Thulium	Tm	69	168.93
Iridium	Ir	77	192.2	Tin	Sn	50	118.69
Iron	Fe	26	55.84	Titanium	Ti	22	47.90
Krypton	Kr	36	83.80	Tungsten	W	74	183.85
Lanthanum	La	57	138.91	Uranium	U	92	238.03
Lawrencium	Lr	103	[257]	Vanadium	V	23	50.94
Lead	Pb	82	207.2	Xenon	Xe	54	131.30
Lithium	Li	3	6.93	Ytterbium	Yb	70	173.04
Lutetium	Lu	71	174.97	Yttrium	Y	39	88.90
Magnesium	Mg	12	24.31	Zinc	Zn	30	65.37
Manganese	Mn	25	54.93	Zirconium	Zr	40	91.22
Mendelevium	Md	101	[256]				

For radioactive elements the mass of the longest-lived isotope is given in brackets.

Table 3. Shortened Periodic Table of the Elements.

1.Period	H							He
	1.Group	2.Group	3.Group	4.Group	5.Group	6.Group	7.Group	8.Group
2.Period	3 Li	4 Be	5 B	6 C	7 N	8 O	9 F	10 Ne
3.Period	11 Na	12 Mg	13 Al	14 Si	15 P	16 S	17 Cl	18 Ar
4.Period	19 K	20 Ca	31 Ga	32 Ge	33 As	34 Se	35 Br	36 Kr
5.Period	37 Rb	38 Sr	49 In	50 Sn	51 Sb	52 Te	53 I	54 Xe
6.Period	55 Cs	56 Ba	81 Tl	82 Pb	83 Bi	84 Po	85 At	86 Rn
7.Period	87 Fr	88 Ra						

sub-group elements

Table 4. The Long Periodic Table of the Elements.

	1a	2a	3b	4b	5b	6b	7b	8b			1b	2b	3a	4a	5a	6a	7a	8a 0
1. Per.	1 H																	2 He
2. Per.	3 Li	4 Be											5 B	6 C	7 N	8 O	9 F	10 Ne
3. Per.	11 Na	12 Mg											13 Al	14 Si	15 P	16 S	17 Cl	18 Ar
4. Per.	19 K	20 Ca	21 Sc	22 Ti	23 V	24 Cr	25 Mn	26 Fe	27 Co	28 Ni	29 Cu	30 Zn	31 Ga	32 Ge	33 As	34 Se	35 Br	36 Kr
5. Per.	37 Rb	38 Sr	39 Y	40 Zr	41 Nb	42 Mo	43 Tc	44 Ru	45 Rh	46 Pd	47 Ag	48 Cd	49 In	50 Sn	51 Sb	52 Te	53 I	54 Xe
6. Per.	55 Cs	56 Ba	57 La 58–71	72 Hf	73 Ta	74 W	75 Re	76 Os	77 Ir	78 Pt	79 Au	80 Hg	81 Tl	82 Pb	83 Bi	84 Po	85 At	86 Rn
7. Per.	87 Fr	88 Ra	89 Ac 90–103															

Lanthanides	58 Ce	59 Pr	60 Nd	61 Pm	62 Sm	63 Eu	64 Gd	65 Tb	66 Dy	67 Ho	68 Er	69 Tm	70 Yb	71 Lu
Actinides	90 Th	91 Pa	92 U	93 Np	94 Pu	95 Am	96 Cm	97 Bk	98 Cf	99 Es	100 Fm	101 Md	102 No	103 Lw

Table 5. The Electrochemical Series.

Li	⇌	$Li^{\oplus}$	+	$e^{\ominus}$	− 3.05 volt
K	⇌	$K^{\oplus}$	+	$e^{\ominus}$	− 2.92 volt
Ba	⇌	$Ba^{2\oplus}$	+	$2e^{\ominus}$	− 2.90 volt
Sr	⇌	$Sr^{2\oplus}$	+	$2e^{\ominus}$	− 2.89 volt
Ca	⇌	$Ca^{2\oplus}$	+	$2e^{\ominus}$	− 2.76 volt
Na	⇌	$Na^{\oplus}$	+	$e^{\ominus}$	− 2.71 volt
Mg	⇌	$Mg^{2\oplus}$	+	$2e^{\ominus}$	− 2.34 volt
Al	⇌	$Al^{3\oplus}$	+	$3e^{\ominus}$	− 1.67 volt
Cr	⇌	$Cr^{2\oplus}$	+	$2e^{\ominus}$	− 0.91 volt
Zn	⇌	$Zn^{2\oplus}$	+	$2e^{\ominus}$	− 0.76 volt
Fe	⇌	$Fe^{2\oplus}$	+	$2e^{\ominus}$	− 0.44 volt
Co	⇌	$Co^{2\oplus}$	+	$2e^{\ominus}$	− 0.28 volt
Ni	⇌	$Ni^{2\oplus}$	+	$2e^{\ominus}$	− 0.25 volt
Pb	⇌	$Pb^{2\oplus}$	+	$2e^{\ominus}$	− 0.13 volt
H_2	⇌	$2H^{\oplus}$	+	$2e^{\ominus}$	± 0.00 volt
Cu	⇌	$Cu^{2\oplus}$	+	$2e^{\ominus}$	+ 0.34 volt
$2I^{\ominus}$	⇌	I_2	+	$2e^{\ominus}$	+ 0.53 volt
Ag	⇌	$Ag^{\oplus}$	+	$e^{\ominus}$	+ 0.80 volt
Hg	⇌	$Hg^{2\oplus}$	+	$2e^{\ominus}$	+ 0.85 volt
$2Br^{\ominus}$	⇌	Br_2	+	$2e^{\ominus}$	+ 1.06 volt
Pt	⇌	$Pt^{2\oplus}$	+	$2e^{\ominus}$	+ 1.2 volt
$2Cl^{\ominus}$	⇌	Cl_2	+	$2e^{\ominus}$	+ 1.36 volt
Au	⇌	$Au^{3\oplus}$	+	$3e^{\ominus}$	+ 1.50 volt
$2F^{\ominus}$	⇌	F_2	+	$2e^{\ominus}$	+ 2.65 volt

Index

The structure of this book is such that each program should be studied in a continuous manner. The index is intended only to allow the user a quick repetition of particular subjects. The **boldface** numbers refer to the programs, the normal print to the sections.